# Sharjah International Conference on Physics of Advanced Materials 2023



Journal of Physics: Conference Series Volume 2751

Sharjah, United Arab Emirates
25-27 April 2023

ISBN: 978-1-7138-9737-8
ISSN: 1742-6588

**Printed from e-media with permission by:**

Curran Associates, Inc.
57 Morehouse Lane
Red Hook, NY 12571

CURRAN ASSOCIATES INC. proceedings .com

**Some format issues inherent in the e-media version may also appear in this print version.**



Institute of Physics
Dirac House, Temple Back
Bristol BS1 6BE UK

Phone: 44 1 17 929 7481
Fax: 44 1 17 920 0979

techtracking@iop.org

**Additional copies of this publication are available from:**

Curran Associates, Inc.
57 Morehouse Lane
Red Hook, NY 12571 USA
Phone: 845-758-0400
Fax: 845-758-2633
Email: curran@proceedings.com
Web: www.proceedings.com

# TABLE OF CONTENTS

## Preface

The Sharjah International Conference on Physics of Advanced Materials (SICPAM) was held on April 23–25, 2023, on the beautiful campus of the University of Sharjah. It brought together more than 120 scientists from 21 countries. The list of invited speakers included leading scientists, editors of reputable journals, young researchers, and students in the field of material science from Europe, the USA and Asia.

Materials have been the key behind major technological breakthroughs throughout history. Their properties can be engineered and optimized to develop specific applications. The transformation of iron into steel ignited the second industrial revolution and the development of railways before the end of the nineteenth century. Similarly, semiconductors enabled the design of solid-state devices that gave birth to modern electronics and the digital revolution (the third industrial revolution).

Today, we are witnessing the birth of the fourth industrial revolution with the fast development of generative AI and the production of the first versions of quantum computers. Quantum materials are emerging as the building blocks of quantum technologies.

The conference covered a wide range of materials and some of the advanced techniques that are used to investigate their properties, as well as the computational methods that are emerging as very powerful tools for material design and screening for diverse applications, including energy conversion and storage, gas sensing and biosensing, water treatment and desalination, etc.

The conference was an enriching experience and an excellent opportunity for many scientists to develop connections and possible collaborations. We would like to express our high appreciations to the University of Sharjah for hosting the conference and members of the scientific and organizing committees who worked hard to make the event a great success that will remain in our memory for a long time.

Our sincere thanks go to the sponsors who generously supported the conference, namely: Sharjah Electricity and Water Authority (SEWA), Sharjah Broadcasting Authority, Sharjah National Oil Corporation (SNOC) and Dubai Electricity and Water Authority (DEWA).

Finally, I would like to address my deepest thanks to the co-Editors who worked tirelessly to ensure that all submitted papers are thoroughly reviewed and the referees' concerns are properly addressed.

Prof. Nouar Tabet, Chairman of the Scientific Committee

## Editors

Prof. Hamid M.K. Al Naimiy, University of Sharjah.

Prof. Yousef Haik, University of Sharjah

Prof. Nouar Tabet, University of Sharjah

Prof. Hussain Alawadhi, University of Sharjah

Prof. Atta Gaffar Attaelmanan, University of Khorfakkan

Prof. Kais Daoudi, University of Sharjah

Prof. Nasser Hamdan, American University of Sharjah

Prof. Mounir Kaidi, University of Sharjah

Prof. Tahar Laoui, University of Sharjah

Prof. Bashir Suleiman, University of Sharjah

Dr Hussein M. Elmehdi, University of Sharjah

Dr Rachik Soualah, University of Sharjah

**Scientific Committee**

Prof. Nouar Tabet, University of Sharjah, UAE, Chairman

Prof. Bashir Suleiman, Department of Applied Physics & Astronomy, University of Sharjah, UAE, Vice Chairman

Prof. Mounir Gaidi, Department of Applied Physics & Astronomy, University of Sharjah, UAE, Coordinator

Prof. Christophe Ballif, Photovoltaics and Thin Film Electronics Laboratory, EPFL, Switzerland

Prof. Zahid Hussain, Former Group Leader at Advanced Light Source, Lawrence Berkeley National Laboratory, USA

Prof. Abdul Ghani Olabi, Sustainable Energy & Power Systems Research Centre, University of Sharjah, UAE

Dr Aaesha Abdulla Alnuaimi, Director of Solar Innovation Center, DEWA, UAE

Prof. Hussain Alawadhi, Center for Advanced Materials Research, University of Sharjah, UAE

Prof. Elhadj Dogheche, Institute of Electronics, Microelectronics and Nanotechnology, CNRS, France

Prof. Tahar Laoui, Mechanical and Nuclear Engineering Department, University of Sharjah, UAE

**Organizing Committee**

Dr. Hussein M. Elmehdi, University of Sharjah, UAE, Chairman

Fatima Sajwani, University of Sharjah, UAE, Vice Chairman

Dr Yassir A. Abdu, Department of Applied Physics and Astronomy, University of Sharjah, UAE

Dr Mohammad Azeem, Department of Applied Physics and Astronomy, University of Sharjah, UAE

Dr Rachik Soualah, Department of Applied Physics and Astronomy, University of Sharjah, UAE

Dr. Anissa M. Bettayeb, Head of Academic Computing Services Section, University of Sharjah, UAE

Nawal Nayfeh, Department of Applied Physics and Astronomy, University of Sharjah, UAE

Racha Al Khoury, Department of Applied Biology, University of Sharjah, UAE

Omar Adwan, Department of Applied Physics and Astronomy, University of Sharjah, UAE

Khalid Al Raboy, Director of Public Relations Department, University of Sharjah, UAE

Journal of Physics: Conference Series **2751** (2024) 011002 doi:10.1088/1742-6596/2751/1/011002

# Peer Review Statement

All papers published in this volume have been reviewed through processes administered by the Editors. Reviews were conducted by expert referees to the professional and scientific standards expected of a proceedings journal published by IOP Publishing.

- **Type of peer review:** Undefined
- **Conference submission management system:** Morressier
- **Number of submissions received:** 42
- **Number of submissions sent for review:** 37
- **Number of submissions accepted:** 31
- **Acceptance Rate (Submissions Accepted / Submissions Received × 100):** 73.8
- **Average number of reviews per paper:** 2
- **Total number of reviewers involved:** 24
- **Contact person for queries:**
  **Name:** Nouar Tabet
  **Email:** ntabet@sharjah.ac.ae
  **Affiliation:** University of Sharjah

Journal of Physics: Conference Series **2751** (2024) 012001 doi:10.1088/1742-6596/2751/1/012001

# Linear vs. Non-linear Behaviour in Ion Irradiation Nanostructuring of Nickel and Silicon Surfaces

**Mogtaba B Mekki[1,2] and Michael R C Hunt[1]**

[1] Department of Physics, Durham University, Durham DH1 3LE, U.K.

[2] Department of Physics, College of Engineering and Physics, King Fahd University of Petroleum and Minerals, Dhahran 31261, Saudi Arabia

**Email: mogtaba.b.mekki@durham.ac.uk , m.r.c.hunt@durham.ac.uk**

**Abstract**. Spontaneous nanometre-scale quasi-periodic ripple-like structures are formed at the surface of polycrystalline Ni films and Si(111) single crystal wafers by irradiation with a broad $Ar^+$ ion beam at room temperature and studied with Atomic Force Microscopy as a function of fluence. The development of these structures can be reproduced by numerical solution of a continuum equation describing the evolution of surface morphology under ion irradiation, using realistic coefficients derived from material properties. In particular, we demonstrate that differences observed in pattern formation on the two surfaces under the conditions studied, such as wavelength stability and exponential growth of interface width for the Ni surfaces compared with wavelength coarsening and interface width saturation on Si(111), can be understood in terms of a cross-over between linear and non-linear behaviours.

## 1. Introduction

Irradiation of a solid surface by a beam of ions with sufficiently high kinetic energy can lead to the removal of atoms from its surface, a process known as ion beam sputtering. Under certain conditions, it has been found that ion irradiation can lead to the formation of self-organized regular patterns such as nano-ripples (one-dimensional quasi-periodic nanostructures), nano-dots, and nano-holes (zero-dimensional nanostructures) [1-3]. Such self-organized structures can be used for a range of applications, including templates for the growth of metal nanowire and nanodot arrays [4, 5], a bioactive surfaces [6], patterns for evaluation of scanned probe microscopy tips [7], luminescent emitters [8], and highly optically absorbing/anti-reflective surfaces [9, 10]. Producing nanostructures by this approach has several advantages in comparison with focused ion beam patterning [11], electron beam lithography and optical lithography [12] due to its low cost and flexibility. In particular, ion-beam induced nanopatterning involves a single step and only requires a broad ion beam, with the result that alignment and focusing are not significant issues.

Unfortunately, there are some drawbacks associated with this process, such as a lack of predictability as to which conditions are appropriate for the formation of a desired structure of selected pitch/wavelength, as our understanding of the mechanism of self-organisation is far from complete. An in-depth grasp of how to optimise primary experimental parameters such as the ion energy, incidence angle, dose, flux, sample temperature, sample type (i.e. amorphous, polycrystalline or crystalline) and sample rotation during irradiation is a key requirement for future application. Such knowledge requires continued development of the theoretical basis for the formation process of these induced self-organized surfaces.

Significant theoretical efforts have been expended to explain ion bombardment-induced periodic nanostructures on various substrates. The first successful theoretical approach for the understanding of induced self-organized ripple formation was formulated by Bradley and Harper [13] (known as the BH continuum model). The linear BH partial differential equation describes the development of self-organised ripple structures as resulting from the balance between two competing mechanisms: sputter-induced surface roughening, which can be understood as producing a negative effective surface tension, and smoothing due to an isotropic thermal surface relaxation originating from thermally-induced diffusion. Although this linear continuum theory was successful in the prediction of many features of ripple formation, it has several shortcomings: (1) it does not predict the formation of ripples at low temperature, where thermal diffusion is negligible (e.g., on most surfaces at room temperature); (2) it does not correctly foretell the surface evolution at long time scales, in particular the experimentally observed power-law coarsening of ripple wavelength; (3) it does not predict the saturation of the interface width, $w$, at high ion fluence; (4) it does not predict the observed rotation of ripple orientation by 90° at high ion dose and temperature; (5) and it cannot explain the formation of nano-dots (quantum dots) and nano-holes [7, 8, 14-16].

To address these failings several theoretical models have been developed which extend the BH model [17-20]. In particular, Makeev, Cuerno and Barabasi (MCB) [20] derived a continuum equation to describe the self-organisation process by considering the microscopic elementary processes occurring during sputter-induced surface morphology evolution which led to inclusion of higher-order derivatives of the local surface height. This model successfully explained the formation of quasi-periodic structures at temperatures below which the thermal surface relaxation mechanism can play a sufficient role to counterbalance roughening through introducing a non-thermal effective ion-induced diffusion [21]. In this model, the temperature-independent fourth derivative terms in the description of the rate of evolution of the surface topography (i.e., local surface height) play the same role of surface smoothing as conventional diffusion but without involving any mass transport. The continuum equation developed by Makeev and co-authors addressed many of the limitations of the BH continuum equation such as successfully describing the rotation of ripple orientation at high incident beam angle and high fluence, ripple saturation, and ripple formation at room temperature.

Although considered for the case of nanodot formation [22], to the best of our knowledge there is currently no direct comparative work between the results of experiment and theory regarding ion irradiation-induced ripple formation presented in the literature, which we address in this work. The experimentally-observed evolution of the surface topography of polycrystalline Ni thin films and Si(111) surfaces under inert gas ion irradiation as a function of ion fluence are compared with the MCB model, which we solve numerically. In particular, the coefficients for the various terms in the MCB equation are derived either from experimentally determined parameters or the results of simulations of the sputtering process using the Stopping and Range of Ions in Matter (SRIM) packages [23]. Calculations of interface width and wavelength show the same general trends as the experimental data. Moreover, we find that, over the range of irradiation conditions studied the experimental results, particularly ripple wavelength and interface width, can be modelled in purely linear terms for the case of Ni but require the inclusion of non-linear terms if the behaviour of the Si surface is to be correctly described. These differences can be explained by the values of the characteristic "cross-over time", $\tau_y$ , between linear and nonlinear regimes which we find numerically for both materials from the Makeev model. It would be expected that the higher surface erosion rate and lower ion penetration depth in Ni compared with Si would favour non-linearity in the former case, however the difference in incident ion beam angle between experiments on the two surfaces is found to dominate, indicating the delicate balance of factors governing ion-induced pattern formation and the need to carefully consider all experimental variables when comparing measurements to theoretical predictions in these systems.

## 2. Methods

Samples consisting of virgin Si(111) single crystal wafers and 100 nm thick polycrystalline Ni films deposited by thermal evaporation on Si(111) substrates were cut to $1\times1$ cm$^2$ and introduced into a home-built ultra-high vacuum (UHV) system (base pressure $1\times10^{-7}$ mbar) at Durham University where they were irradiated with a defocussed $Ar^+$ beam produced by a cold cathode ion source (Vacuum Generators AG21) such that the ion beam uniformly illuminated the surface of the samples. Ion bombardment was undertaken at ion energy, $E_{ion}$, of 5.0 keV, ion flux, $f$, of $2.1\times10^{13}$ ions cm$^{-2}$ s$^{-1}$ and an incident angle, $\theta$, of 75° with respect to the surface normal for the Ni surfaces. For Si(111) the irradiation conditions were $E_{ion}$ = 3.8 keV, $f$ = $1.6\times10^{13}$ ions cm$^{-2}$ s$^{-1}$ and $\theta$ = 60°. In both cases, a range of ion fluences was studied by varying the total irradiation time at fixed fluence. The chamber pressure typically increased to $1\times10^{-5}$ mbar during ion bombardment and the current incident upon the sample was measured directly using a Keithley 445 electrometer.

After ion bombardment samples were removed from the UHV system and imaged, in Durham, by Atomic Force Microscopy (AFM) under ambient atmosphere using a Veeco NanoScope IIIa scanned probe microscope in tapping mode with sharp tapping etched silicon probes (TESP). Care was taken to image all samples at a range of scan angles, scan speeds and feedback loop conditions to confirm that all images of surface structures observed reflected the topography of the surfaces rather than experimental artefacts. Multiple different areas on each samples were imaged to ensure unformity. After imaging, analysis of the surface topography of both Ni and Si(111) samples was undertaken using Gwyddion [24] from which a number of key parameters were extracted, such as interface width and average surface height.

Surface structural evolution was explored using a special case of the MCB model. The linear erosion rate expected for ion sputtering of a planar surface, $-v_0$, along with first and third-order terms which contribute only to the motion of the quasi-periodic ripples parallel to the surface, but do not otherwise influence surface morphology [20], are neglected. The neglect of these terms is equivalent to a co-ordinate transformation into a frame of reference moving at the same velocity as the eroding surface along the (global) surface normal and at the velocity of propagation of the bombardment-induced surface features in the plane perpendicular to the normal and leads to equation 1, which is similar to the noiseless anisotropic Kuramoto-Sivashinsky equation [25]:

$$\frac{\partial h(x,y,t)}{\partial t} = v_x \frac{\partial^2 h}{\partial x^2} + v_y \frac{\partial^2 h}{\partial y^2} - D_{xx}\frac{\partial^4 h}{\partial x^4} - D_{xy}\frac{\partial^4 h}{\partial x^2 \partial y^2} - D_{yy}\frac{\partial^4 h}{\partial y^4} + \frac{\lambda_x}{2}\left(\frac{\partial h}{\partial x}\right)^2 + \frac{\lambda_y}{2}\left(\frac{\partial h}{\partial y}\right)^2 \quad (1)$$

where $v_x$, and $v_y$ are effective surface tension coefficients in the $x$ and $y$ directions respectively. $D_{xx}$ and $D_{yy}$ are the orthogonal effective ion-induced surface diffusion coefficients and $D_{xy}$ is the diagonal effective ion-induced surface diffusion coefficient. $\lambda_x$, $\lambda_y$ are lateral coefficients in $x$ and $y$ directions respectively. Values for these parameters were determined for the specific cases of Ni and Si using results determined from SRIM [23] simulations and the defintions of Makeev, Cuerno and Barabasi [20] to determine sputter yield, average ion penetration depth, longitudinal and transverse ion straggles.

For negative values of $v_x$, $v_y$ and positive values of $D_{xx}$, $D_{yy}$ periodic ripples will form with wavevector oriented along the $x$ or $y$ direction. The characteristic wavelength, $\ell$ of these ripples is given by:

$$\ell = \frac{2\pi}{k_{max}} = 2\pi\sqrt{\frac{2\times\min(D_{xx}, D_{yy})}{\max|v_x, v_y|}} \quad (2)$$

where $k_{max}$ is the fastest growing wavevector.

Numerical integration was performed using Exponential Time Differencing (ETD) with Runge–Kutta accuracy of order 4 (ETDRK4). Discretization of equation 1 was performed using the Spectral Fourier Method. Since equation 1 is categorized as a stiff semi-linear partial differential equation, we choose a small temporal increment $\Delta t$ = 0.01 to get a stable numerical solution and impose periodic boundary conditions $h(x, y, t) = h(x + L_x, y, t) = h(x, y + L_y, t)$, where the domain size is $L_x \times L_y = 180\times180$ nm$^2$. The initial surface was selected to be a (Gaussian) randomly rough surface with a local surface height

Journal of Physics: Conference Series **2751** (2024) 012001 doi:10.1088/1742-6596/2751/1/012001

uniformly distributed between [-0.04, 0.04] nm with a zero average. The computational codes were written in Python and run on the Hamilton supercomputer at Durham University.

## 3. Results and Discussion

AFM topographs illustrating the evolution of the surfaces of the polycrystalline Ni thin film and Si(111) surface with ion fluence are presented in Figure 1 and Figure 2. The direction of the incident ion beam, projected onto the sample surface, is illustrated by an arrow in each case. For both surfaces the formation of one-dimensional quasi-periodic structures – or ripples – is clearly apparent. Two possible ripple orientations can be predicted theoretically [13, 20] known as "parallel mode" and "perpendicular mode" ripples, depending on the orientation of the ripple wavevector with respect to the projection of the ion beam direction onto the surface. Defining the direction of the wave-vector of the ripple structures, $k_{max}$, at right angles to the average direction of the "wave-fronts" it can be seen that perpendicular-mode ripples form in both cases.

As fluence increases the wavelength, $\ell$, of the ripples formed on the Ni film surface initially remains constant, as shown in Figures 1(a-c) and Figure 3(a), followed by a sudden and abrupt increase in value, Figure 1(d) which is accompanied by a reduction in interface width, Figure 3(c), which otherwise increases exponentially. Examination of the thin film samples post-irradiation indicated that at fluences beyond approximately $5\times10^{16}$ ions cm$^{-2}$ the 100 nm thick Ni film was completely eroded and the transition in wavelength and interface width reflects the formation of periodic structures in the Si(111) substrate supporting the Ni film. In contrast to the constant wavelength of the self-organised ripples formed on polycrystalline Ni, $\ell$ is seen to change significantly during irradiation of the Si(111) surface, coarsening as a function ion fluence, Figure 2(a-d) and Figure 3(b). The variation of wavelength with ion fluence can be fitted with a power-law dependence $\ell \sim \phi^n$ (red curve, Figure 3(b)) with a coarsening exponent $n \approx 0.2$. Likewise, the evolution of the interface width on the bare Si(111) surface (Figure 3(d)) is very different to the exponential increase observed for polycrystalline Ni and is observed to saturate at large fluence. A somewhat similar saturation of interface width by Ziberi and co-workers [14] for irradiation of Si(100) surfaces. The similarity between the behaviours of the (100) and (111) surfaces is to be expected due to amorphization under ion bombardment, in comparison with the facet-dependent structural evolution seen on some metal surfaces [1]. Note that we required a rather higher range of fluence to observe structures on the Si(111) surface in comparison with the polycrystalline Ni films and the interface width remains small.

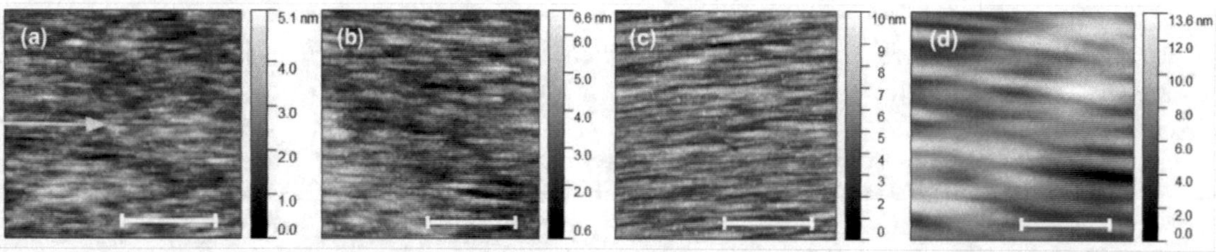


**Figure 1.** AFM topographs of pattern formation in polycrystalline Ni films on Si(111) produced under $Ar^+$ ion bombardment ($E_{ion}$ = 5.0 keV, ion flux $f = 2.1\times10^{13}$ ions cm$^{-2}$ s$^{-1}$, incident angle $\theta = 75°$, at room temperature) as a function of fluence: (a) $1.1\times10^{16}$ ions cm$^{-2}$, (b) $1.6\times10^{16}$ ions cm$^{-2}$, (c) $3.5\times10^{16}$ ions cm$^{-2}$, and (d) $1.2\times10^{17}$ ions cm$^{-2}$ (at which point the Ni is removed and the Si substrate exposed). The scale bar in each image is 400 nm in length. The projection of the ion beam onto the target surface is indicated by the yellow arrow in panel (a).

The behaviour of the surface topography of the polycrystalline Ni film under ion irradiation, at least until the Si(111) substrate is reached, is consistent with the linear BH model which predicts a constant ripple wavelength and exponentially increasing interface width [13]. However, the ripple coarsening observed for irradiation of the bare Si(111) surface is similar to that observed for medium energy ion irradiation of Si(100) in the work of Datta and Chini [26] and requires the inclusion of non-linear terms in the description of surface evolution under irradiation, as does the saturation of interface width shown in Figure 3(d).

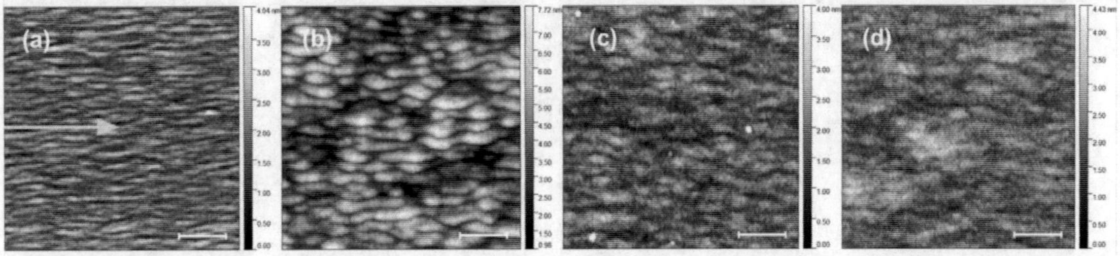


**Figure 2.** AFM topographs of pattern formation on Si(111) due to $Ar^+$ ion bombardment ($E_{ion}$ = 3.8 keV, ion flux $f$ = 1.6×10$^{13}$ ions cm$^{-2}$ s, incident angle $\theta$ = 60°, at room temperature) as a function of fluence: (a) 1.7×10$^{17}$ ions cm$^{-2}$, (b) 2.3×10$^{17}$ ions cm$^{-2}$, (c) 3.5×10$^{17}$ ions cm$^{-2}$ and (d) 4.6×10$^{17}$ ions cm$^{-2}$. The scale bar in each image is 400 nm in length. The projection of the ion beam onto the target surface is indicated by the yellow arrow in panel (a).

To explain the differences in the evolution of wavelength and interface width between the two materials, and in particular to understand the differences in the relative importance of non-linear terms between the two systems the time-evolution of the local surface height, $h(x,y,t)$ was determined by numerical integration of equation 1 using parameters derived for both the experimentally studied systems, Ni and Si, as described above. The sputtering yields for Ni and Si surfaces as a function of incident ion beam energy and angle, obtained by carrying out SRIM simulations [23], were used to produce erosion rate factors, $F$, which are corrected from those presented by Makeev *et al.* [20] to accurately describe the variation of sputter yield at high angles found from the SRIM simulations. The full set of parameters used for equation 1, derived for each of the materials, is shown in Table 1.

**Table 1.** Calculated coefficients used in the numerical solution of the MCB model.

| | Ni | Si |
|---|---|---|
| Target density, $\rho_N$ (atoms/cm$^3$) | 9.1×10$^{22}$ | 5.0×10$^{22}$ |
| Average ion penetration depth[a], $a$ (nm) | 3.3 | 7.6 |
| Sputtering yield[a], $Y$(atoms/ion) | 13.15 | 5.90 |
| Corrected erosion rate factor[b], $F$(nm/s) | 0.0303 | 0.0189 |
| Effective surface tension coefficient in the $x$-direction, $\nu_x$ (nm$^2$/s) | 0.0663 | -0.0068 |
| Effective surface tension coefficient in the $y$-direction, $\nu_y$ (nm$^2$/s) | -0.0031 | -0.0148 |
| Effective ion surface diffusion coefficient in the $x$-direction, $D_{xx}$ (nm$^4$/s) | -0.2598 | 0.1065 |
| Effective ion surface diffusion coefficient in the $y$-direction, $D_{yy}$ (nm$^4$/s) | 0.0023 | 0.0282 |
| Effective ion surface diffusion coefficient in the $xy$-direction, $D_{xy}$ (nm$^4$/s) | -0.0775 | 0.1899 |
| Lateral $x$-direction coefficient, $\lambda_x$ (nm/s) | 0.0375 | 0.0112 |
| Lateral $y$-direction coefficient, $\lambda_y$ (nm/s) | -0.0011 | -0.0069 |

[a] Obtained from SRIM simulations, [b]correction based on SRIM results

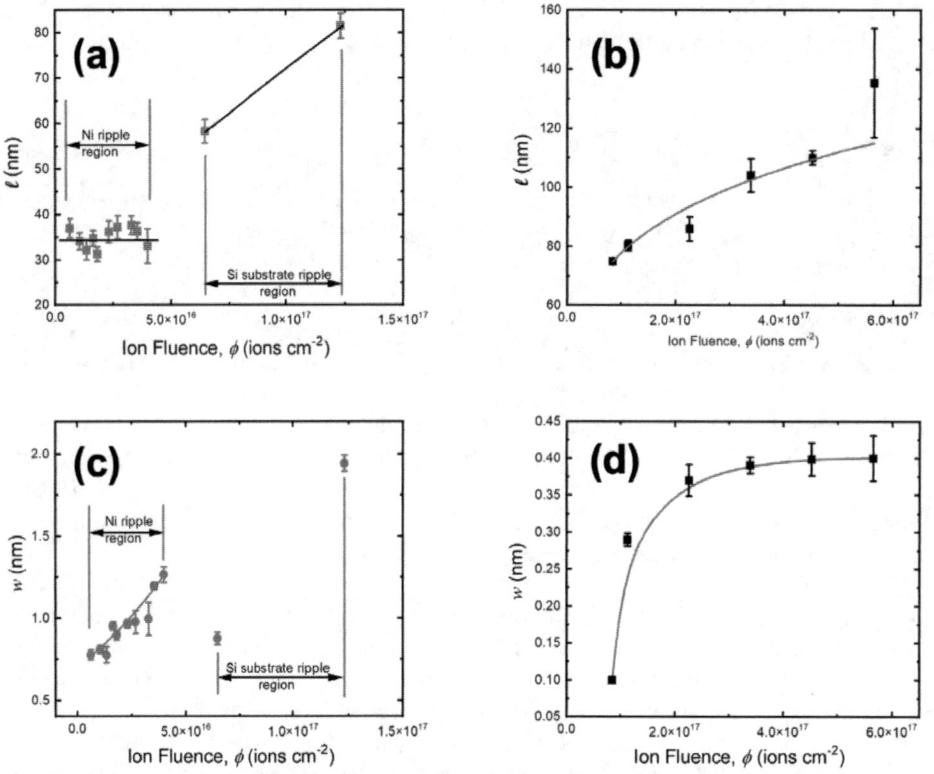


**Figure 3.** Evolution of ripple wavelength for (a) polycrystalline Ni and (b) Si(111) surfaces and interface width, $w$, for (c) polycrystalline Ni and (d) Si(111) as a function of ion fluence. Note the sudden transition in both wavelength and interface width in (a) and (c) at the point the Ni film is completely eroded and the Si substrate reached. The ripple wavelength on Si(111) shows a power law dependence $\ell \sim \phi^n$, with $n \approx 0.2$, as shown by the red line in (b). In the early stages of ripple formation on polycrystalline Ni the increase in interface width increases exponentially, as shown by the fitted line in (c). The line in (d), indicating saturation of the interface width on Si(111) is a guide to the eye. Irradiation conditions were: ion energy $E_{ion} = 5.0$ keV, ion flux $f = 2.1\times10^{13}$ ions cm$^{-2}$ s$^{-1}$, incident angle $\theta = 75°$ and room temperature for Ni; ion energy $E_{ion} = 3.8$ keV, ion flux $f = 1.6\times10^{13}$ ions cm$^{-2}$ s$^{-1}$, ion angle $\theta = 60°$ and room temperature for Si(111).

Numerical solutions for surface topography as a function of iteration (equivalent to increasing fluence) for the Ni surface are presented in Figure 4 and those for Si(111) in Figure 5. For the Ni surface, the ripple wavelength, $\ell$, remains constant even after approximately 10,000 iterations, the increasing number of computational cycles up to this point leading only to more clearly-defined and ordered structures. Figure 6(a) shows, for the Ni surface the evolution of the ripple wavelength with computational iterations for selected iteration times. The wavelength of the ripple features is initially constant then, after a certain computational time begins to coarsen. This behaviour can be understood in terms of a regime of linear behaviour below a "cross-over time", $\tau_y \sim 1 \times 10^6$ iterations (or equivalently, for fixed flux, a cross-over fluence, $\phi_c$), in which the contribution of non-linear terms in equation (1) are negligible, followed by a regime dominated by non-linear terms for $t > \tau_y$ [27]. However, in our experiment the latter case can only be reached at fluences beyond that at which we completely erode the Ni film, Figure 1 and Figure 3(a).

The behaviour of the interface width, *w*, and the average interface height, $\bar{h}(t)$, is shown in Figure 6(c). $\bar{h}(t)$ is defined as:

$$\bar{h}(t) = \frac{1}{NM}\sum_{y=0}^{y=N-1}\sum_{x=0}^{x=M-1}(h(x,y,t) - h_{planar}(t)) \quad (3)$$

where $h(x,y,t)$ is the local interface height at (*x*,*y*) and time *t* and $h_{planar}(t)$ is the (decreased) height which would be expected due to sputter erosion of a uniform planar surface under the same irradiation conditions. Below $\tau_y$ the interface width, *w*, is seen to increase exponentially and $\bar{h}(t)$ remains zero, before *w* saturates and $\bar{h}(t)$ decreases rapidly at larger numbers of iterations. The decrease in $\bar{h}(t)$ can be understood in terms of an enhancement of erosion rate in comparison with a planar surface when non-linear terms in the MCB equation become dominant and the surface coarsens. Such variations in sputter yield have been observed theoretically by Makeev and Barabasi for both rough surfaces [28] and those with ion irradiation-induced periodic structures [29].

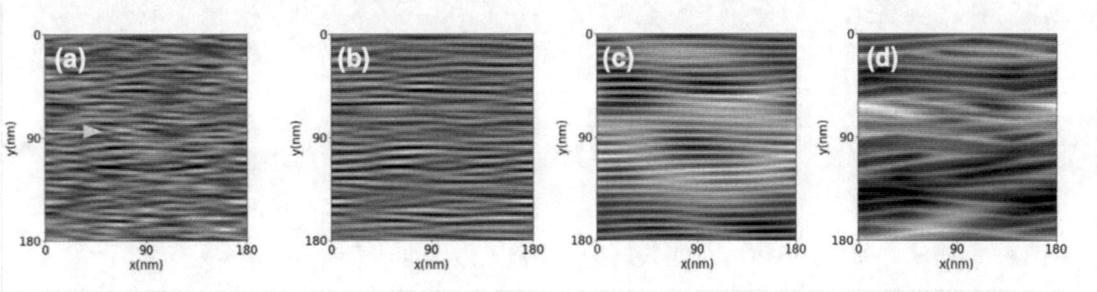


**Figure 4.** Simulated surface evolution during ion bombardment of Ni, under the same conditions as those employed experimentally, shown at selected computational iterations (cycles): (a) $5.0\times10^4$, (b) $2.0\times10^5$, (c) $1.0\times10^6$, and (d) $1.9\times10^6$ iterations. The projection of the ion beam onto the target surface is indicated by the arrow in (a) and is the same in all panels.

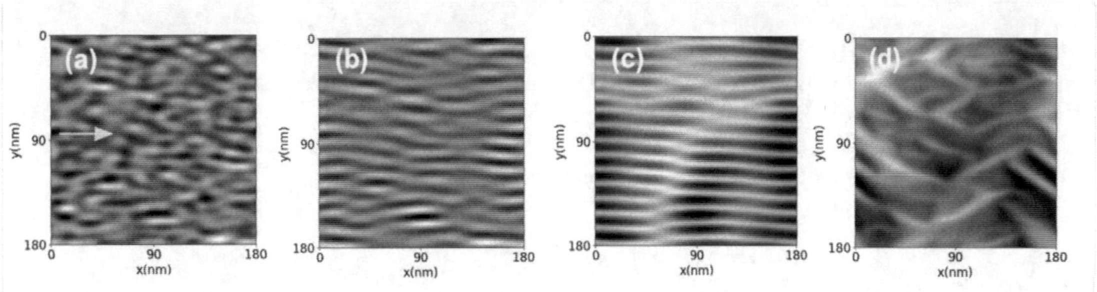


**Figure 5.** Simulated surface evolution during ion bombardment of Si, under the same conditions as those employed experimentally, shown at selected computational iterations (cycles): (a) $5.0\times10^4$, (b) $2.0\times10^5$, (c) $6.0\times10^5$, and (d) $1.9\times10^6$ iterations. The projection of the ion beam onto the target surface is indicated by the arrow in (a) and is the same in all panels.

A similar computational analysis of the evolution of the Si(111) surface, Figures 6(b) and 6(d), shows qualitatively similar behaviour of $\ell$, *w* and $\bar{h}(t)$, with the key difference that the cross-over time between the dominance of linear and non-linear terms occurs earlier at $\tau_y{\sim}6 \times 10^5$ iterations. The earlier cross-over time for the Si(111) surface coupled with the higher range of fluences employed experimentally

($1.7\times10^{17}$ ions cm$^{-2}$ to $4.6\times10^{17}$ ions cm$^{-2}$ for Si(111) compared with $1.1\times10^{16}$ ions cm$^{-2}$ to $3.5\times10^{16}$ ions cm$^{-2}$ for polycrystalline Ni) explains the significant differences in the behaviour of interface width and wavelength observed.

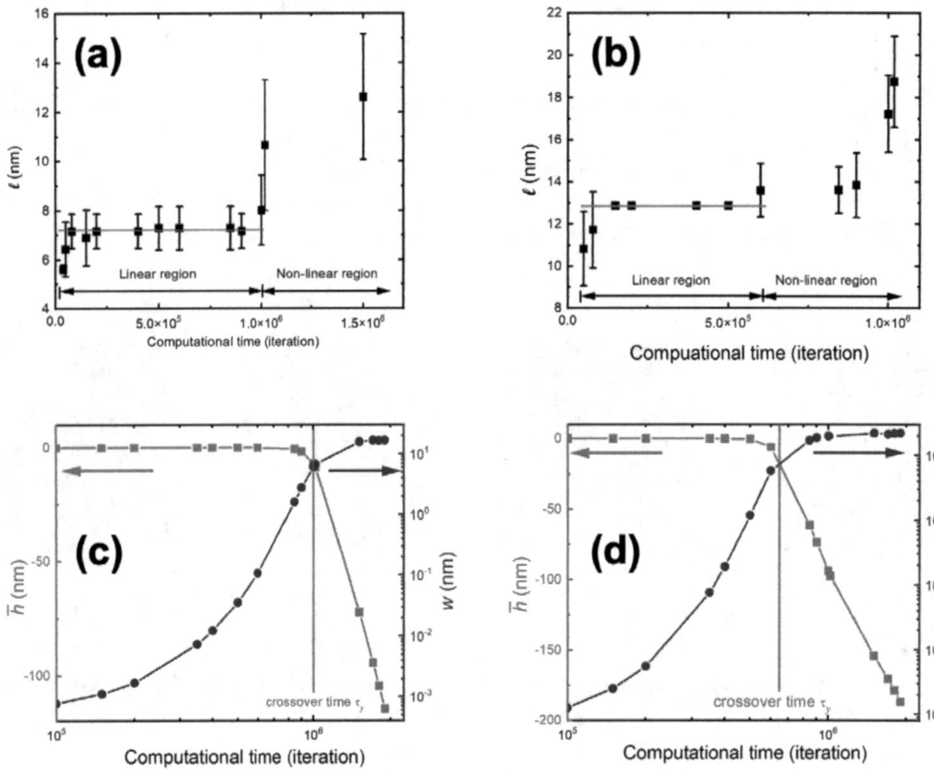


**Figure 6.** Results from simulations of surface evolution: Ripple wavelength as a function of computational time (iterations) for (a) Ni and (b) Si; Variation of mean surface height, $\bar{h}(t)$, and interface width, $w$, for (c) Ni and (d) Si.

By considering equation (1), the cross-over time can be defined as [27]:

$$\tau_y \sim \frac{D_{yy}}{\nu_y^2} \ln\left(\frac{\nu_y}{\lambda_y}\right) = \frac{a}{F} Z(\theta, a, a_\alpha, a_\beta) \quad (4)$$

where $D_{yy}$, $\nu_y$, $a$ and $F$ are defined in Table 1 and $Z$ is a function of $a$, ion beam incidence angle, $\theta$, and $a_\alpha$, $a_\beta$ which are the ratios of the longitudinal and lateral ion straggles to the average ion range, respectively. From the examination of the material-dependent terms $a$ and $F$ it might be initially thought that the cross-over time for Si should be greater than that of Ni due to the larger average ion penetration depth and smaller erosion rate factor (Table 1). This would certainly be the case if the experiments were performed with identical incident ion beam angles. However, the difference of 15° between the two sets of measurements and the resulting change in $Z$ is enough to reverse the order at which the cross-over time is reached, indicating the delicate dependence of ion beam-induced pattern formation on the conditions of irradiation as well as the intrinsic properties of the surfaces and the ions incident upon them.

A comparison of the experimentally determined values of ripple wavelength for the polycrystalline Ni surface, Figure 3(a) with the linear region of the computationally-determined wavelength for the same surface, Figure 6(a) shows that the latter is systematically smaller by a factor of ~5. Likewise, a comparison of the experimental interface width obtained on the silicon surface at saturation, Figure 3(d), with the computationally determined value, Figure 6(d) shows that the latter is a factor of ~50 smaller. Given the explicit use of materials parameters to determine the coefficients of equation (1) the disagreement between the calculated evolution of the silicon and nickel surfaces under ion bombardment and those observed experimentally point to either inaccuracies in determining the ion straggle when using TRIM, as suggested by Facsko and co-workers [22], or additional processes which have not been explicitly incorporated into the description modelled by equation (1). We can expect that our neglect of thermally induced diffusion at room temperature is unlikely to impact significantly in describing surface evolution at room temperature. However, the current model neglects the influence of viscous surface relaxation and flow [20, 30, 31] and ion-induced mass transport [32] which, through changing the balance between competing processes of roughening and smoothing which lead to ripple formation, would influence the theoretically determined values for ripple wavelength and interface width.

## 4. Conclusions

Experimental observations of the formation of quasi-periodic patterns (ripples) due to $Ar^+$ ion irradiation of the surfaces of two different materials, polycrystalline Ni and Si(111), show significant differences. In particular, ripples formed on polycrystalline Ni display a wavelength which is constant with fluence and an interface width (amplitude) which increases exponentially, whereas the structures formed on Si(111) coarsen with increase irradiation and show amplitude saturation.

Numerical simulations performed by solving the continuum equation for surface evolution under ion irradiation derived by Makeev, Cuerno and Barabasi (MCB) using realistic coefficients derived from the physical properties of the two materials and the results of calculated ion irradiation parameters show a good agreement with experimental trends. The simulations indicate that, under the conditions studied, the differences between experimentally observed pattern formation on polycrystalline Ni and Si(111) surfaces can be understood in terms of differing values of the cross-over time from behaviour dominated by linear to that controlled by non-linear terms in the MCB equation [20, 27]. Moreover, our results indicate the subtle interplay between materials properties, ion beam and incident beam angle in determining the fluence (or time) evolution of ripple structures under ion bombardment.

Numerical modelling using the MCB equation with coefficients derived for specific physical systems and experimental validation as reported in this work paves the way for quantitative predictions of the evolution of ripple-like surface structures produced by ion irradiation. However, it is clear from the differences between experimentally and theoretically determined values of wavelength and interface width that further work is required to fully capture all contributions to morphological evolution in a quantitative continuum model. Fully understanding the conditions for the formation of specific surface structures with selected amplitude and wavelength should ultimately enable ion bombardment to be used as an effective tool for the application-specific tailoring of surfaces.

**Acknowledgement** Mogtaba B Mekki acknowledges the support provided for this work by Durham University, UK and would like to acknowledge KFUPM, Saudi Arabia. Both authors thank Prof. Del Atkinson for the use of facilities for growth of Ni films.

## References

[1] Chan W L and Chason E 2007 *J. Appl. Phys.* **101** 121301.
[2] Aste T and Valbusa U 2005 *New J. Phys.* **7** 122.
[3] Valbusa U, Boragno C and Mongeot F 2002 *J. Phys. Cond. Mat.* **14** 8153-75
[4] Toma A, Chiappe D, Massabò D, Boragno C and Buatier de Mongeot F 2008 *Appl. Phys. Letts.* **93** 163104.

[5] Oates T W H, Keller A, Noda S and Facsko S 2008 *Appl. Phys. Letts.* **93** 063106.
[6] Dell'Anna R, Masciullo C, Iacob E, Barozzi M, Giubertoni D, Böttger R, Cecchini M and Pepponi G 2017 *RSC Adv.* **7** 9024-30
[7] Frost F, Hirsch D and Schindler A 2001 *Appl. Surf. Sci.* **179** 8-12
[8] Facsko S, Dekorsy T, Koerdt C, Trappe C, Kurz H, Vogt A and Hartnagel H 1999 *Science* **285** 1551-3
[9] Bhowmik D and Karmakar P 2018 *AIP Conf. Proc.* **1953** 100071.
[10] Datta D P and Som T 2021 *Solar Energy* **223** 367-75
[11] Xu Z, Fu Y, Han W, Wei D, Jiao H and Gao H 2016 *Curr. Nanosci.* **12** 696-711
[12] Knight S, Prabhu V M, Burnett J H, Liddle J A, Soles C L and Diebold A C, *Nanoelectronics Lithography*, in *Handbook of Nanophysics*, Sattler K D, Editor. 2011, CRC Press: Boca Raton, U.S.A. p. 19-1 - 35.
[13] Bradley R M and Harper J M E 1988 *J. Vac. Sci. Technol. A* **6** 2390-5
[14] Ziberi B, Frost F, Höche T and Rauschenbach B 2005 *Phys. Rev. B* **72** 235310.
[15] Frost F, Schindler A and Bigl F 2000 *Phys. Rev. Letts.* **85** 4116-9
[16] Brown A-D and Erlebacher J 2005 *Phys. Rev. B* **72** 075350.
[17] Shipman P D and Bradley R M 2011 *Phys. Rev. B* **84** 085420.
[18] Castro M, Muñoz-García J, Cuerno R, Hernández M d M G and Vázquez L 2007 *New J. Phys.* **9** 102.
[19] Vogel S and Linz S J 2005 *Phys. Rev. B* **72** 035416.
[20] Makeev M A, Cuerno R and Barabási A L 2002 *Nucl. Instrum. Meth. B* **197** 185-227
[21] Makeev M A and Barabási A-L 1997 *Appl. Phys. Letts.* **71** 2800-2
[22] Facsko S, Kurz H and Dekorsy T 2001 *Phys. Rev. B* **63** 165329.
[23] Ziegler J F, Ziegler M D and Biersack J P 2010 *Nucl. Instrum. Meth. B* **268** 1818-23
[24] Nečas D and Klapetek P 2012 *Cent. Eur. J. Physics* **10** 181-8
[25] Cuerno R and Barabasi A L 1995 *Phys. Rev. Letts.* **74** 4746-9
[26] Datta D P and Chini T K 2007 *Phys. Rev. B* **76** 075323.
[27] Park S, Kahng B, Jeong H and Barabási A L 1999 *Phys. Rev. Letts.* **83** 3486-9
[28] Makeev M and Barabasi A L 2004 *Nucl. Instrum. Meth. B* **222** 316-34
[29] Makeev M A and Barabási A-L 2004 *Nucl. Instrum. Meth. B* **222** 335-54
[30] Umbach C C, Headrick R L and Chang K C 2001 *Phys. Rev. Letts.* **87** 246104
[31] Mayer T M, Chason E and Howard A J 1994 *J. Appl. Phys.* **76** 1633-43
[32] Carter G and Vishnyakov V 1996 *Phys. Rev. B* **54** 17647-53

# Investigating antitumor therapeutic efficacy using magnetic hyperthermia of $Fe_3O_4$ nanoparticles

**Venkatesha Narayanaswamy[1], Jayalakshmi Jagal[1], Imaddin A. Al-Omari[2], Hafsa Khurshid[3], Mohamed Haider[1,4], Ihab M. Obaidat[5] and Bashar Issa[1,6,7*]**

[1]Research Institute of Medical & Health Sciences, University of Sharjah, Sharjah 27272, UAE.
[2]Department of Physics, Sultan Qaboos University, P.O. Box 36, Muscat PC 123, Sultanate of Oman.
[3]Department of Physics, College of Sciences, University of Sharjah, Sharjah, P.O. Box 27272, UAE.
[4]Department of Pharmaceutics and Pharmaceutical Technology, College of Pharmacy, University of Sharjah, Sharjah 27272, UAE.
[5]Department of Physics, College of Sciences, University of Sharjah, Sharjah P.O. Box 27272, UAE.
[6]Department of Medical Diagnostic Imaging, College of Health Sciences, University of Sharjah, Sharjah, P.O. Box 27272, UAE.
[7]Department of Biomedical Engineering, Faculty of Engineering and Natural Sciences, Istinye University, Istanbul, 34010, Turkey
*Correspondence: bissa@sharjah.ac.ae

**Abstract:** Four sets of $Fe_3O_4$ nanoparticles were synthesized using a one-pot hydrothermal method using ethylene glycol (EG) and water reaction mixture. The EG coated $Fe_3O_4$ nanoparticles were characterized using XRD (X-ray diffraction) and DLS (dynamic light scattering). The nanoparticles hydrodynamic sizes are in the range 60-100 nm and showed good dispersibility in water and cellular medium without any aggregation. The specific absorption rate (SAR) under an alternating magnetic field (AMF) was measured to evaluate the magneto hyperthermic effect of the nanoparticles under field amplitudes of 51.85, 43.87, 35.89, and 27.92 kA/m and field frequencies in the range 384.5-167.30 kHz. The nanoparticles synthesized using 30% EG showed the highest SAR value of 372.95 W/g, whereas the particles synthesized using 70% EG showed a SAR value of 217.12 W/g. SAR values show a linear dependency on the AC frequency whereas the field amplitude displayed a nonlinear effect on the SAR. Breast cancer cells (MDA) treated with the nanoparticles for 30 minutes under an AMF of frequency of 384.5 kHz and strength of 24.67 kA/m showed a 50% decrease in cell viability. The cellular viability further decreased to 25% after 24 h treatment under the AMF which is remarkable for the therapeutic application of the nanoparticles. The apoptotic cell death showed a dependency on the AMF frequency and strength combinations.

## 1. Introduction

The use of magnetic nanoparticles (MNPs) in an alternating magnetic field (AMF) to generate heat and selectively destroy cancer cells through hyperthermia is a potential cancer treatment method [1,2]. Due to their high biocompatibility and minimal collateral damage to healthy tissue, $Fe_3O_4$ magnetic nanoparticles have drawn the most attention among different magnetic nanoparticles [3]. Magnetic nanoparticles are also examined for drug delivery and targeting tumor cites using an external magnetic field, biosensors to detect specific biomolecules, contrast agents in MRI allowing for enhanced imaging of specific tissues, separation and purification of biomolecules and cells, magnetic scaffolds for tissue engineering, and magnetic cell labeling [4–8]. Magnetic nanoparticles generate localized heating in tissues under an AMF [9]. The technique has been explored as a potential treatment for cancer, as it has been shown to induce tumor cell death by hyperthermia-mediated apoptosis and necrosis [1]. The saturation magnetization, shape, Neel relaxation, and Brownian relaxation of the nanoparticles are some of the factors that affect the magnetic heating capabilities of the particles [10,11]. Synthetic conditions are used to optimize the properties by employing different synthesis methods like organometallic decomposition, hydrothermal, and microwave assisted synthesis [12,13]. Synthesizing high efficiency nanoparticles for magnetic heating is essential to develop therapeutic agents for magnetic hyperthermia. Designing appropriate nanoparticles for the treatment of malignancies requires a greater understanding of the cell death mechanism triggered via hyperthermia. Numerous studies have shown that magnetic hyperthermia causes apoptosis in cancer cells by activating caspase, upregulating pro-apoptotic proteins, and downregulating anti-apoptotic BCL-2 proteins, as well as by heating cancer cells to high temperatures [14]. Furthermore, it has been demonstrated that magnetic hyperthermia causes oxidative stress, which produces reactive oxygen species (ROS), which then causes cancer cells to undergo apoptosis [15,16]. This study is intended to synthesize the nanoparticles with high heating efficiency using the hydrothermal method with EG acting as a co-solvent and surfactant. Furthermore, nanoparticles efficiency is examined for the killing of cancer cells with respect to frequency and field amplitude of AMF. The cell death mechanism is investigated using the flow cytometry and MTT method with two different AMF parameters.

## 2. Experimental

### 2. 1 Materials and Methods

$Fe_3O_4$ nanoparticles were synthesized by a one-step hydrothermal method using a mixture EG and water solvent. 2.8 g of anhydrous iron chloride $FeCl_3$ (17.28 mmol) and iron chloride tetrahydrate $FeCl_2$. $4H_2O$ (8.64 mmol) was dissolved in 100 ml of water and EG reaction mixture. Four sets of reaction mixture are used with EG concentration 10, 30, 50, and 70 %. They are labeled as S1, S2, S3, and S4, respectively. To obtain a homogenous solution, the reaction mixture of salts and solvent is stirred at room temperature for 20 minutes. 1 N NaOH solution is added dropwise under constant stirring to adjust the pH to 12-13, and the precipitate produced is stirred continuously for 20 minutes. The reaction mixture is transferred to a Teflon-lined autoclave, which is sealed and heated in an oven at 180 ºC for 6 h. After the autoclave cooled to room temperature, the produced nanoparticles were separated with an external magnet and washed using distilled water. Using an IR lamp, the nanoparticles are dried before being used for further characterization and magnetic hyperthermia investigations to kill cancer cells.

### 2.2 Nanoparticles Characterization and Cell Culture

The diffraction pattern is obtained using an X-ray powder diffractometer (Bruker D8) with a Cu Ka radiation source (wavelength=1.5406 A, generator voltage=40 kV, and generator current=40 mA). The powder sample was scanned at a rate of 0.02º/0.1 s throughout a 2-theta range of 25–70º. The hydrodynamic diameter and zeta potential of EG coated nanoparticles are determined using the Anton Paar Litesizer DLS 500 using forward scattering angle. Cells from the MDA-MB-231 human breast cancer cell line (ATCC® HTB-26, ATCC, Manassas, VA, USA) were grown in RPMI and Mccoy's media conditions, which were both supplemented with 10% FBS and 1% penicillin/streptomycin, respectively. For further investigation, all cell lines were cultured at 37 oC with 5% CO2 and humidified air. After being washed with PBS and transferred to a 15-ml centrifuge tube containing medium at each time point, the cells were resuspended in 500 ml of binding buffer before being mixed with 5 µl of Annexin V-FITC and 5 µl of PI. Using BD Biosciences, San Jose, California, USA, flow cytometric analysis was carried out after 15 minutes.

## 3. Results and Discussions

*3.1 XRD*

Figure 1 displays the XRD patterns of the nanoparticles synthesized with varied EG percentages in the reaction mixture from the hydrothermal treatment. The ferrite nanoparticles characteristic Bragg diffraction peaks had 2-theta values of 30.19° (220), 35.57° (311), 43.22° (400), 53.47° (422), 57.10° (511), and 62.70° (440). These peaks are associated with the cubic spinel structure of ferrite and the Fd3m (227) space group, and the lattice parameters a = b = c = 8.3724 Å. The observed lattice parameters are consistent with the previously reported lattice parameter of $Fe_3O_4$ nanoparticles and were determined by multiple peak fitting approach utilizing Jade XRD. [13]. As indicated in Figure 1, there were no additional peaks of any other impurity phase in the XRD patterns. FWHM determined from the highest intensity peak (311) is used to determine the average crystallite sizes of the produced nanoparticles by the Debye-Scherrer formula [17]. The average sizes are in the range of 15-17 nm and depend on the percentage of EG and water mixture used for the synthesis. EG acts as a solvent and surfactant for controlling particle size which is evident from the decreasing particle size with the increase in EG percentage in reaction mixture. The average size and lattice parameters of the four sets of nanoparticles are listed in Table 1.

Journal of Physics: Conference Series **2751** (2024) 012002 doi:10.1088/1742-6596/2751/1/012002

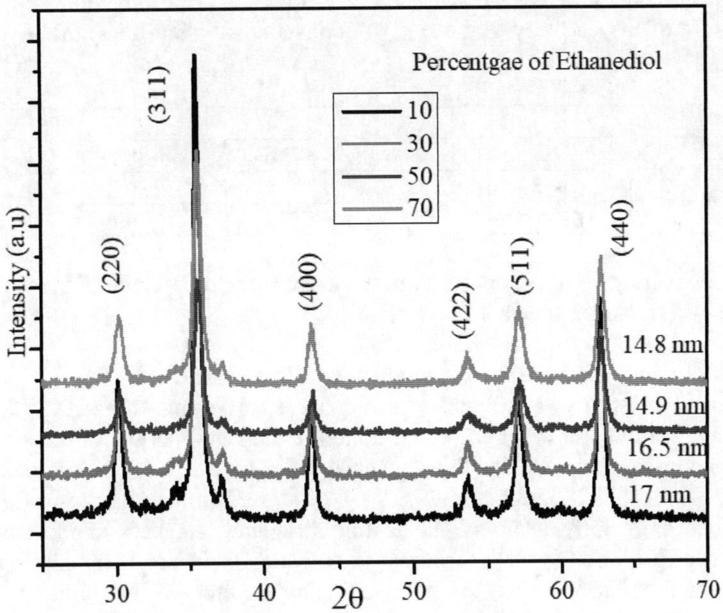


**Figure 1**. XRD patterns of the nanoparticles synthesized using hydrothermal method with ethylene glycol percentage 10, 30, 50, and 70% in the reaction mixture.

*3.2 Hydrodynamic diameter and Zeta potential using DLS.*

The dispersed nanoparticle hydrodynamic diameter and zeta potential are determined using the Anton Paar Litesizer with forward scattering angle. Figure 2(a) and (b) shows the nanoparticle zeta potential and hydrodynamic size. EG coating of the nanoparticles is responsible for their extremely high hydrodynamic diameter, which is in the range of 60-120 nm compared to the crystallite sizes determined by XRD. Table 1 includes the synthetic conditions, average crystalline sizes, hydrodynamic diameter, and zeta potential of the nanoparticles. The nanoparticle in water forms a stable dispersion, as shown by a positive zeta potential value at neutral pH.

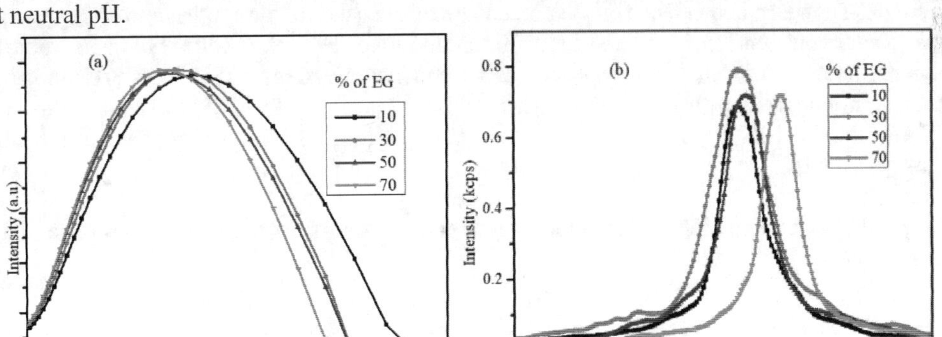


**Figure 2**. (a) Hydrodynamic diameter of $Fe_3O_4$ nanoparticles obtained using dynamic light scattering method. (b) Zeta potential of nanoparticles at neutral pH.

| Sample | Percentage of EG | Average crystallite size (nm) from XRD | Lattice constant (A) | Hydrodynamic diameter (nm) | Zeta potential (mV) |
|---|---|---|---|---|---|
| S1 | 10 | 17 | 8.3724 | 95.2 | 22.3 |
| S2 | 30 | 16.5 | 8.3736 | 86.2 | 22.5 |
| S3 | 50 | 14.9 | 8.3724 | 80.3 | 23.3 |
| S4 | 70 | 14.8 | 8.3724 | 79.6 | 28.6 |

**Table 1**. The average crystallite size, lattice constant, hydrodynamic diameter, and zeta potential of $Fe_3O_4$ nanoparticles.

*3.3 Magnetic hyperthermia measurements of nanoparticles dispersions*
The nanoparticles are dispersed in water by probe sonication for 10 minutes to obtain the water dispersion of nanoparticles. This is important in order to determine the magneto thermic effect using nanoScale Bio magnets hyperthermia instrument. AMF with frequencies of 384.50, 330.55, 304.75, and 167.30 kHz and field amplitudes of 51.85, 43.87, 35.89, and 27.92 kA/m is used to acquire the heating profiles of the nanoparticles. In order to precisely determine the heating efficiency, adiabatic conditions were maintained for the calorimetric tests. Using a thermal probe, the rise in water dispersion temperature is measured. SAR values for the dispersion of nanoparticles are obtained using the slope of the initial heating profile. The SAR values are given by equation 1[18]:

$$SAR\ (W/g) = \frac{C}{m_{MNP}} \frac{dT}{dt} \tag{1}$$

$m_{MNP}$ is the mass (in g) of the MNPs in the water dispersion, C (4184 $Jkg^{-1}K^{-1}$) is the specific heat capacity of water, and dT/dt is the initial slope of the temperature-time curve derived from AMF treatment. Figure 3(a) displays the heating profiles of S2 (30% EG) nanoparticles produced under a field amplitude of 27.92 kA/m and different frequencies. Figure 3(b) displays the SAR values obtained for nanodispersions of 1 and 2 mg/ml as a function of EG %. S2 nanoparticles showed the highest SAR value indicating high thermal conversion ability among the four set of nanoparticles. S2 nanoparticles (30% EG) showed the highest SAR value of 372.95 W/g whereas S4 (70%) had a SAR value of 217.12 W/g under AMF. The SAR values of the nanoparticles showed a dependence on the percentage of EG used for the synthesis. The heating profiles are obtained under constant frequency and field amplitude range. The SAR values obtained from the four sets of nanoparticles at constant field amplitude and frequency are shown in Figures 3(c) and (d). SAR values increase nonlinearly with AMF strength and almost linearly with AMF frequency. Based on the linear response theory (LRT), where SAR increases linearly with AMF frequency and quadratically with AMF intensity, the results slightly deviate from those anticipated [19]. The role of dipolar interactions at this particle concentration could be the cause of this deviation from the LRT. When compared to several papers that synthesis nanoparticles utilizing the coprecipitation approach, the SAR values observed in this study are significantly higher [20,21].

*3.4 MTT Cytotoxicity Assay*
Since S2 (30%) nanoparticles showed high thermal conversion ability, invitro studies for the killing of cancer cells are examined with two different sets of AMF parameters (AMF1: 24.67 kA/m-384.50 kHz and AMF2:47.74 kA/m-167.5 kHz) which generate a saturation temperature of 46 °C for 1 mg/ml nano dispersion. The effect of magnetic field frequency and field amplitude parameters, which result in the same saturation temperature, on the mechanism of cell death is examined using these parameter combinations. Four set of cells were utilized to examine the magnetic cytotoxicity: Cells from the control group (C) were incubated at 37 °C throughout the experiment. One group of cells (F) was subjected to a magnetic field without of nanoparticles to examine the effects of the magnetic field. Another group of cells (P) were treated

without the presence of a field in order to study the effects of the nanoparticles alone. For the AMF treatment cells (P+F) were treated with 1 mg/ml of S2 nanoparticles for 30 minutes exposure time and 46 °C saturation temperature.

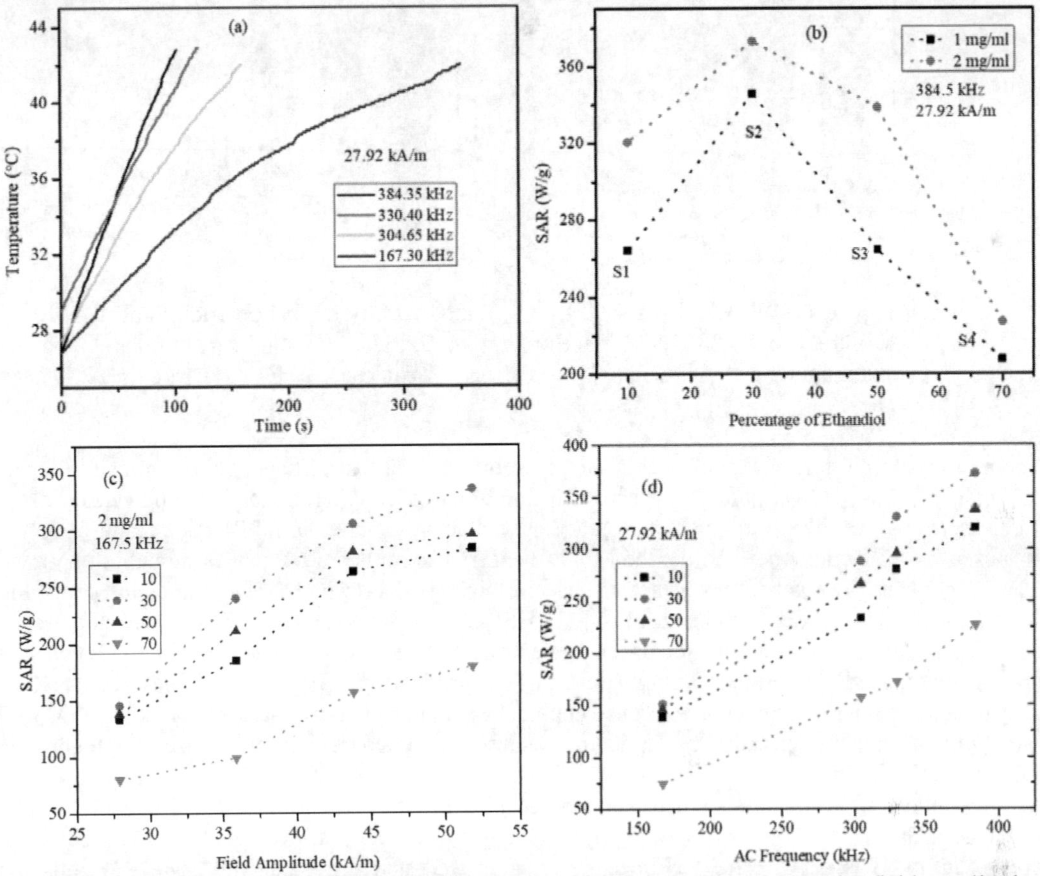


**Figure 3**. (a) Heating profiles obtained from the S2 nano dispersion under field amplitude 27.92 kA/m and various frequencies. (b) SAR values of the S1-S4 nanoparticles obtained under AMF (384.5 kHz and 27.92 kA/m), (c) Field amplitude dependent SAR values at 167.5 kHz, (d) AC frequency dependent SAR of S1-S4 nanoparticles at 27.92 kA/m.

The cytotoxicity of EG coated nanoparticles treated with two sets of AMF settings at a concentration of 1 mg/ml is determined using the MTT assay. MDA-MB-231 that was treated with AMF treatment was seeded in triplicate on a 96-well tissue culture plate (Corning, Sigma-Aldrich Co., St. Louis, MO, USA) at a density of 5000 cells per well. The absorbance of the MTT assay was calculated at 570 nm using the ELISA plate reader SynergyTM HTX microplate reader (BioTek, Winooski, VT, USA) used to evaluate cell viability. Figure 4 illustrates the cytotoxicity of 1 mg/ml nanoparticles treated with AMF (24.67 kA/m-384.50 kHz) reduced the viability of MDA-MB-231 cells by about 48 and 75% after 0 and after 24 h, respectively. The results of the cell proliferation experiment for cells treated with AMF showed a significant difference for various time points for cell lines compared to the control group with a p value of 0.0001. It was shown that there was no magnetic field toxicity when the viability of cancer cells treated with the magnetic field alone was comparable to the viability of the control at various time points.

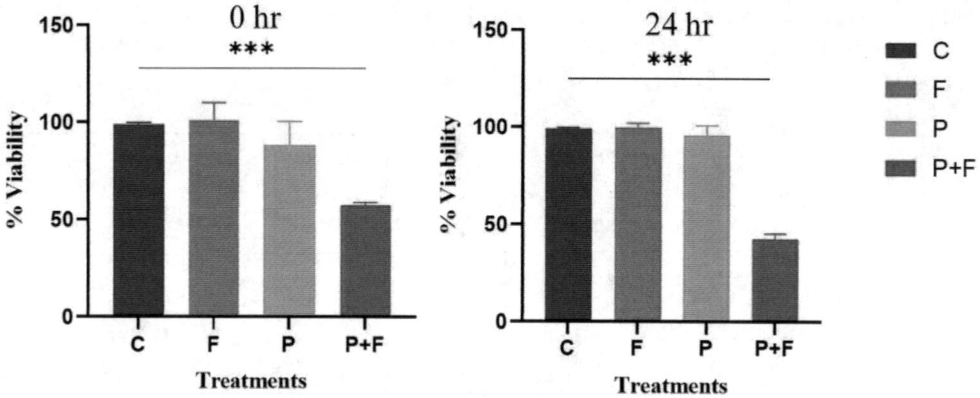


**Figure 4.** Cell viability of MDA cells treated with $Fe_3O_4$ nanoparticles under AMF with frequency of 384.5 kHz and strength of 24.67 kA/m obtained after 0 and 24 hr treatment (C-Cells, F-Cells+AMF, P-Cells+Particles, & P+F-ells+Particles+AMF).

*3.5 Apoptosis*

In response to numerous physiological and pathological stimuli, apoptosis, a type of programmed cell death, occurs. It is distinguished by separate morphological and biochemical alterations, including cell shrinkage, chromatin condensation, DNA fragmentation, and caspase enzyme activation [22]. As a result of cellular stress caused by magnetic hyperthermia, the MNPs' heat may stimulate a number of signalling pathways that cause apoptosis. The activation of heat shock proteins (HSPs), which are essential for controlling cellular stress responses, is one such mechanism [23]. HSPs can accelerate protein breakdown, stop proteins from misfolding, and control the function of apoptotic proteins like caspases. Additionally, oxidative stress brought on by magnetic hyperthermia might result in the production of reactive oxygen species (ROS), which can harm cellular components and cause apoptosis. The mitogen-activated protein kinase (MAPK) pathway, which can cause apoptosis by activating caspases and encouraging the release of cytochrome c from the mitochondria, is one of the signalling pathways that ROS can activate [24,25]. The apoptosis of cancer cells involves early apoptosis, late apoptosis, and necroptosis. Using a flow cytometer from BD Biosciences in San Jose, California, the apoptotic and necrotic cell death of MDA-MB-231 cells after treatment with two sets of AMF field and frequency parameters is determined. MDA-MB-231 cells were exposed to magnetic nanoparticles and AMF for 0 hours, and the apoptosis rate was approximately 48%. within 24 hours, it increased to 75%, as illustrated in Figure 5. Nanoparticles are biocompatible and don't have any inherent toxicity, as evidenced by the fact that cells treated with the field (F) and particle (P) alone have minimal apoptosis. The percentage of early and late apoptosis obtained from the flow cytometry are listed in Table 2. From the apoptosis data it is clear that two sets of AMF parameters used cause apoptosis in different amounts: AMF1 (24.67 kA/m-384.50 kHz) shows high rate of late apoptosis compared to little early apoptosis effect, whereas AMF2 (47.74 kA/m-167.5 kHz) conditions show significant apoptosis both early and late. Total apoptosis seems to be the same for both sets of parameters. The apoptotic cell death showed a dependency on the AMF frequency and field strength combinations though the cells are subjected to similar thermal treatment of 46 °C for 30 min. This can be related to the impact of the field parameters on the uptake of nanoparticles by cells, ROS produced as a result of thermal heating, and the response of the cell thermal shock proteins, which are essential factors for cell death. Understanding the effect of field parameters on the cell death pathways will help to determine the differential effect of modulating reactive oxygen species and heat shock proteins response to the AMF parameters. These studies will help in designing the in vivo magnetic hyperthermal treatment of cancer tumors. In order to understand the impact of AMF settings on the gene and shock protein expressed by the nanoparticles, additional studies are required. Overall, these studies suggest that magnetic hyperthermia induce apoptosis in cancer cells, which may contribute to its potential as a therapeutic technique for cancer treatment.

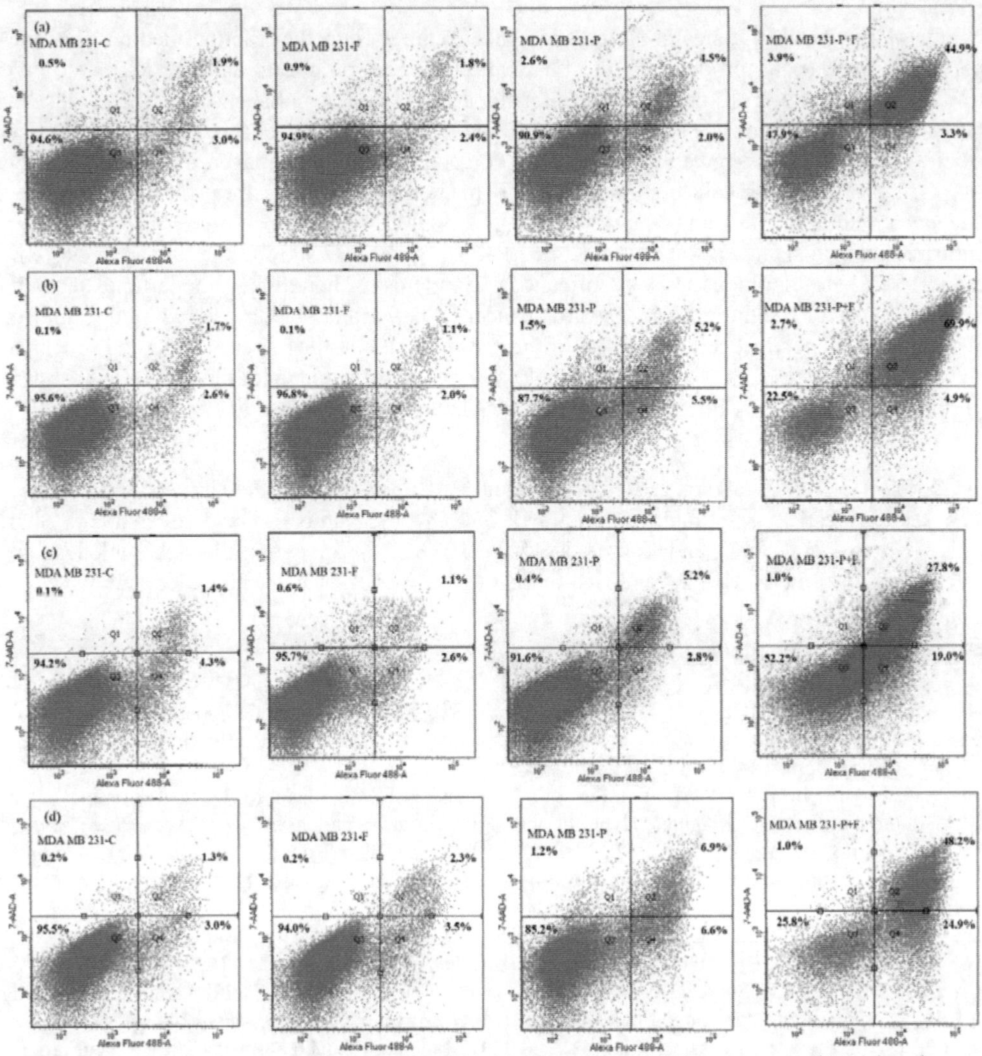

**Figure 5.** Apoptosis assay of MDA-MB-231 (a) 0 hr after AMF1 treatment, (b) 24 hr after AMF1 treatment, (c) 0 hr after AMF2 treatment, and (d) 24 hr after AMF2 treatment.

| | Time after AMF treatment (hr) | Field amplitude (kA/m) | Field frequency (kHz) | % of early apoptosis | % of late apoptosis | % of total apoptosis |
|---|---|---|---|---|---|---|
| AMF1 | 0 | 24.67 | 384.50 | 3.3 | 44.9 | 48.2 |
| | 24 | 24.67 | 384.50 | 4.9 | 69.9 | 74.8 |
| AMF2 | 0 | 47.74 | 167.5 | 19 | 27.8 | 46.8 |
| | 24 | 47.74 | 167.5 | 24.9 | 48.2 | 73.1 |

**Table 2**. Apoptosis of MDA-MB-231 cells subjected to two distinct sets of frequencies and field strengths.

## 4. Conclusions

Ethylene glycol and water mixtures in various percentages are used as the reaction medium in a one-step hydrothermal technique to produce $Fe_3O_4$ nanoparticles. The produced nanoparticles had average crystalline diameters between 15 and 17 nm. The nanoparticles' zeta potential shows that their dispersions are stable in aqueous medium. The proportion of EG and water employed for the synthesis determines the nanoparticles' ability for heat conversion; the 30% EG reaction mixture displayed the highest SAR value, measuring 372.95 W/g. SAR values show a linear dependency on the AC frequency whereas the field amplitude has a nonlinear effect on the SAR values. The two sets of AMF parameters used for the invitro treatment causes apoptosis in a different rate, 24.67 kA/m-384.50 kHz shows high late apoptosis whereas 47.74 kA/m-167.5 kHz conditions show significant early apoptosis. Though the nanoparticles showed high SAR and high magnetic heating toxicity towards cancer cells. Further studies are required to understand the field parameters dependent cellular uptake of the nanoparticles on the effect of ROS and shock protein response for thermal heating. These studies will help to improve the potential of the magnetic nanoparticles for the designing the in vivo treatment of cancer tumors using AMF.

**Acknowledgments:** The Collaborative Research Grant #2101050262 from the University of Sharjah and Sharjah Research Academy is gratefully acknowledged by the authors. The University of Sharjah's Competitive Grant 2201110363 and Targeted Grant 2101110252 are gratefully acknowledged by the authors. Research Institute of Medical and Health Sciences (RIMHS) is also acknowledged by the writers.

## References

[1] Mahmoudi K, Bouras A, Bozec D, Ivkov R and Hadjipanayis C 2018 Magnetic hyperthermia therapy for the treatment of glioblastoma: a review of the therapy's history, efficacy and application in humans *International Journal of Hyperthermia* **34** 1316–28.

[2] Chang D, Lim M, Goos J A C M, Qiao R, Ng Y Y, Mansfeld F M, Jackson M, Davis T P and Kavallaris M 2018 Biologically Targeted Magnetic Hyperthermia: Potential and Limitations *Frontiers in Pharmacology* **9.**

[3] Pradhan P, Giri J, Samanta G, Sarma H D, Mishra K P, Bellare J, Banerjee R and Bahadur D 2007 Comparative evaluation of heating ability and biocompatibility of different ferrite-based magnetic fluids for hyperthermia application *Journal of Biomedical Materials Research Part B: Applied Biomaterials* **81B** 12–22.

[4] Issa B, Qadri S, Obaidat I M, Bowtell R W and Haik Y 2011 PEG Coating Reduces NMR Relaxivity of $Mn_{0.5}Zn_{0.5}Gd_{0.02}Fe_{1.98}O_4$ Hyperthermia Nanoparticles *J Magn Reson Imaging* **34** 1192–8.

[5] Joshy K S, Augustine R, Mayeen A, Alex S M, Hasan A, Thomas S and Chi H 2020 $NiFe_2O_4$/poly (ethylene glycol)/lipid–polymer hybrid nanoparticles for anti-cancer drug delivery *New J. Chem.* **44** 18162–72.

[6] Anik M I, Hossain M K, Hossain I, Mahfuz A M U B, Rahman M T and Ahmed I 2021 Recent progress of magnetic nanoparticles in biomedical applications: A review *Nano Select* **2** 1146–86.

[7] Cardoso V F, Francesko A, Ribeiro C, Bañobre-López M, Martins P and Lanceros-Mendez S 2018 Advances in Magnetic Nanoparticles for Biomedical Applications *Advanced Healthcare Materials* **7** 1700845.

[8] Martins P M, Lima A C, Ribeiro S, Lanceros-Mendez S and Martins P 2021 Magnetic Nanoparticles for Biomedical Applications: From the Soul of the Earth to the Deep History of Ourselves *ACS Appl. Bio Mater.* **4** 5839–70.

[9] Dennis C L and Ivkov R 2013 Physics of heat generation using magnetic nanoparticles for hyperthermia *Int J Hyperthermia* **29** 715–29.

[10] Fabris F, Lima E, Biasi E D, Troiani H E, Mansilla M V, Torres T E, Pacheco R F, Ibarra M R, Goya G F, Zysler R D and Winkler E L 2019 Controlling the dominant magnetic relaxation mechanisms for magnetic hyperthermia in bimagnetic core–shell nanoparticles *Nanoscale* **11** 3164–72.

[11] Gavilán H, Simeonidis K, Myrovali E, Mazarío E, Chubykalo-Fesenko O, Chantrell R, Balcells L, Angelakeris M, Morales M P and Serantes D 2021 How size, shape and assembly of magnetic nanoparticles give rise to different hyperthermia scenarios *Nanoscale* **13** 15631–46.

[12] F. Hasany S, H. Abdurahman N, R. Sunarti A and Jose R 2013 Magnetic Iron Oxide Nanoparticles: Chemical Synthesis and Applications Review *Current Nanoscience* **9** 561–75.

[13] Ma M, Zhang Y, Guo Z and Gu N 2013 Facile synthesis of ultrathin magnetic iron oxide nanoplates by Schikorr reaction *Nanoscale Res Lett* **8** 16.

[14] Singh R, Letai A and Sarosiek K 2019 Regulation of apoptosis in health and disease: the balancing act of BCL-2 family proteins *Nat Rev Mol Cell Biol* **20** 175–93.

[15] Chen F, Wang C-C, Kim E and Harrison L E 2008 Hyperthermia in combination with oxidative stress induces autophagic cell death in HT-29 colon cancer cells *Cell Biol Int* **32** 715–23.

[16] Cellai F, Munnia A, Viti J, Doumett S, Ravagli C, Ceni E, Mello T, Polvani S, Giese R W, Baldi G, Galli A and Peluso M E M 2017 Magnetic Hyperthermia and Oxidative Damage to DNA of Human Hepatocarcinoma Cells *Int J Mol Sci* **18** 939.

[17] Patterson A L 1939 The Scherrer Formula for X-Ray Particle Size Determination *Phys. Rev.* **56** 978–82.

[18] Andreu I and Natividad E 2013 Accuracy of available methods for quantifying the heat power generation of nanoparticles for magnetic hyperthermia *International Journal of Hyperthermia* **29** 739–51.

[19] Hergt R and Dutz S 2007 Magnetic Particle Hyperthermia – Biophysical Limitations of a Visionary Tumour Therapy *Journal of Magnetism and Magnetic Materials* **311** 187–92.

[20] Narayanaswamy V, Al-Omari I A, Kamzin A S, Issa B, Tekin H O, Khourshid H, Kumar H, Mallya A, Sambasivam S and Obaidat I M 2021 Specific Absorption Rate Dependency on the Co2+ Distribution and Magnetic Properties in $Co_xMn_{1-x}Fe_2O_4$ Nanoparticles *Nanomaterials* **11** 1231.

[21] Narayanaswamy V, Sambasivam S, Saj A, Alaabed S, Issa B, Al-Omari I A and Obaidat I M 2021 Role of Magnetite Nanoparticles Size and Concentration on Hyperthermia under Various Field Frequencies and Strengths *Molecules* **26** 796.

[22] Zimmermann K C and Green D R 2001 How cells die: Apoptosis pathways *Journal of Allergy and Clinical Immunology* **108** S99–103.

[23] Court K A, Hatakeyama H, Wu S Y, Lingegowda M S, Rodríguez-Aguayo C, López-Berestein G, Ju-Seog L, Rinaldi C, Juan E J, Sood A K and Torres-Lugo M 2017 HSP70 Inhibition Synergistically Enhances the Effects of Magnetic Fluid Hyperthermia in Ovarian Cancer *Mol Cancer Ther* **16** 966–76.

[24] Belhadj Slimen I, Najar T, Ghram A, Dabbebi H, Ben Mrad M and Abdrabbah M 2014 Reactive oxygen species, heat stress and oxidative-induced mitochondrial damage. A review *International Journal of Hyperthermia* **30** 513–23.

[25] Kassis S, Grondin M and Averill-Bates D A 2021 Heat shock increases levels of reactive oxygen species, autophagy and apoptosis *Biochimica et Biophysica Acta (BBA) - Molecular Cell Research* **1868** 118924.

# Generation of Nanoparticles from FeCu Powder Using Femtosecond Laser

M. AlAwadhi[a,b*], A. Yassine [a,c]

[a] *Department of Physics, American University of Sharjah, 26666 Sharjah, United Arab Emirates*

[b] *Dubai Police G. H. Q, Dubai, United Arab Emirates*

[c] *College of Education, American University in Emirates, Dubai, United Arab Emirates*

**Abstract.** The usage of optical force techniques such as lasers, provides contactless, and non-mechanical, microcrystal manipulation. Nanocrystalline materials are commonly produced using methods such as mechanical alloying, electrodeposition... (etc.). Bimetallic alloys and nanostructures containing Iron and Copper have been utilized for environmental solutions due to their exceptional electric and magnetic properties. The purpose of this study is to investigate the physical and optical properties of Iron Copper (FeCu) nanoparticles that are generated using femtosecond laser from $Fe_{50}Cu_{50}$ powder. Ethanol and vacuum mediums are utilized as alternative ablation mediums. Characterization of the material has been conducted through XRD, EDX and Absorption Spectroscopy. The resulted tests revealed a consistency in the iron and copper content, indicating that the material did not undergo any oxidation during the ablation processes. The absorption spectroscopy results revealed that the ablated nanoparticles resonated at a different wavelength than the bulk material, demonstrating that there was a shift in the optical properties. Additionally, the materials generated from ablation in Ethanol resulted in smaller grains compared to Vacuum ablation. Moreover, the chemical homogeneity was retained with no changes compared to the starting micron size particles which were utilized as a target for the ablation process.

***Keywords—Iron-Copper; Nanoparticles; Laser-ablation; femtosecond laser***

* Corresponding author.
*E-mail address:* g00082783@aus.edu (M. AlAwadhi).

## 1.Introduction

Due to their cost-effectiveness and high engineerability, Iron alloy systems are among the most extensively used systems in practical applications. For instance, it has been reported that FeCu nanoparticles have shown promising solution for wastewater treatment and dye removal [1] due to the high removal capacities of iron material, and the exceptional antibacterial properties of copper material [2]. The ability of Cu to protect corrosive materials such as iron promotes the utilization of the bimetallic nanoparticles FeCu for environmental applications [3].

Traditionally, these alloys can be produced through different techniques to obtain different type and quantities of materials, one of which is [4].

Iron-Copper alloys (FeCu) have been previously investigated and presented intriguing magnetic and electron transport properties in bulk scale [5]. On the other hand, the lack of compatibility between Iron and Copper has rendered these more common techniques ineffective. The challenge of limited miscibility between Iron and Copper resulted in finding innovative methods to alloy these two elements [6,7]. Methods such as electrodeposition [8], magnetron-sputtering [9], and mechanical alloying [10], were previously used alloying iron and copper. Mechanical alloying, a solid-state alloying method is often used to produce uniform and homogenous materials, by relying on the repeated process of cold-welding, fracturing, and rewelding of the material in a high-energy ball mill [11–14]. Although this method is often utilized to produce nanocrystalline materials (1-100*nm*) [11], the production of these material can take long periods of time (up to 26h) [15].

The aim of this study is to produce FeCu nanoparticles using laser ablation in ethanol and pulsed laser deposition (PLD) from Iron Copper alloys and to investigate their physical and optical properties through different characterization techniques and compare them with the bulk material.

### *1.1. Laser Ablation Mediums*

Laser-metal interaction mechanisms mainly depend on the power density, the laser pulse duration, average power, and peak power [16]. The combination of these properties and the use of a suitable ablation medium allows for a smoother ablation process.

Laser irradiation and Laser induced fragmentation were developed to produce nanostructures, however the Femtosecond Laser Ablation in liquid technique is the most feasible process, as it is an innovative method that occurs at once, in a short period of time, non-polluting and with minimal involvement of chemical materials [17–19].

The laser ablation technique in liquid is proved to generate stable nanoparticles with tunable size and shape [20]. For instance, the optical properties of Gold and Copper nanoparticles were tuned to be widely applicable in biomedical engineering fields and photovoltaics [21,22].

Nanostructures are generated after immersing the target in liquid and heating it to its boiling point with laser pulse. Plasma cloud filled with vapor atoms is formed, expanded, and condensed in the liquid. The produced nuclei agglomerate with each other by weak forces and acquire specific features that depend on laser parameters and liquid medium [23,24]. For instance, Aluminium nanoparticles showed three different sizes after pulse laser ablation in three different liquid media [24].

The properties of the ablated metallic nanostructures (size and shape) through femtosecond laser are highly dependent on the energy, pulse duration, wavelength of the laser used and the surrounding liquid medium in which ablation is performed [25,26]. It has been reported that ZnO nanoparticles presented different size with different optical properties with changing the pulse energy due to the local rise of temperature at the surface of the nanoparticles which lead to the melting and fusion of small nanostructures into bigger ones [27,28].

Research shows that nanostructures stability depends on liquid media. For example, GaAs nanoparticles were stable for several months when they are ablated in ethanol compared to water, and water with Ammonium Sulfide solution in which the nanostructures visibly aggregated [29].

Since iron-copper alloys undergo oxidation in water medium during dispersion [30], alternative ablations mediums must be investigated. Gündoğdu et al. [31] investigated the use of ethanol as an ablation medium for femtosecond laser ablation processes, and it has been concluded that the ablated material maintained its stability during the ablation process when utilizing the specific medium of ablation.

On the other hand, it has been recorded that pulsed laser deposition using femtosecond laser is an efficient technique to generate nanoparticles in vacuum on a substrate with an average size of 10 nm [32] as the ultrashort pulses do not affect the generated nanostructures and transform the targeted material to plasma state due to the high temperature. This technique relies on a vacuum chamber that is enclosed to prevent impurities from interacting with the material during the ablation process. A vacuum pump is connected to the chamber and used to theoretically remove any gases and background air, or as much as possible [33], the fact that creates an ideal condition for testing the properties of the nanoparticles, as it prevents the interference of background gases [34,35].

# 2. Methodology

## *2.1.Sample Preparation*

Iron Copper alloy was prepared using mechanical alloying method with a Fritsch Pulverisette 7 Premium line planetary ball mill. An equiatomic alloy was produce by grinding 2.5g of Iron powder, and 2.5g of copper powder, in a tungsten carbide grinding bowl. Eight 10mm stainless steel grinding balls were used to grind the mixture into an alloy at 600 rpm for 10 hours, resulting in the $Fe_{50}Cu_{50}$ alloy. 1g of the resulting powder was then pressed using a hydraulic press into a 1cm diameter palette to be used the target during the ablation process. Two palettes were created using the same powder source and each was used individually for the two methods of ablation (Ethanol Laser Ablation and Vacuum Laser Ablation).

## *2.2.Ethanol Laser Ablation*

The FeCu target was ablated using femtosecond laser. The pulse duration was 40 fs. The sample were covered with ethanol (10 ml) to prevent simultaneous oxidation. Rb - YAG laser with wavelength of 1030 nm were used for the ablation technique, with pulse energy of 10 W and repetition rate of 50 kHz and laser fluence of 20.408 *J/cm²*. The Ablation process took 3 minutes, and the laser beam spot size was 50mm. The parameters mentioned above are summarized in the Table 1 below.

The fluence was calculated using equation (1).

$$Fluence\ \left({}^{J}/_{cm^2}\right) = \frac{Laser\ Pulse\ Energy\ (J)}{Focal\ Spot\ Area\ (cm^2)} \qquad (1)$$

**Table 1.** Experimental Parameters for Ethanol Laser Ablation

| Sample | Pulse Duration *(fs)* | Wavelength *(nm)* | Laser Fluence *(J/cm²)* | Pulse Energy *(W)* | Repetition Rate *(kHz)* | Ablation time *(min)* | Ethanol volume *(ml)* |
|---|---|---|---|---|---|---|---|
| **Vacuum Ablation** | 40 | 1030 | 26 | 10 | 50 | 3 | 10 |

Silicon and glass substrates were used to collect FeCu nanoparticles through evaporation for characterization through SEM and EDX and optical absorption spectroscopy to test the morphology, chemical composition, and optical properties of the nanomaterials respectively.

## *2.3.Vacuum Laser Ablation*

Samples were loaded 24 hours prior to the ablation process into the set up and were vacuum pumped up to a pressure of $2.87\times10^{-5}$ mbar to avoid contamination of the material during the experiment. The sample is heated for 2 min prior to the experiment and placed in the vacuum chamber. 1 cm from the source, glass and silicon substrates were placed to collect the nanoparticles for characterization through SEM, EDX and XRD. The vacuum chamber consists of a rotating holder to avoid formation of craters on the surface. The sample is targeted by Rb – YAG femtosecond laser with wavelength 1030 nm, pulse energy

10 W, repetition rate of 50 kHz and spot size of 70 $\mu m$ for 10 min. The parameters mentioned above are summarized in the Table 2 below.

**Table 2.** Experimental Parameters for Vacuum Laser Ablation

| Sample | Pulse Duration *(fs)* | Wavelength *(nm)* | Laser Fluence *(J/cm²)* | Pulse Energy *(W)* | Repetition Rate *(kHz)* | Ablation time *(min)* | Vacuum Pressure *(mBar)* |
|---|---|---|---|---|---|---|---|
| **Vacuum Ablation** | 125 | 1030 | 20.408 | 10 | 50 | 10 | $2.87 \times 10^{-5}$ |

## 3. Results and Discussion

### *3.1. Vacuum Ablation*

Scanning Electron Microscopy (SEM), Energy Dispersive X-ray (EDX) and X-ray diffraction (XRD) characterization tests were conducted on the sample deposited on the Si wafer in vacuum. The results obtained from these tests can be seen in the figures below.

Figure 3 shows a live EDX map of a single FeCu particle which displays the material distribution of a single particle. As can be noted from the live map, there is a separation of the alloy in the material. Additionally, Table 3 indicates that there is a drop in the copper content while as the iron content remains at approximately 50%.

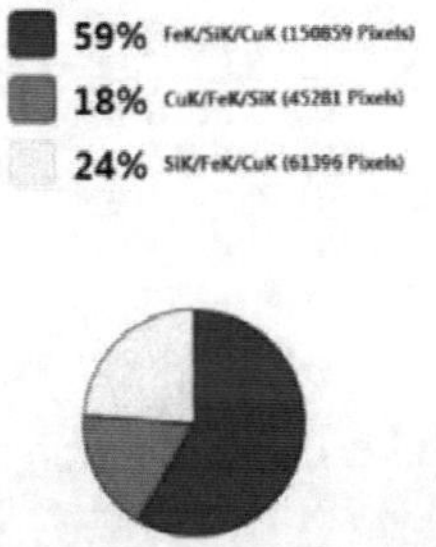


Table 3: Live MAP of a $Fe_{50}Cu_{50}$ single particle

| **Material** | **Fe** | **Cu** | **Si** |
|---|---|---|---|
| (%wt.) | 59 | 18 | 24 |

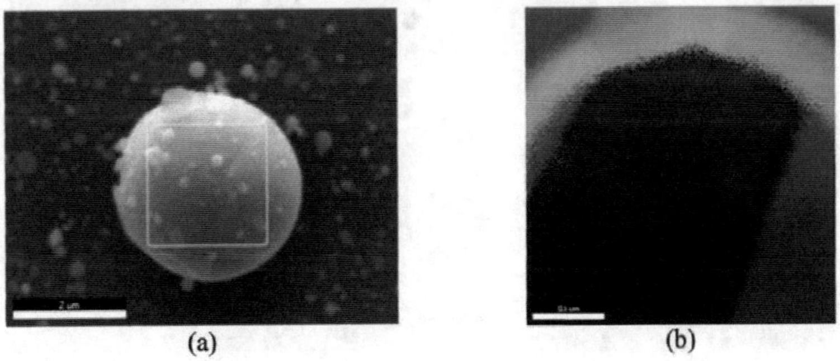


*Figure 1: (a) SEM of a Single $Fe_{50}Cu_{50}$ Particle (b) Live MAP of that same particle*

Journal of Physics: Conference Series **2751** (2024) 012003 doi:10.1088/1742-6596/2751/1/012003

This can also be proven by the detailed EDX map of that same particle in Figure 3 (b and d), which show that atomic separation in a clear manner. Moreover, the map analysis presented in Table 4, confirms that atomic weight shift that was seen in the previous figure. In both figures the Si content can be disregarded, since the targeted substrate is Silicon wafer, so the high silicon weight percentage is expected.

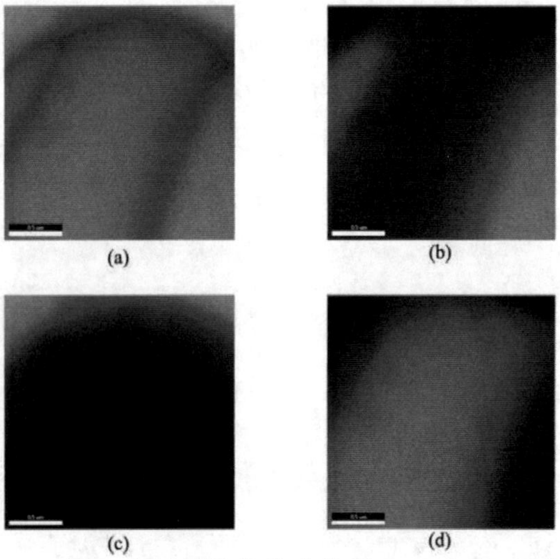
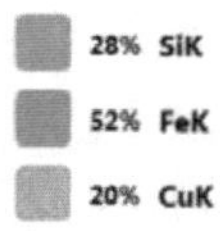


Table 4: MAP of a $Fe_{50}Cu_{50}$ single particle

| Material | Fe | Cu | Si |
|---|---|---|---|
| (%wt.) | 52 | 28 | 20 |

*Figure 2: (a) MAP of a Single $Fe_{50}Cu_{50}$ particle showing (b) Iron content (c) Copper content (d) Silicon content*

Additionally, SEM images were taken at different magnifications. Figure 4 shows the FeCu coated silicon wafer, at different magnifications at the following horizontal field width: 6.16 $\mu m$, 71.1 $\mu m$, 91.9mm, 248mm, 484mm. As can be clearly seen, the particle selected for the analysis, shows larger diameter compared to the particles observed in the surrounding.

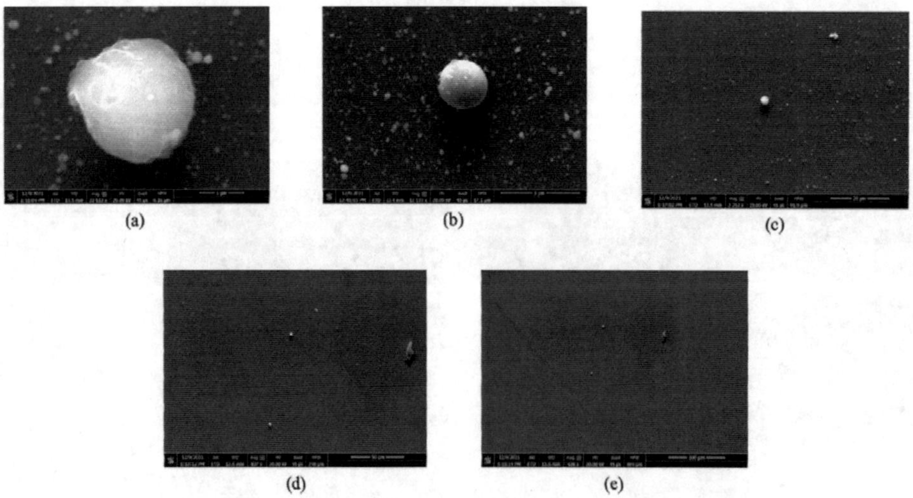

*Figure 3: SEM Images of the $Fe_{50}Cu_{50}$ coated silicon wafer at different magnifications, the Horizontal field width of each is as follows (a) 6.16$\mu m$, (b)17.1$\mu m$, (c) 91.9$\mu m$, (d) 248$\mu m$, (e) 484$\mu m$*

Additionally, a thorough EDX analysis was conducted, and presented a slightly contradictory piece of information regarding the weight and atomic content of copper presented in the previous figures. The data found in Table 5, extracted from the particle seen in Figure 6 can be contributed towards the fact that the selected particle is an agglomerate, which will not properly reflect individual particles.

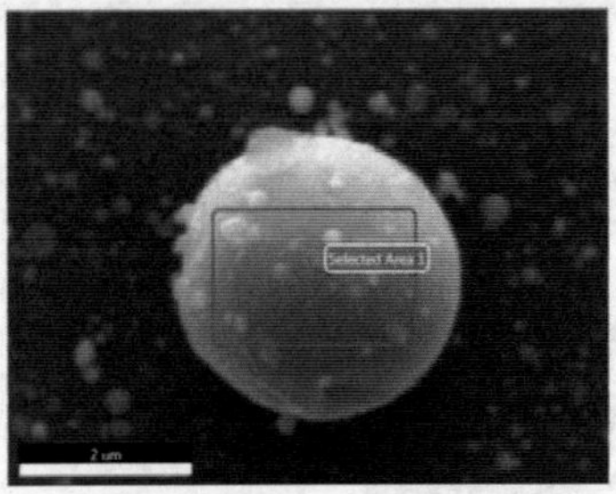


Table 5: EDX of a $Fe_{50}Cu_{50}$ single particle

| Material | Weight (%) | Atomic (%) |
|---|---|---|
| Silicon (Si) | 7.56 | 14.6 |
| Iron (Fe) | 55.61 | 53.98 |
| Copper (Cu) | 36.83 | 31.42 |

*Figure 4: EDX Results of a single $Fe_{50}Cu_{50}$ Particle*

This data was confirmed from the XRD in Figure 7. The results obtained show a shoulder peak, which indicate a dual phase (phase separation) compared to the original material that will be discussed later. This ablation process resulted in the separation of the alloy into Fe and Cu alloy as proven by the figures presented.

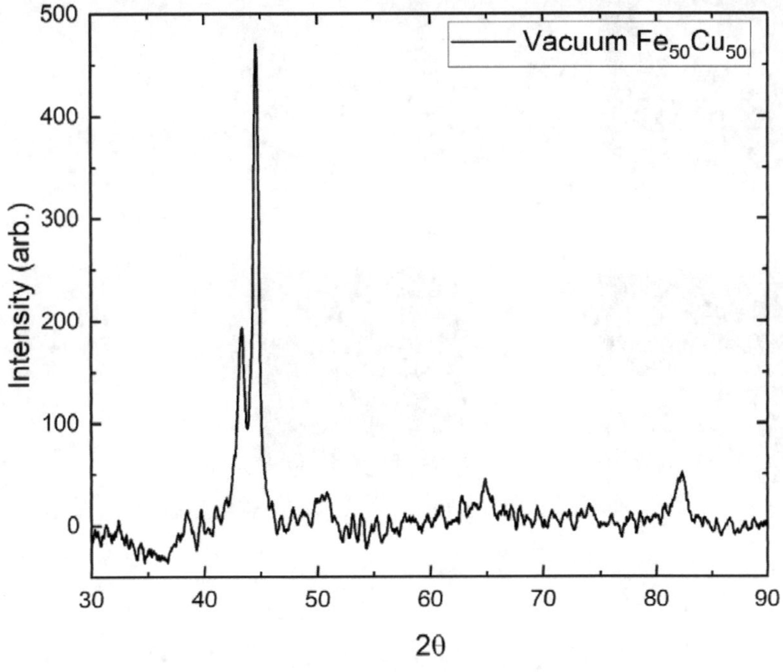


*Figure 5: XRD of Vacuum Ablated Sample*

### *3.2.* Original Sample

The Original sample (ball milled sample) characterization was conducted to compare it to the ablated samples. As can be seen in Figure 8, SEM images of the bulk sample display the powder at different magnifications, the view field of each is as follows 31.1µm, 425µm, 2.9mm. The size of the particles observed are considered massive compared to the laser ablated particles.

*Figure 6: SEM Images of the $Fe_{50}Cu_{50}$ Original sample at different magnifications, the view field of each is as follows (a) 31.1µm, (b) 425µm, (c)2.9mm*

EDX and MAP analysis was conducted to analyse the distribution of the materials in the alloy. Figure 9 indicate a homogenous distribution in the material, with a 50-50 weight composition as shown in Table 6.

Table 6: MAP of Ball Milled $Fe_{50}Cu_{50}$

| Target Composition | Fe wt% | Cu wt% |
|---|---|---|
| $Fe_{50}$-$Cu_{50}$ | 51.4 | 48.6 |

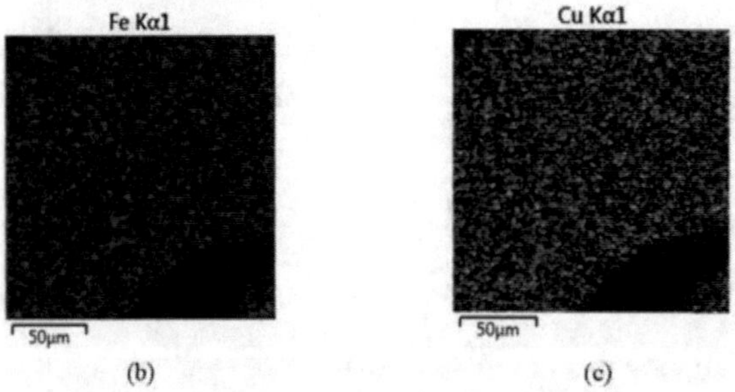


*Figure 7: EDX MAP Results of a $Fe_{50}Cu_{50}$ of the Original Sample*

Additionally, the alloying success was confirmed by the single (FCC) phase observed in the XRD of the original sample. The ball milled sample resulted in a single phase, face-centered cubic, 50-50 iron-copper alloy.

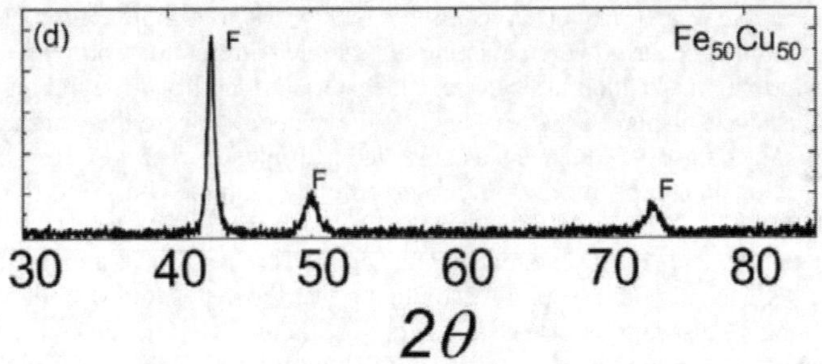


*Figure 8: XRD of Original Sample*

## 3.3. Laser Ablation in Ethanol

As conducted on the previous samples, SEM and EDX characterization tests were completed on the sample ablated in ethanol by femtosecond laser. The results of the tests can be seen in the figures below.

Unlike the vacuum ablated samples, ethanol ablated samples display even smaller sized particles (approximately 0.06mm), as seen in Figure 11. This could be attributed to a number of reasons. For starters, the ablation medium of both samples was different. In the vacuum sample, the target and the wafer were placed in a vacuum chamber of approximated $2.87 \times 10^{-5}$mBar. On the other hand, the ethanol ablation sample was completely submerged into the fluid, not allowing any external contamination to the process.

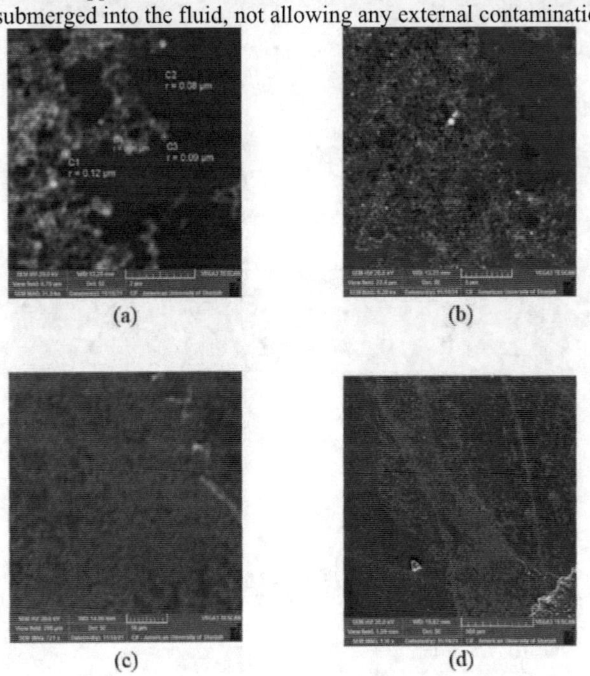


*Figure 9: SEM Images of the $Fe_{50}Cu_{50}$ Ethanol Ablated sample at different magnifications, the view field of each is as follows (a) 13.25µm, (b) 22.4µm, (c) 228µm, (d) 1.59mm*

Homogeneity in the sample was clearly observed in the EDX-MAP as seen in Figure 12. Based on the MAP figures, no separation occurred when ablating the sample with femtosecond laser in ethanol, the uniformity of the distribution is a good indicator of the material's stability throughout this process. Additionally, the EDX analysis displayed the resilience of the material. Where the weight percentage of iron and copper maintained the 50-50 balance, as observed in Table 7.

On the other hand, high Silicon compositions were observed in this sample compared to the ethanol. This could be attributed to two different factors. The first, being that the SEM used when analyzing this sample has lower power, and thereby a smaller view field. This leads to the inclusion of more silicon gaps into the EDX analysis. The second reason being that the deposition method (sputtering) resulted in a non-homogenous distribution of the sample on the silicon wafer, where on the other hand, the vacuum deposition provides a homogenous and effective coating method. The combination of these two reasons, which is the situation that is reported, resulted in this significantly high silicon content compared to that of the vacuum sample.

(a)

Table 7: MAP of Ethanol Ablated $Fe_{50}Cu_{50}$

| Fe wt% | Cu wt% | Si wt% |
|---|---|---|
| 2.9 | 2.4 | 94.7 |

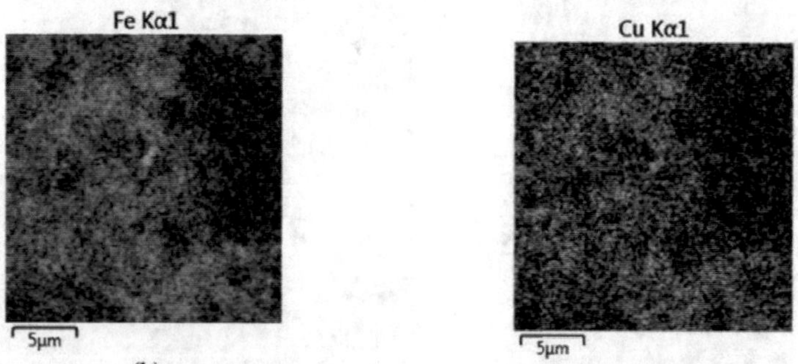


(b) (c)

*Figure 10: EDX Results of a $Fe_{50}Cu_{50}$ of the Ethanol Ablated Sample*

The absorption spectrum of FeCu nanoparticles ablated in ethanol were compared with the absorption spectrum of the bulk material. As it is observed in Figure 13, The graph shows an intense peak at approximately 230 nm for the nanostructures ablated in ethanol, in addition to the observed range that falls between 193.792nm and 1051.827nm.

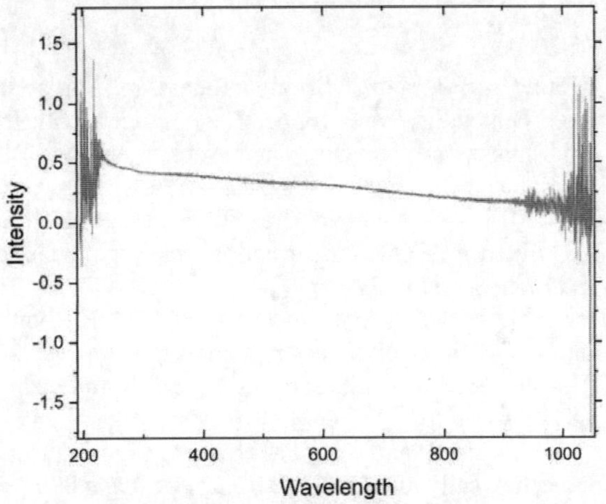


*Figure 11: Absorption spectrum of Ethanol Ablated $Fe_{50}Cu_{50}$ nanoparticles*

On the other hand, the absorption spectrum of the parent material FeCu Alloy [36] clearly shows an edge at 500 nm, and that the range of the spectrum falls between 450 and 900 nm [33]. The difference in the absorption peak indicates that the surface plasmons of the nanoparticles resonate at an energy that is different than the plasmons resonance energy of the bulk material. The plasmonic nanoparticles of FeCu resonate at a wavelength in the ultra-violet region unlike the surface plasmons of the bulk material that shows its resonance in the visible region. The shift in the wavelength indicates that the original material and the nanoparticles have different optical properties, molecular structures, electronic properties, and interatomic distances.
This Analysis confirmed that the ablated nanostructures are tuned in terms of size to exhibit optical and electronic properties that are different than the bulk material.

# 4. Conclusion

In conclusion, two effective laser ablation methods were investigated, femtosecond laser ablation in Ethanol, and laser ablation in vacuum chamber. A comparison of the samples generated was conducted with the original ball milled powder. It was noted that the materials generated from the same source material using ablation in Ethanol (0.06mm) resulted in smaller grains compared to Vacuum ablation (4-5mm). Additionally, the EDX of both samples showed that the nanoparticles generated using ethanol maintained their 50-50 ratio, compared to that of the vacuum ablated and deposited samples which exhibited a reduction in copper. Moreover, the distribution of the sample in EDX MAP of the vacuum indicate a separation of the material. On the other hand, the EDX-MAP of the ethanol sample showed more even distribution. The XRD of the vacuum ablated samples pointed towards a dual phase material, where twin peaks were observed in the graph, confirming the separation of the iron-copper separation and the results obtained from the EDX-MAP. It is also important to note that the vacuum ablation required a 24-hour prior sample preparation to achieve near–vacuum pressure, and the process took up to 10 minutes. Whereas the ethanol ablation process lasted approximately 3minutes and required to prior sample preparation. The samples obtained from ethanol ablation showed more promise in the ablation of $Fe_{50}Cu_{50}$ alloys as the process is more efficient, generated better results, and less time consuming. Further investigation into the parameters of the ablation process is required to obtain optimum results for both mediums of ablation.

Journal of Physics: Conference Series **2751** (2024) 012003 doi:10.1088/1742-6596/2751/1/012003

# 5. References:

[1] Goodman Dlamini N, Kotze Basson A and Srirama Rajasekhar Pullabhotla V 1125 A Comparative Study between Bimetallic Iron@copper Nanoparticles with Iron and Copper Nanoparticles Synthesized Using a Bioflocculant: Their Applications and Biosafety **8** 1125

[2] Sepúlveda P, Rubio M A, Baltazar S E, Rojas-Nunez J, Sánchez Llamazares J L, Garcia A G and Arancibia-Miranda N 2018 As(V) removal capacity of FeCu bimetallic nanoparticles in aqueous solutions: The influence of Cu content and morphologic changes in bimetallic nanoparticles *J Colloid Interface Sci* **524** 177–87

[3] Chowdhury M N K, Beg M D H, Khan M R and Mina M F 2013 Synthesis of copper nanoparticles and their antimicrobial performances in natural fibres *Mater Lett* **98** 26–9

[4] Ferry M 2016 Alloys: Iron *Reference Module in Materials Science and Materials Engineering* (Elsevier) pp 1–11

[5] Awadhi M Al, Egilmez M, Abuzaid W and Alami A H 2022 Magnetic properties and non-fermi liquid behaviour in mechanically alloyed FeCu *J Alloys Compd* **890** 161812

[6] Alami A H and Hawili A A 2020 Fe-Cu Metastable Alloy *Reference Module in Materials Science and Materials Engineering* (Elsevier)

[7] Chien C L, Liou S H, Kofalt D, Yu W, Egami T, Watson T J and McGuire T R 1986 Magnetic properties of FexCu100-x solid solutions *Phys Rev B* **33** 3247–50

[8] Barthem V M T S, Noce R D, Macedo W A A and Givord D 2009 Magnetic properties of electrodeposited Fe-poor Fe-Cu alloys *Brazilian Journal of Physics* **39** 182–5

[9] Chien C L, Liou S H, Kofalt D, Yu W, Egami T, Watson T J and McGuire T R 1986 Magnetic properties of FexCu100-x solid solutions *Phys Rev B* **33** 3247–50

[10] Martínez-Blanco D, Gorria P, Pérez M J, Blanco J A and Gonzaléz M A 2007 Low temperature neutron diffraction and magnetization of Fe25Cu75 solid solutions *J Non Cryst Solids* **353** 859–61

[11] Suryanarayana C 2001 Mechanical alloying and milling *Prog Mater Sci* **46** 1–184

[12] Suryanarayana C 2019 Mechanical Alloying: A Novel Technique to Synthesize Advanced Materials *Research* **2019** 1–17

[13] Rabiee M, Mirzadeh H and Ataie A 2017 Processing of Cu-Fe and Cu-Fe-SiC nanocomposites by mechanical alloying *Advanced Powder Technology* **28** 1882–7

[14] Benjamin J S 1970 Dispersion Strengthened Superalloys by Mechanical Alloying

[15] Zaluski L, Zaluska A and Ström-Olsen J O 1995 Hydrogen absorption in nanocrystalline Mg2Ni formed by mechanical alloying *J Alloys Compd* **217** 245–9

[16] Chun M K and Rose K 1970 Interaction of high-intensity laser beams with metals *J Appl Phys* **41** 614–20

[17] Mangababu A, Sarang Dev G, Chandu B, Bharati M S S, Debashish P, Venugopal Rao S and Nageswara Rao S V S 2020 Fabrication and characterization of GaAs nanoparticles achieved using femtosecond laser ablation *Mater Today Proc* **33** 2385–9

[18] Barcikowski S, Devesa F and Moldenhauer K Impact and structure of literature on nanoparticle generation by laser ablation in liquids

[19] Chinnathambi S, Chen S, Ganesan S and Hanagata N 2014 Silicon quantum dots for biological applications *Adv Healthc Mater* **3** 10–29

[20] Zeng H, Du X W, Singh S C, Kulinich S A, Yang S, He J and Cai W 2012 Nanomaterials via Laser Ablation/Irradiation in Liquid: A Review *Adv Funct Mater* **22** 1333–53

[21] Gezgin S Y, Kiliç H Ş, Kepceoğlu A, Bayir S, Nalbantoğlu I E and Toprak A 2016 Plasmonic tuning of gold doped thin films for layers of photovoltaic devices *AIP Conf Proc* **1722** 200013

[22] Gezgin S Y, Kepceoğlu A, Kılıç H Ş, Gezgin S Y, Kepceoğlu A and Kılıç H Ş 2017 An experimental investigation of localised surface plasmon resonance (LSPR) for Cu nanoparticles depending as a function of laser pulse number in Pulsed Laser Deposition *AIPC* **1815** 030020

[23] Ghorbani H R 2014 A review of methods for synthesis of Al nanoparticles *Oriental Journal of Chemistry* **30** 1941–9

[24] Yang G W 2007 Laser ablation in liquids: Applications in the synthesis of nanocrystals *Prog Mater Sci* **52** 648–98

[25] Oseguera-Galindo D O, Martínez-Benítez A, Chávez-Chávez A, Gómez-Rosas G, Pérez-Centeno A and Santana-Aranda M A 2012 Effects of the confining solvent on the size distribution of silver NPs by laser ablation *Journal of Nanoparticle Research* **14** 1–6

[26] Tsuji T, Iryo K, Watanabe N and Tsuji M 2002 Preparation of silver nanoparticles by laser ablation in solution: influence of laser wavelength on particle size *Appl Surf Sci* **202** 80–5

[27] Al-Nassar S I, Hussein F I and Ma A K 2019 The effect of laser pulse energy on ZnO nanoparticles formation by liquid phase pulsed laser ablation *Journal of Materials Research and Technology* **8** 4026–31

[28] Zhang D, Gökce B and Barcikowski S 2017 Laser Synthesis and Processing of Colloids: Fundamentals and Applications *Chem Rev* **117** 3990–4103

[29] Arola E, Dahl J, Tuominen M, Laukkanen P, Niemi T and Salminen T 2012 Single-step fabrication of luminescent GaAs nanocrystals by pulsed laser ablation in liquids *Optical Materials Express, Vol. 2, Issue 6, pp. 799-813* **2** 799–813

[30] Swiatkowska-Warkocka Z 2021 Bimetal CuFe Nanoparticles—Synthesis, Properties, and Applications *Applied Sciences* **11** 1978

[31] Gündoğdu Y, Kepceoğlu A, Yiğit Gezgin S, Küçükçelebi H and Şükür Kılıç H 2019 Femtosecond Laser Ablation Synthesis of Nanoparticles and Nano-Hybrides in Ethanol Medium *Mater Today Proc* **18** 1803–10

[32] Liu B, Hu Z, Che Y, Chen Y and Pan X 2007 Nanoparticle generation in ultrafast pulsed laser ablation of nickel *Appl Phys Lett* **90** 044103

[33] Hamad A H, Khashan K S and Hadi A A, Laser Ablation in Different Environments and Generation of Nanoparticles, in "*Applications of Laser Ablation - Thin Film Deposition, Nanomaterial Synthesis and Surface Modification",* Chapter: 8, pp.177-196, Edited by Dongfang Yang, DOI 10.5772/62615, INTECHOPEN, 2016.

[34] Amoruso S, Ausanio G, Bruzzese R, Vitiello M and Wang X 2005 Femtosecond laser pulse irradiation of solid targets as a general route to nanoparticle formation in a vacuum *Phys Rev B Condens Matter Mater Phys* **71** 033406

[35] Cheung J and Horwitz J 1992 Pulsed Laser Deposition History and Laser-Target Interactions *MRS Bull* **17** 30–6

[36] Alami A H and Hawili A A 2020 Synthesis, characterization and applications of FeCu alloys *Applied Surface Science Advances* **1** 100027

Journal of Physics: Conference Series **2751** (2024) 012004 doi:10.1088/1742-6596/2751/1/012004

# Oxidative Stripping of Starch-AgNPs for Nanomolar Indirect Electrochemical Sensing of BSA Using an L-Cysteine Modified Glassy Carbon Electrode

Rania Ziad[a,b], Abdel-Nasser Kawde[a], Ayman AbdelHamid[a], Kais Daoudi[b,c]

[a]Pure and Applied Chemistry Group, Department of Chemistry, College of Sciences, University of Sharjah, P. O. Box 27272, United Arab Emirates.

[b]Center for Advanced Materials Research, Research Institute of Sciences and Engineering, University of Sharjah, P. O. Box 27272, Sharjah, United Arab Emirates.

[c]Department of Applied Physics and Astronomy, College of Sciences, University of Sharjah, P. O. Box 27272, United Arab Emirates.

Email: akawde@sharjah.ac.ae

**Abstract**. In this study, an L-cystine-modified glassy carbon electrode (GCE) was developed as an electrochemical platform based on AgNP oxidative stripping for nanomolar indirect detection of bovine serum albumin (BSA) biomacromolecules. The Poly-L-Cys was formed by L-cysteine (L-Cys) monomers electro-polymerization on the GCE surface. L-Cys pending groups interacted with AgNPs through Ag-S chemical bonds while it interacted with BSA through van der Waals forces and hydrogen bonding. To characterize the bare GCE and GCE/L-Cys, cyclic voltammetry and electrochemical impedance spectroscopy were used. The oxidative stripping of AgNPs by the addition of BSA was monitored using linear sweep stripping analysis.

## 1. Introduction

Genetic information encoded in nucleic acids has enormously fascinated researchers and caught their interest in unraveling its mysteries. Proteins, which are the molecular expression of genetic information, are the major constituent of biological activities. As a vital component of the cell, proteins also play an important role in biological processes. These can be released in pathological conditions, which can be termed protein biomarkers. The sensing of these proteins can be vital in disease diagnosis and treatment. Quantitative analysis of proteins has real applications in pharmacology, clinical medicine, biotechnology, and food nutrition [1, 2]. In vertebrate blood, serum albumin is the most abundant protein. It has a major role in the circulation system mainly in transporting and deporting many proteins and different compounds. They maintain plasma osmotic pressure while also exerting a great role in scavenging free oxygen radicals [2, 3].

Bovine serum albumin (BSA) is a typical serum protein, which has many physiological functions such as binding, transportation, and delivery of different molecules like fatty acids, bilirubin, porphyrins, and steroids [4]. It is one of the dietary proteins, which can be absorbed from food. By consuming bovine milk and meat, humans also get BSA. It is reported that exposure to pathogenic BSA can result in diseases such as membranous nephropathy [5], mad cow disease [6], insulin-dependent diabetes mellitus [7], and Creutzfeldt-Jakob disease [8]. As a result, more attention is being paid to the detection of BSA in immunological and bio-analytical studies, such as fluorimetric determination [9, 10], quartz crystal microbalance [11], electrochemical polarography [12], and infrared spectroscopy [13]. The major drawbacks of these analytical methods are that they necessitate costly equipment and complicated procedures and are associated with limited selectivity. Electrochemical methods, on the other hand, have benefits over conventional techniques as electrochemical biosensors facilitate simple and rapid sensing at low cost [14]. Metal nanoparticles such as Pt, Ag, and Au were extensively employed for fabricating electrochemical sensors due to the intrinsic characteristics existing in these nanoparticles, mainly the electrocatalytic properties, small size, mass transfer, electrochemical conductivity, in addition to their compatibility with living organisms [15, 16].

Modified electrodes fabricated by electro-polymerization of amino acids have been of interest recently due to the simplicity of their preparation, displaying thin films at the modified surface with pending selective functional groups originating from the R-side chains of the amino acids, which can help in forming multilayers functionalized with very selective arm groups [17]. In addition to the amine ($-NH_2$) and carboxyl (-COOH) terminal functional groups existing in all amino acids, L-Cys has the advantageous thiol functional group (-SH), which makes it a special monomer for modifying electrochemical sensors either using chemical or electrochemical modifications [18]. Benefiting from the above-mentioned properties, the naturally occurring zwitterion (L-Cys) is a very promising linker that could be used to stabilize the noble metal nanoparticles, such as AgNPs at the surface of an electrochemical platform and help therefore in indirect electrochemical detection of biomolecules using oxidative stripping voltammetry [19–21].

Electrochemical sensing of BSA was achieved in this work by altering the surface of GCE with L-cysteine while using Starch-AgNPs (SAgNPs) as the electroactive species. In this study, we focused on the electrochemical detection of BSA, with a particular emphasis on the complexation of BSA with SAgNPs. The electrochemical detection protocol is based on the stripping of $Ag^0$, which is present in nanoparticles and bound to BSA at the surface of an L-Cysteine-modified glassy carbon electrode (GCE/L-Cys).

## 2. Experimental part

### *2.1. Instrumentation*
A three-electrode electrochemical cell was used in all experiments, employing bare GCE and GCE/L-Cys working electrodes, KCl-saturated Ag/AgCl as reference electrode, and a Pt wire as counter electrode. A CHI1232 electrochemical workstation was used to run all voltammetry techniques and record the electrochemical measurements, while a Zahner Zennium Pro electrochemical workstation was employed to run electrochemical impedance spectroscopy (EIS).

### *2.2. Working electrode preparation (GCE/L-Cys)*
5.0 mM of L-cystine was dissolved in 0.3 M phosphate buffer solution (PBS) at pH = 9.0, and its electrochemical polymerization on the surface of GCE was done using cyclic voltammetry (CV) with three successive cycles at the potential range of -0.8 V to +1.8 V and a scan rate of 20 mV/s. To remove the residual physically adsorbed L-Cys, the GCE/L-Cys electrode surface was rinsed several times with deionized water. CV and EIS were used to characterize both sensors, the bare GCE and GCE/L-Cys modified sensor, using 5.0 mM equimolar solution of ferro/ferricyanide $[Fe(CN)_6]^{3-/4-}$ in 0.1 M KCl.

Linear sweep voltammetry (LSV) was used to evaluate the electrocatalytic activity of bare GCE and GCE/L-Cys electrodes towards the sensing of BSA, where different concentrations of BSA were spiked into a 2.0 mL cell containing 1.0 mL of SAgNPs and 1.0 mL of citrate buffer (pH=4), based on which a calibration curve was then established.

### *2.3. Preparation of BSA/SAgNPs*

SAgNPs were synthesized according to an experimental procedure reported in our earlier work [22]. The obtained SAgNPs were centrifuged for 10 min at 4000 rpm before use. The absorbance spectra of SAgNPs and BSA/SAgNPs were obtained using UV-vis spectrometry and were used to determine the stoichiometry of complexation between BSA and SAgNPs. All analyses were performed in a microplate reader at a wavelength range of 200–800 nm. BSA concentrations of 0.0, 45.5, 83.3, 115.4, 142.9, 333.3, 375.0, 444.4, and 500.0 nM were analyzed by spiking a constant volume of 100 μL of SAgNP with increasing volumes of BSA. $\Delta\lambda/\Delta\lambda_{max}$ was plotted against the concentration of BSA to find the complexation stoichiometry of BSA and SAgNPs, where $\Delta\lambda$ is the shift in the peak maximum and $\Delta\lambda_{max}$ is the maximum absorbance shift.

## 3. Results and discussion

Figure 1a depicts a plot of $\Delta\lambda/\Delta\lambda_{max}$ vs. various BSA concentrations in nM. When the BSA concentration was increased, the $\Delta\lambda/\Delta\lambda_{max}$ began to rise until it reached a constant value of 1.0. The optimal BSA/SAgNPs stoichiometry was 83.3 nM, after which the value of $\Delta\lambda/\Delta\lambda_{max}$ stabilized at 1.0. Following that, a stock working solution of BSA/SAgNPs was prepared by mixing 1000 μL of SAgNPs with 200 μL of BSA and ultrasonicating the solution for 5 minutes. The obtained BSA/SAgNPs solution was then used for constructing an LSV calibration curve.

The electrochemical activities of unmodified GCE and GCE/L-Cys electrodes were studied using CV in PBS (pH=7) as the blank and 5.0 mM $K_3Fe(CN)_6/K_4Fe(CN)_6$ in 0.1 M KCl. The CVs of both sensors, the unmodified GCE and GCE/L-Cys, were carried out at different scan rates ranging from 5.0 mV/s to 200 mV/s, as shown in Figures 1b,c. Both Figures show CVs of bare GCE and GCE/L-Cys electrodes with reversible anodic and cathodic processes taking place at the surface of the sensor for the redox couple $K_3Fe(CN)_6/K_4Fe(CN)_6$. The CVs of the electrodes in the PBS buffer (pH=7) revealed no electrochemical activity. The electrochemical surface areas of the sensors were determined using the Randles-Sevcik equation (1), given below.

$$I_p = 2.69 \times 10^5 v^{1/2} n^{3/2} C D^{1/2} A \quad (1)$$

Where A is the electrochemical active surface in $cm^2$, D is the diffusion coefficient of the $[Fe(CN)_6]^{4-}$ which is subjected to oxidation at the surface of the electrode taken as $6.31\times10^{-6}$ $cm^2s^{-1}$, C is the concentration of the same ion, taken as 5.0 mM, n represents the number of electrons involved in the ferro/ferricyanide redox reaction, taken as 1 $e^-$, and $v$ is the scan rate of the reaction and is represented in V/s.

A voltammogram overlay for GCE and GCE/L-Cys electrodes is depicted in Figures 1b and 1c. The electroactive surface areas were calculated using the CVs shown in Figures 1b-c, and are shown in Figure 1d as a bar graph. The GCE active electrochemical area was determined to be $6.66\times10^{-3}$ $cm^2$ while a surface area of $4.44\times10^{-3}$ $cm^2$ was found for GCE/L-Cys. The non-conductive nature of L-Cys caused the decrease in surface area.

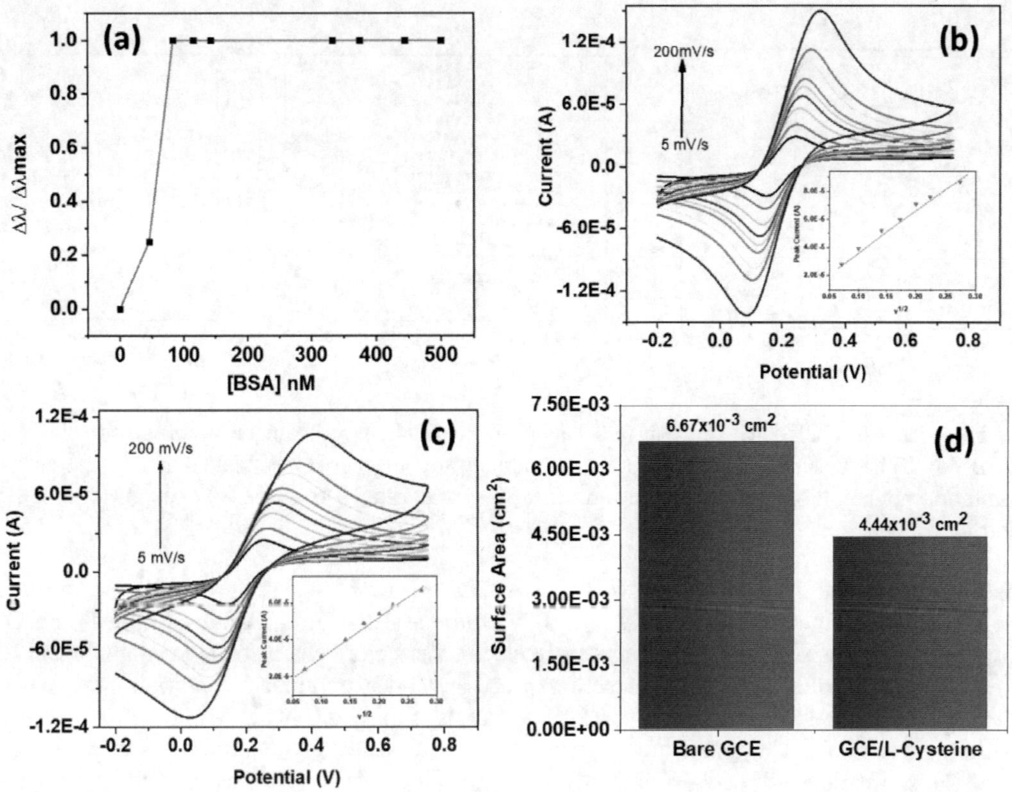


**Figure 1.** (a) $\Delta\lambda/\Delta\lambda_{max}$ plot vs. BSA concentrations ranging from 0 to 500 nM. (b, c) Overlay of CV at various scan rates ranging from 5.0 $mVs^{-1}$ to 200 $mVs^{-1}$, used for the determination of the electroactive surface area for (b) unmodified GCE and (c) GCE/L-Cys. (d) Surface area calculated for bare GCE and GCE/L-Cys.

The oxidative stripping of SAgNPs in the presence of BSA was monitored using LSV. Figure 2a depicts LSVs of various solutions at the surface of GCE/L-Cys in citrate buffer (pH=4.0) over a potential range of -0.50–0.50 V. In the absence of SAgNPs, no anodic peak was observed for GCE/L-Cys in both the blank and BSA solutions, as shown in Figure 2a. When AgNPs were added to the buffer solution, a clearly defined oxidation peak was observed at +0.16 V, confirming the stripping of SAgNPs on the electrode's surface. The addition of BSA to the Ag-containing solution improved the current response. This improvement supported the formation of a BSA/SAgNPs complex on the surface of the GCE electrode.

The influence of the concentration of BSA was studied and a calibration curve was established (Figure 2b-c). Different concentrations of BSA were spiked into a 2.0 mL cell containing 1.0 ml of SAgNPs and 1.0 ml of citrate buffer (pH=4.0). The concentration of BSA was varied from 0.0 to 45.5 nM, and the relative peak current was measured by linear sweep anodic stripping voltammetry (LSASV) using GCE/L-Cys electrode as shown in Figure 2b. There was a linear increase in the peak current with BSA concentration from 1.4 μA to 23 μA, with a limit of detection of 4.0 nM and a limit of quantification of 13.5 nM, as shown in Figure 2c.

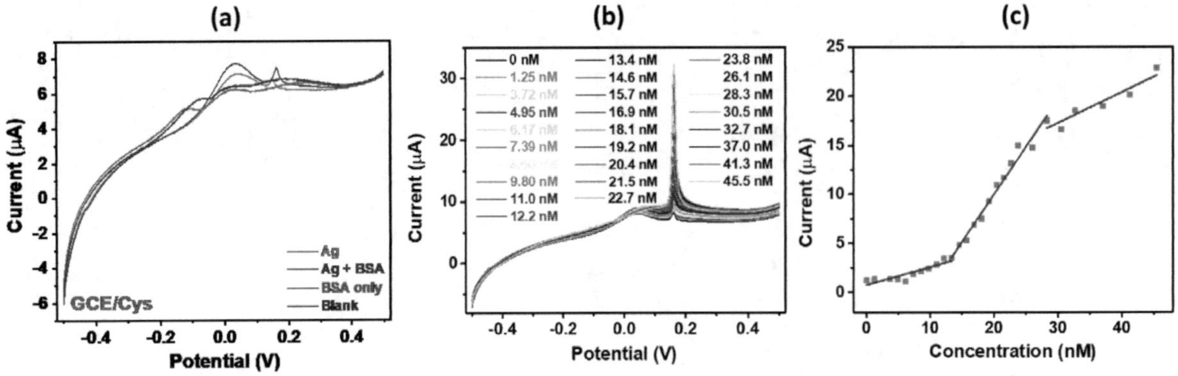


**Figure 2.** (a) LSV of GCE/L-Cys in citrate 0.1 M buffer (pH=4.0) in the absence and presence of BSA and SAgNPs. (b) LSASV of BSA (0 nM to 45.5 nM) in citrate buffer (pH = 4.0) at GCE/L-Cys electrode. Accumulation potential: -0.5 V; accumulation time: 30 s; scan rate: 100 mV/s. (c) Peak current vs. BSA concentration.

## 4. Conclusion

For the first time, the mechanism of BSA sensing with the assistance of SAgNPs stripped on a GCE modified by L-Cys was studied. The sensors were electrochemically characterized using CV and EIS, and LSASV was used for the BSA detection process. Under optimized conditions, the electro-polymerization of the selected monomer (L-Cys) yields a polymer with a high affinity for Ag nanoparticles and BSA. This phenomenon was used to fabricate a voltammetric sensor for BSA detection. This was done by using Starch-AgNPs as an electroactive species to modify a GCE by immobilizing L-Cys on its surface. L-Cys interacted with BSA via van der Waals weak interactions as well as strong hydrogen bonding leading to a reduction in the interaction with competitive water molecules. Moreover, the thiol group of L-Cys may have interacted strongly with AgNPs via Ag-S covalent bond leading to the production of more sensitive analytical results. LSV was used to detect BSA on the surface of GCE/L-Cys indirectly by stripping the Ag metal in its nanoparticle form (SAgNPs). The results revealed that an enhancement in the stripping oxidation peak of Ag upon the addition of BSA was obtained only in the case of LSV, due to the ability to obtain relevant current responses within a few milliseconds after the electrode was stimulated.

## 5. Declaration of Competing Interest

No known conflict or competing interest

## 6. Acknowledgments

The authors acknowledge the research fund provided by the Department of Graduate studies, The research Institute of Sciences and Engineering, University of Sharjah, Graduate students support grant dated 14/06/2022 and student ID: U20104420, and the Competitive Research Project (V.C.R.G/ R. 438/2021).

## References

[1] Doumas, B.T., Peters, T.: Serum and urine albumin: a progress report on their measurement and clinical significance. Clinica Chimica Acta. 258, 3–20 (1997). https://doi.org/10.1016/S0009-8981(96)06446-7

[2] Dockal, M., Carter, D.C., Rüker, F.: Conformational Transitions of the Three Recombinant Domains of Human Serum Albumin Depending on pH. Journal of Biological Chemistry. 275, 3042–3050 (2000). https://doi.org/10.1074/JBC.275.5.3042

[3] Qin, W., Dan, W., Bin, D., Zaijun, L., Yanqiang, H.: A spectrophotometric method for

determination of total proteins in cow milk powder samples using the o-nitrophenylfluorone/Mo(VI) complex. Journal of Food Composition and Analysis. 19, 76–82 (2006). https://doi.org/10.1016/J.JFCA.2005.04.003

[4] Babu, E., Muthu Mareeswaran, P., Singaravadivel, S., Bhuvaneswari, J., Rajagopal, S.: A selective, long-lived deep-red emissive ruthenium(II) polypyridine complexes for the detection of BSA. Spectrochim Acta A Mol Biomol Spectrosc. 130, (2014). https://doi.org/10.1016/j.saa.2014.04.060

[5] Debiec, H., Lefeu, F., Kemper, M.J., Niaudet, P., Deschênes, G., Remuzzi, G., Ulinski, T., Ronco, P.: Early-Childhood Membranous Nephropathy Due to Cationic Bovine Serum Albumin. New England Journal of Medicine. 364, (2011). https://doi.org/10.1056/nejmoa1013792

[6] Hillerton, J.E.: Bovine Spongiform Encephalopathy: Current Status and Possible Impacts. J Dairy Sci. 81, (1998). https://doi.org/10.3168/jds.s0022-0302(98)75869-2

[7] Persaud, D.R., Barranco-Mendoza, A.: Bovine serum albumin and insulin-dependent diabetes mellitus: Is cow's milk still a possible toxicological causative agent of diabetes?, (2004)

[8] Brown, D.R.: Neurodegeneration and prion disease. (2005)

[9] Zhang, W., He, X.W., Chen, Y., Li, W.Y., Zhang, Y.K.: Molecularly imprinted polymer anchored on the surface of denatured bovine serum albumin modified CdTe quantum dots as fluorescent artificial receptor for recognition of target protein. Biosens Bioelectron. 31, 84–89 (2012). https://doi.org/10.1016/J.BIOS.2011.09.042

[10] Yang, Y.Q., He, X.W., Wang, Y.Z., Li, W.Y., Zhang, Y.K.: Epitope imprinted polymer coating CdTe quantum dots for specific recognition and direct fluorescent quantification of the target protein bovine serum albumin. Biosens Bioelectron. 54, 266–272 (2014). https://doi.org/10.1016/J.BIOS.2013.11.004

[11] Lin, T.Y., Hu, C.H., Chou, T.C.: Determination of albumin concentration by MIP-QCM sensor. Biosens Bioelectron. 20, 75–81 (2004). https://doi.org/10.1016/J.BIOS.2004.01.028

[12] Luo, D., Lan, J., Zhou, C., Luo, C.: Polarographic Behavior of Co(II)-BSA or -HSA Complex in the Presence of a Guanidine Modifier. Anal Chem. 75, (2003). https://doi.org/10.1021/ac0300643

[13] Wittemann, A., Ballauff, M.: Secondary structure analysis of proteins embedded in spherical polyelectrolyte brushes by FT-IR spectroscopy. Anal Chem. 76, (2004). https://doi.org/10.1021/ac0354692

[14] Vestergaard, M., Kerman, K., Tamiya, E.: An overview of label-free electrochemical protein sensors, (2007)

[15] Hamidi, H., Haghighi, B.: Fabrication of a sensitive amperometric sensor for NADH and H2O2 using palladium nanoparticles-multiwalled carbon nanotube nanohybrid. Materials Science and Engineering C. 62, (2016). https://doi.org/10.1016/j.msec.2016.01.058

[16] Saha, K., Agasti, S.S., Kim, C., Li, X., Rotello, V.M.: Gold nanoparticles in chemical and biological sensing, (2012)

[17] Zhang, D., Li, L., Ma, W., Chen, X., Zhang, Y.: Electrodeposited reduced graphene oxide incorporating polymerization of l-lysine on electrode surface and its application in simultaneous electrochemical determination of ascorbic acid, dopamine and uric acid. Materials Science and Engineering: C. 70, 241–249 (2017). https://doi.org/10.1016/J.MSEC.2016.08.078

[18] Zhang, K., Luo, P., Wu, J., Wang, W., Ye, B.: Highly sensitive determination of Sunset Yellow in drink using a poly (l-cysteine) modified glassy carbon electrode. Analytical Methods. 5, (2013). https://doi.org/10.1039/c3ay40873f

[19] Duan, D., Yang, H., Ding, Y., Ye, D., Li, L., Ma, G.: Three-dimensional molecularly imprinted electrochemical sensor based on Au NPs@Ti-based metal-organic frameworks for ultra-trace detection of bovine serum albumin. Electrochim Acta. 261, (2018). https://doi.org/10.1016/j.electacta.2017.12.146

[20] Sangili, A., Kalyani, T., Chen, S.M., Rajendran, K., Jana, S.K.: Label-free electrochemical immunosensor based on L-cysteine-functionalized AuNP on reduced graphene oxide for the detection of dengue virus E-protein in dengue blood serum. Compos B Eng. 238, (2022). https://doi.org/10.1016/j.compositesb.2022.109876

[21] Fathalipour, S., Pourbeyram, S., Sharafian, A., Tanomand, A., Azam, P.: Biomolecule-assisted synthesis of Ag/reduced graphene oxide nanocomposite with excellent electrocatalytic and antibacterial performance. Materials Science and Engineering: C. 75, 742–751 (2017). https://doi.org/10.1016/J.MSEC.2017.02.122

[22] Elgamouz, A., Bajou, K., Hafez, B., Nassab, C., Behi, A., Haija, M.A., Patole, S.P.: Optical sensing of hydrogen peroxide using starch capped silver nanoparticles, synthesis, optimization and detection in urine. Sensors and Actuators Reports. 2, (2020). https://doi.org/10.1016/j.snr.2020.100014

# Enhancing aqueous Zn-Mn battery performance using $Na^+$ ion conducting ceramic membrane

**S.T. Senthilkumar**[1], **Hussain Alawadhi**[1,2] **and Anis Allagui**[1,3]

[1]Center for Advanced Materials Research, Research Institute of Sciences and Engineering, University of Sharjah, P.O. Box 27272, Sharjah, United Arab Emirates
[2]Department of Applied Physics & Astronomy, University of Sharjah, P.O. Box 27272, Sharjah, United Arab Emirates
[3]Department of Sustainable and Renewable Energy Engineering, University of Sharjah, P.O. Box 27272, Sharjah, United Arab Emirates

senkumphy@gmail.com, halawadhi@sharjah.ac.ae and aallagui@sharjah.ac.ae

**Abstract**. The low cell voltage and capacity of conventional Zn-$MnO_2$ batteries often result in limited energy density. In this study, we assembled a Zn-$MnO_2$ battery based on the acid-alkaline electrolyte decoupled concept and reversible $MnO_2/Mn^{2+}$ deposition/dissolution chemistry to increase the cell voltage and capacity. We used a $Na^+$ ion conducting NASICON ceramic membrane in the battery to decouple the acid and alkaline electrolytes effectively. The assembled Zn-$MnO_2$ battery demonstrated a cell voltage of 2.43 V and a coulombic efficiency (CE) of 90% at a current density of 0.2 mA/cm$^2$. It also exhibited excellent rechargeability with continuous charge and discharge cycles. After 100 cycles, the battery exhibits a capacity of 0.36 mAh with a maximum CE of 93.42%. A study on the battery's self-discharge performance showed that a maximum of 71% capacity could be recovered from the charged battery after 24 hours of rest. Finally, we fabricated a pouch-type acid-alkaline electrolyte decoupled Zn-$MnO_2$ battery and examined its feasibility.

## 1. Introduction

Our modern civilization and industrialization heavily rely on natural resources such as fossil fuels and coal-based components for electricity production, resulting in alarming global warming due to greenhouse gas emissions, particularly carbon dioxide ($CO_2$) [1–3]. To reduce $CO_2$ emissions from the electricity production sector, several countries have extensively deployed solar and wind power as clean, sustainable, and green energy technologies. However, their intermittent nature poses a challenge that requires the integration of large-scale electrical energy storage (EES) systems [4]. Electrochemical rechargeable batteries are considered a potential candidate for deployment in large-scale EES systems [4]. Therefore, there is a competition to develop rechargeable batteries by adopting new chemistry and architecture, with a focus on aqueous electrolytes [5,6] to meet the requirements for large-scale EES applications, such as easy manufacturing, transportability, environmentally friendly, reliable energy storage, and longer lifespan.

As of now, various aqueous batteries based on monovalent ($Na^+$, $Li^+$, and $H^+$) [7,8] and multivalent ($Zn^{2+}$, $Mg^{2+}$, and $Al^{2+}$) [9] chemistries have emerged for deployment in large-scale EES. Among the various aqueous batteries, the Zn-$MnO_2$ battery has recently received tremendous attention for large-scale stationary EES because its major components (i.e., Zn, $MnO_2$, and electrolyte) are cheaper, user-friendly and safer [10]. Additionally, the high ionic conductivity of aqueous electrolytes makes this system more advantageous than flammable organic electrolyte systems, such as Li-ion batteries [5,6]. Zn offers a highly reversible electrochemical redox reaction ($Zn/Zn^{2+}$) with a redox potential of -0.76 V (vs. standard hydrogen electrode (SHE)) and a theoretical specific capacity of 820 mAh/g [11]. $MnO_2$ also offers a theoretical capacity of 308 mAh/g for one-electron processes and 616 mAh/g for two-electron processes [11]. However, in the conventional Zn-$MnO_2$ battery, the $Zn^{2+}$ ion insertion/deinsertion process at $MnO_2$ supports only a one-electron transfer process, which not only limits the capacity (308 mAh/g) but also limits the cell voltage (~1.2-1.4 V) [10]. Therefore, the conventional Zn-$MnO_2$ battery can only achieve the practical energy density of ~369-430 Wh/kg.

Cui et al. [12] recently proposed a $MnO_2/Mn^{2+}$ deposition/dissolution chemistry with a two-electron reaction that theoretically allows for a high capacity of 616 mAh/g and a redox potential of 1.224 V (vs. SHE). This chemistry has been used to develop a Zn-$MnO_2$ battery with a theoretical cell voltage of 1.99 V [13,14], which, in conjunction with $MnO_2$'s capacity (616 mAh/g), could achieve a theoretical energy density of 1226 Wh/kg. However, the use of an acidic electrolyte to ensure $MnO_2/Mn^{2+}$ dissolution in the Zn-$MnO_2$ battery poses significant challenges. For instance, the Zn metal may undergo self-dissolution in acidic electrolytes, leading to hydrogen evolution and associated corrosion [15]. Besides, the operation of those batteries above 2 V during charging could cause additional corrosion at the Zn anode and hydrogen evolution. Subsequently, the resulting hydrogen evolution reactions can reduce the proton content in the electrolyte, which could affect the reversibility of the $MnO_2/Mn^{2+}$ reaction. Overall, this circumstance could reduce the battery's cycle life [10].

Conversely, it is acknowledged that using alkaline electrolytes can provide better stability to the Zn anode and enable it to have a high redox potential (-1.119 V (vs. SHE)) through the reversible $Zn/Zn(OH)_4^{2-}$ redox reaction. Considering this, a recent approach involved combining the Zn anode immersed in an alkaline electrolyte with the $MnO_2$ cathode (or carbon cathode electrode) immersed in an acidic electrolyte within the same battery system, separated by selective ion exchange polymer membranes [16,17]. This alkaline-acid electrolyte decoupling strategy extended the theoretical cell voltage to 2.423 V [18] and energy density to 1492 Wh/kg for the Zn-$MnO_2$ battery. However, the use of polymer-based ion-exchange membranes (For instance, Nafion) to decouple the alkaline and acid electrolytes is known to be costly, and their porous nature does not provide complete control over the migration of electrolytic ions and water between the alkaline and acidic electrolytes. This could potentially lead to changes in the pH of the electrolytes, which could ultimately limit the overall operation of the battery. Therefore, further modifications are necessary to effectively decouple the alkaline and acid electrolytes in the Zn-$MnO_2$ battery.

Herein, we assembled an acid-alkaline electrolyte decoupled Zn-$MnO_2$ battery using a NASICON ceramic membrane, which can conduct only $Na^+$ ions. We then employed the assembled battery for electrochemical characterizations. As a result, the assembled battery with NASICON yielded a cell voltage, capacity and coulombic efficiency of 2.43 V, and 90%, respectively, at a current density of 0.2 $mA/cm^2$. After 100 cycles, the capacity was 0.36 mAh, and the coulombic efficiency was 93.42%; however, the retained capacity after 100 cycles was 100%, concerning the capacity of the second cycle. Additionally, we fabricated a pouch-type battery and tested its rechargeability. The capacity of 0.28 mAh with 94.33% CE was attained for the pouch-type battery after 100 cycles.

## 2. Experimental methods

### *2.1. Materials*

We have used the following chemicals and materials in our present work, which are commercially accessible. Sodium hydroxide (NaOH (sd fine chemicals)), Zinc acetate ($ZnC_4H_6O_4$ (sd fine chemicals)), Sodium sulfate ($Na_2SO_4$ (sd fine chemicals)), Manganese acetate ($MnC_4H_6O_4$ (sd fine chemicals)), Sulfuric acid ($H_2SO_4$ (sd fine chemicals)) and Zn foil (Sigma-Aldrich). NASICON ($Na_3Zr_2Si_2PO_{12}$) ceramic membrane received from 4TOONE Corporation, South Korea. Carbon cloth from fuel cell stores was used after heat treatment at 500°C for 1 h. All others were used as received without any modification.

### *2.2. Battery assembly*

Using a home-built double-chamber electrochemical cell, we constructed an acid-alkaline electrolyte decoupled Zn-$MnO_2$ battery. In brief, we used three acrylic plates - two with a thickness of 10 mm and one with a thickness of 5 mm. We attached a NASICON ceramic separator to the 5 mm plate and positioned it between the other two 10 mm plates to create two compartments - one for the acid electrolyte and another for the alkaline electrolyte. The zinc (Zn) foil (3 $cm^2$) served as the anode with the alkaline electrolyte, while a carbon cloth (1 $cm^2$) was used as the cathode current collector with the acidic electrolyte. We note that the carbon cloth was used after heat-treated at 500°C for 1 h. The alkaline electrolyte was a mixture of 6 M NaOH and 2 M $ZnC_4H_6O_4$, and the acid electrolyte was a mixture of 1 M $H_2SO_4$, 0.5 M $Na_2SO_4$, and 0.2 M $MnC_4H_6O_4$. The NASICON was exposed to the electrolytes over an area of ~2.25 $cm^2$. The Whatman filter paper (WFP) was placed between the NASICON and Zn foil, as well as between the NASICON and carbon cloth, to avoid direct physical contact. We injected 1.5 mL of acid electrolyte and 1.8 mL of alkaline electrolyte into their respective compartments and used Ti wire with Zn foil and carbon felt, respectively, to make electrical contacts. Finally, we sealed the hole with plastic glue after injecting the electrolytes.

The pouch-type battery was fabricated using a commercial polymer laminated aluminum film and a NASICON ceramic membrane, with a Zn foil as the anode and a graphite foil as the cathode current collector. The same electrolytes that were used in the home-built double-chamber electrochemical cell were also used in the pouch-type battery.

### *2.3. Electrochemical test*

The electrochemical properties of the anodic ($Zn/Zn(OH)_4^{2-}$) reaction on the Zn foil in an alkaline electrolyte (6 M NaOH/2 M $ZnC_4H_6O_4$) and the cathodic ($Mn^{2+}/MnO_2$) reaction on the carbon felt in an acidic electrolyte (1 M $H_2SO_4$/0.5 M $Na_2SO_4$/0.2 M $MnC_4H_6O_4$) were studied using cyclic voltammetry (CV) at a sweep rate of 0.1 mV/s. The CVs were performed individually on a three-electrode system. To examine the anodic reaction, the CV was recorded using Ti wire, Zn foil (1 $cm^2$), and Hg/HgO as the working, counter, and reference electrodes. Then, to study the cathodic reaction, the CV was recorded using carbon cloth (1 $cm^2$), Pt plate (1 $cm^2$), and Ag/AgCl as the working, counter, and reference electrodes, respectively. We have employed the equation, "$E_{SHE} = E_{Hg/HgO} - 0.826 + 0.059$ pH $+ E^o_{Hg/HgO}$" to convert the experimentally measured potential vs. Hg/HgO to SHE and "$E_{SHE} = E_{Ag/AgCl} + 0.118$ pH $+ E^o_{Ag/AgCl}$" to convert the experimentally measured potential vs. Ag/AgCl to SHE. Where $E_{SHE}$ is the converted potential vs. SHE, $E_{Hg/HgO}$ is experimentally measured potential against Hg/HgO, $E^o_{Hg/HgO}$ is electrode potential of the used Hg/HgO electrode, $E_{Ag/AgCl}$ is experimentally measured potential against Ag/AgCl, $E^o_{Ag/AgCl\ standard}$ is electrode potential of the used Ag/AgCl electrode, and pH refers to the pH of the electrolyte used at anode and cathode side.

CV of the assembled battery was recorded at a sweep rate of 0.2 mV/s within a voltage window of 1.7-2.8 V. The constant voltage technique was used to charge the battery to its cut-off capacity of 0.4 mAh. In contrast, the galvanostatic method was used to discharge the battery to its cut-off voltage of 1.9 V. To charge the batteries, a voltage of 2.8 V was used. The self-discharge performance of the charged battery was evaluated by measuring the open-circuit voltage (OCV) under resting conditions for 24 hours at room temperature.

During the testing of the pouch-type battery, a constant charging voltage of 2.8 V was utilized along with a cut-off capacity of 0.3 mAh, while a cut-off voltage of 1.9 V was employed during the galvanostatic discharge.

### *2.4. Characterization*

The CV measurements of the Zn anode and carbon cloth (cathode) electrodes were recorded using the SP 200 potentiostat from BioLogic instruments. Battery characterizations, including cyclic voltammetry

(CV), galvanostatic charge and discharge (GCD), and self-discharge tests, were carried out using the BCS-800 series battery cycler from BioLogic instrument. X-ray diffraction (XRD) patterns were obtained using the Bruker D8 advance, while the Raman spectrum was collected using the Renishaw InVia Raman microscope.

## 3. Results and Discussion

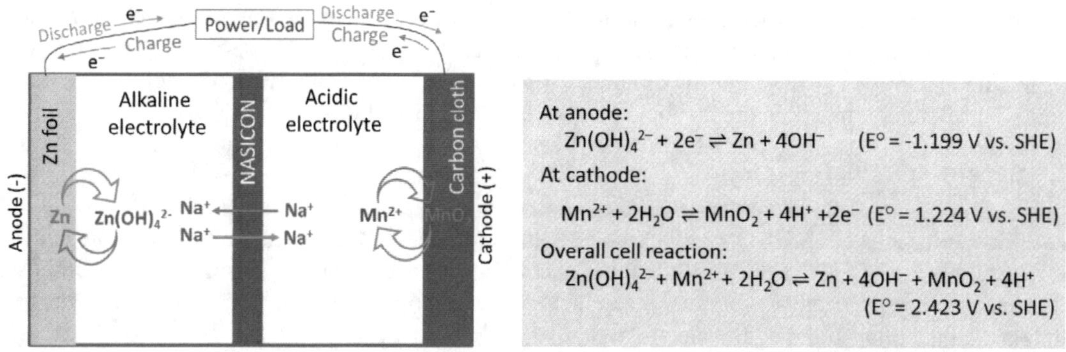


**Figure 1.** Schematic illustration of an acid-alkaline electrolyte decoupled Zn-$MnO_2$ battery that utilizes a NASICON ceramic membrane, and the electrochemical processes involved in this battery.

As shown in Figure 1, the assembled battery consists of a Zn foil anode immersed in an alkaline electrolyte (6 M NaOH/2 M $ZnC_4H_6O$) and a carbon cloth cathode current collector immersed in an acidic electrolyte (1 M $H_2SO_4$/0.5 M $Na_2SO_4$/0.2 M $MnC_4H_6O_4$), separated by a NASICON ceramic membrane. During the battery's charging process, $Mn^{2+}$ ions in the acidic electrolyte oxidize into $MnO_2$ on the carbon cloth's surface, releasing two electrons. These electrons are transferred through the external circuit from the carbon cloth to the Zn anode, resulting in Zn deposition on the Zn anode surface. This deposition occurs due to the reduction of $Zn(OH)_4^{2-}$ in the alkaline electrolyte. The reaction process at the Zn anode and carbon cloth during the battery's discharge process is highly reversible. $Na^+$ ions move between the acid and alkaline electrolyte chambers through the NASICON ceramic membrane during the charge and discharge process to maintain charge balance in the battery.

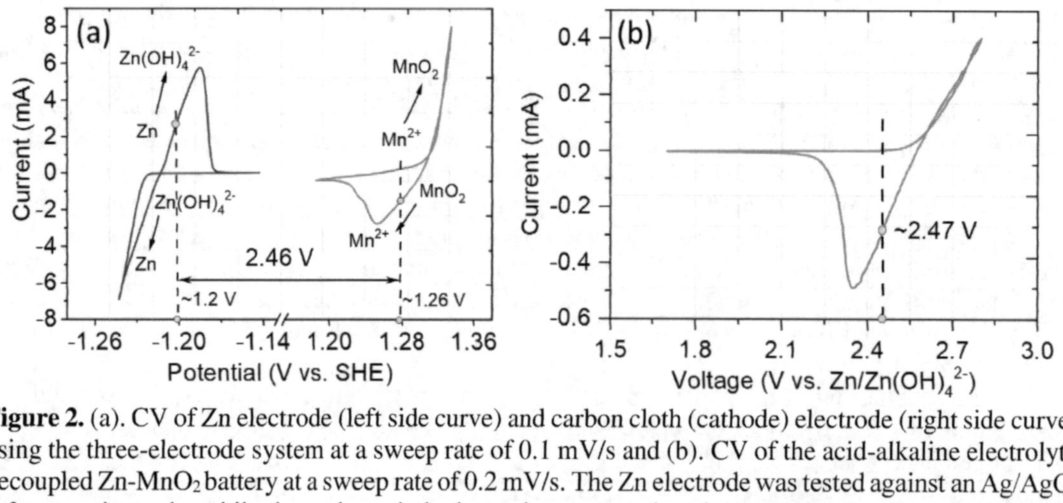


**Figure 2.** (a). CV of Zn electrode (left side curve) and carbon cloth (cathode) electrode (right side curve) using the three-electrode system at a sweep rate of 0.1 mV/s and (b). CV of the acid-alkaline electrolyte decoupled Zn-$MnO_2$ battery at a sweep rate of 0.2 mV/s. The Zn electrode was tested against an Ag/AgCl reference electrode, while the carbon cloth electrode was tested against a Hg/HgO reference electrode. To conveniently estimate the cell voltage from the anode and cathode reaction potentials, the x-axis of the CVs in Figure 2a was converted to a standard hydrogen electrode (SHE).

The redox potential of the $Zn/Zn(OH)_4^{2-}$ reaction at the Zn electrode in 6 M NaOH/2 M $ZnC_4H_6O_4$ electrolyte and the $MnO_2/Mn^{2+}$ reaction on the carbon cloth electrode in 1 M $H_2SO_4$/0.5 M $Na_2SO_4$/0.2 M $MnC_4H_6O_4$ electrolyte was determined using CV measurement before studying the acid-alkaline electrolyte decoupled Zn-$MnO_2$ battery. The CVs recorded at 0.1 mV/s are shown in Figure 2a. The Zn deposition and stripping reactions occur at potentials of -1.22 V and -1.21 V (vs. SHE), respectively, with a relatively small difference in potential, indicating good reversibility. The cathode CV shows an evolution peak for $MnO_2$ deposition starting at ~1.3 V (vs. SHE) and a broad reduction peak corresponding to $MnO_2$ dissolution into $Mn^{2+}$ with the reduction peak potential at ~1.25 V (vs. SHE). Based on the half-wave potential of the CVs, the estimated cell voltage for the assembled Zn-$MnO_2$ battery is ~2.46 V (Figure 2a). The assembled battery was also tested using CV at 0.2 mV/s, and the cell voltage was found to be ~2.47 V (Figure 2b), which matches the theoretical cell voltage of 2.42 V calculated based on the theoretical redox potentials of the $Zn/Zn(OH)_4^{2-}$ and $MnO_2/Mn^{2+}$ reactions (Figure 1).

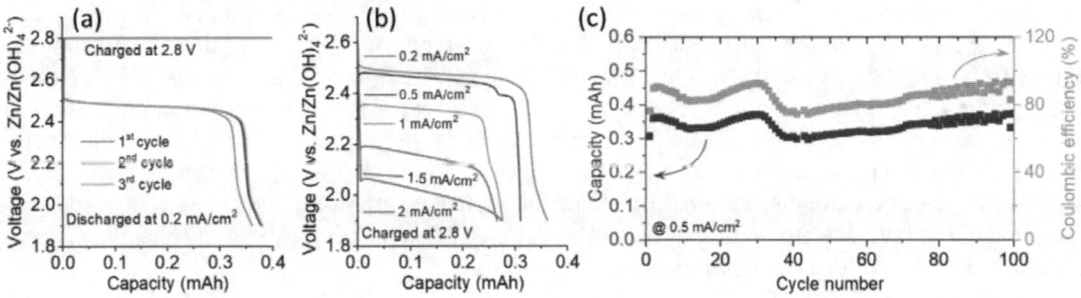


**Figure 3.** Electrochemical performances of the acid-alkaline electrolyte decoupled Zn-$MnO_2$ battery. (a). First three discharge curves at 0.2 mA/cm$^2$. (b). Discharge curves at different current densities, and (c). Cycle performance at 0.5 mA/cm$^2$.

The battery was charged using a constant voltage charging technique until it reached its capacity cut-off of 0.4 mAh and then discharged using the galvanostatic method with a voltage cut-off of 0.9 V. The first three discharge curves of the battery are shown in Figure 3a at 0.2 mA/cm$^2$, and it can be observed that the capacity and CE increased from 0.36 mAh and 90% to 0.38 mAh and 95% after two cycles. The discharge cell voltage of the battery was found to be 2.43 V, close to the theoretical cell voltage (2.42 V) and the estimated cell voltage based on the CV results (~2.46 V). Figure 3b shows discharge curves collected at different current densities after charging the battery to 0.4 mAh at 2.8 V. All discharge curves exhibit clear and flat voltage plateaus, but the capacity decreases from 0.36 to 0.27 mAh as the discharge current density increases from 0.2 to 2 mA/cm$^2$. Even at 2 mA/cm$^2$, the achieved capacity is still 75% of the capacity achieved at 0.2 mA/cm$^2$. Unfortunately, increasing the discharge current density leads to a decrease in the discharge cell voltage from 2.43 V to 2.13 V due to an increase in the battery's internal (or ohmic) resistance and polarization of electrode active materials. The cycle stability of the battery was tested at 0.5 mA/cm$^2$ (Figure 3c). In the first cycle, the capacity and CE were 0.31 mAh and 76.62%, respectively, and in the second cycle, they were 0.36 mAh and 89.84%. However, the capacity of 0.36 mAh, the same as that achieved in the second cycle, was retained after 100 cycles with an improved CE of 93.4%. Therefore, considering the discharge capacity of the second cycle, the achieved capacity retention after 100 cycles is about 100%.

To verify $MnO_2$ deposition/dissolution, Raman spectra and XRD were used to examine the carbon cloth electrode after charge and discharge. The Mn-O band related to $MnO_2$ between ~447-772 cm$^{-1}$ in Raman spectra (Figure 4a), and the diffraction peaks at 36.64° and 40.72° correspond to $MnO_2$ in the XRD pattern have been observed (Figure 4b) after the charge [16]. After discharge, the peaks belonging to $MnO_2$ disappear from the Raman spectra and XRD (Figures 4a and b). These results confirm $MnO_2$ deposition on carbon cloth and its complete dissolution during battery charge and discharge.

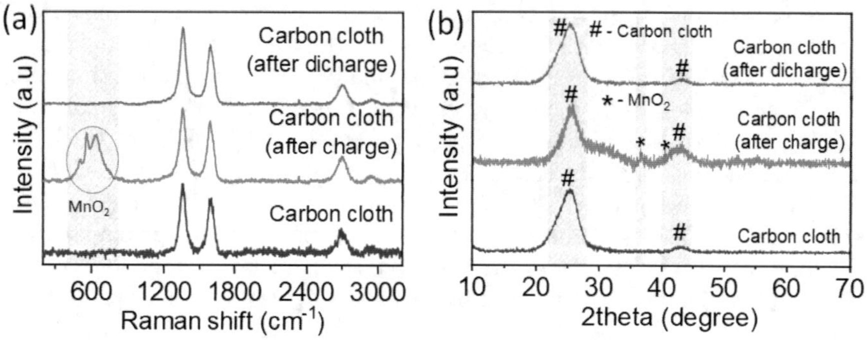


**Figure 4.** Ex-situ studies of carbon cloth electrode after charge and discharge. (a). Raman spectra and (b). XRD pattern.

Studying the self-discharge characteristics of a battery is critically important. As planned, we evaluated the self-discharge performance of a 0.4 mAh charged battery by measuring its open circuit voltage (OCV) under rest conditions for 24 hours. As shown in Figure 5a, a stable OCV voltage was observed throughout the rest period. However, after 24 hours of rest, the battery could only deliver a discharge capacity of 0.23 mAh (Figure 5b), which is 71% of the charged capacity. This indicates the occurrence of self-discharge in the battery. As illustrated in Figure 5c, the self-discharge process could be caused by the exfoliation of cathode active materials from the current collector (i.e., deposited $MnO_2$ on carbon cloth) [19]. Nevertheless, this battery has attractive merits, as >90% of the charged capacity can be recovered when discharged immediately after charging.

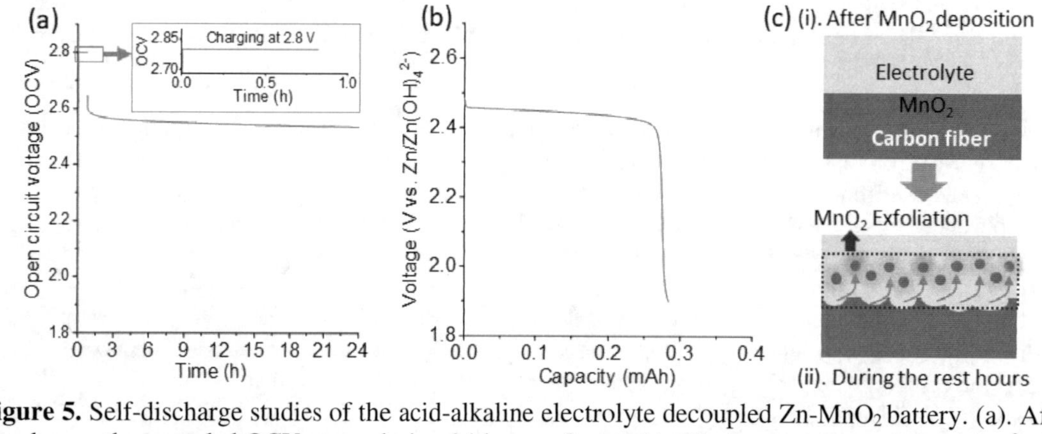


**Figure 5.** Self-discharge studies of the acid-alkaline electrolyte decoupled Zn-$MnO_2$ battery. (a). After the charge, the recorded OCV curve during 24 hours of rest. (b). Discharge curve at 0.2 mA/cm$^2$ after 24 hours of rest and (c). Schematic illustration of $MnO_2$ exfoliation from the electrode. This study was performed after 15 charge and discharge cycles.

We have also conducted additional work on the feasibility of using NASICON ceramic membrane to assemble a pouch-type battery. Notably, the cathode current collector in the assembled pouch cell was graphite foil instead of carbon cloth. Figure 6a shows the results of the first three discharge cycles at 0.05 mA/cm$^2$, where we obtained a capacity of 0.26 mAh with a CE and cell voltage of 86.6% and 2.38 V in the first discharge. In the third discharge cycle, the capacity, CE, and cell voltage were 0.29 mAh, 96.6%, and 2.39 V, respectively. Furthermore, we tested the cycle performance of the fabricated pouch cell at a current density of 0.05 mA/cm$^2$, and the obtained results are presented in Figure 6b. The cycling results demonstrate better cycle stability, with a capacity of 0.28 mAh and a CE of 94.33% after 100 cycles. This outcome suggests the possibility of developing an acid-alkaline electrolyte decoupled pouch-type Zn-$MnO_2$ battery and its practicality, although further research and development are necessary for real-time applications.

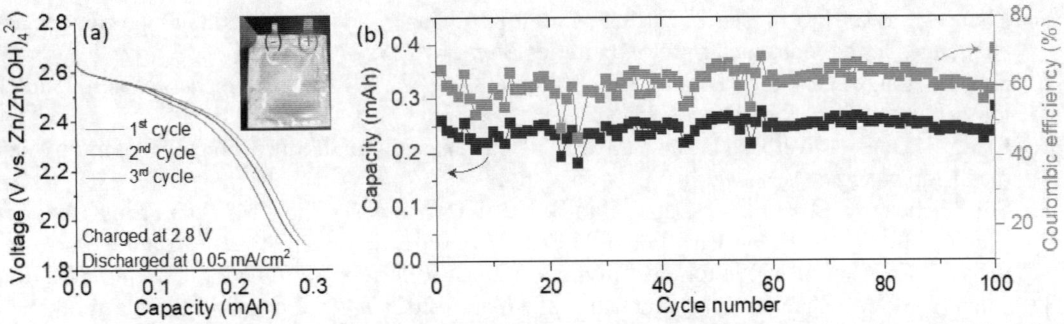


**Figure 6.** Electrochemical performance of the pouch-type acid-alkaline electrolyte decoupled Zn-$MnO_2$ battery at 0.05 mA/cm$^2$. (a) Discharge curves and (b) Cycle performance. The inset in Figure 6a shows the photographic image of the assembled pouch-type acid-alkaline electrolyte decoupled Zn-$MnO_2$ battery.

## 4. Conclusion

In summary, this work demonstrates the successful construction of a Zn-$MnO_2$ battery utilizing the acid-alkaline electrolyte decoupled approach, reversible $MnO_2/Mn^{2+}$ deposition/dissolution chemistry, and a $Na^+$ ion conducting NASICON ceramic membrane. The NASICON ceramic membrane effectively separates the acid and alkaline electrolytes, leading to a cell voltage and a CE of 2.43 V and 90%, respectively, at 0.2 mA/cm$^2$. The assembled battery exhibits remarkable rechargeability, and it was observed that a capacity and CE of 0.36 mAh and 93.42%, respectively, can be maintained even after 100 cycles. The feasibility of the NASICON ceramic membrane is also assessed by fabricating a pouch-type acid-alkaline electrolyte decoupled Zn-$MnO_2$ battery. After 100 cycles, a pouch-type battery achieved a capacity and CE of 0.28 mAh and 94.33%, respectively. Overall, this study emphasizes the potential of utilizing NASICON membranes in acid-alkaline electrolyte decoupled Zn-$MnO_2$ batteries, demonstrating their promising electrochemical performances.

**Acknowledgments**
We thank Prof. Youngsik Kim from Ulsan National Institute of Science and Technology (UNIST) and 4toONE Energy in Ulsan, South Korea, for generously providing the NASICON ceramic membrane used in this study. Additionally, we appreciate the staff members of the advanced materials research laboratory (AMRL) for their support in material characterization.

## References

[1] United Nations 2013 Mitigating Climate Change Through Investments in Fossil Fuel Technologies 75
[2] Hansen J, Kharecha P, Sato M, Masson-Delmotte V, Ackerman F, Beerling D J, Hearty P J, Hoegh-Guldberg O, Hsu S L, Parmesan C, Rockstrom J, Rohling E J, Sachs J, Smith P, Steffen K, Van Susteren L, Von Schuckmann K and Zachos J C 2013 Assessing "dangerous climate change": Required reduction of carbon emissions to protect young people, future generations and nature *PLoS One* **8**
[3] Arfeen Z A, Abdullah M P, Hassan R, Othman B M, Siddique A, Rehman A U and Sheikh U U 2020 Energy storage usages: Engineering reactions, economic-technological values for electric vehicles—A technological outlook *Int. Trans. Electr. Energy Syst.* **30** 1–30
[4] Liu J, Hu C, Kimber A and Wang Z 2020 Uses, Cost-Benefit Analysis, and Markets of Energy Storage Systems for Electric Grid Applications *J. Energy Storage* **32** 101731
[5] Khan T, Garg A K, Gupta A, Madan A K and Jain P K 2023 Comprehensive review on latest advances on rechargeable batteries *J. Energy Storage* **57** 106204
[6] Zhang H, Liu X, Li H, Hasa I and Passerini S 2021 Challenges and Strategies for High-Energy Aqueous Electrolyte Rechargeable Batteries *Angew. Chemie - Int. Ed.* **60** 598–616

[7] Zhou L, Liu L, Hao Z, Yan Z, Yu X F, Chu P K, Zhang K and Chen J 2021 Opportunities and challenges for aqueous metal-proton batteries *Matter* **4** 1252–73
[8] Kim H, Hong J, Park K Y, Kim H, Kim S W and Kang K 2014 Aqueous rechargeable Li and Na ion batteries *Chem. Rev.* **114** 11788–827
[9] Liang Y, Dong H, Aurbach D and Yao Y 2020 Current status and future directions of multivalent metal-ion batteries *Nat. Energy* **5** 646–56
[10] Sun J, Zheng X, Li K, Ma G, Dai T, Ban B, Yuan Y, Wang M, Chuai M, Xu Y, Liu Z, Jiang T, Zhu Z, Chen J, Hu H and Chen W 2023 Scalable production of hydrogen evolution corrosion resistant Zn-Al alloy anode for electrolytic $MnO_2$/Zn batteries *Energy Storage Mater.* **54** 570–8
[11] Chen C, Shi M, Zhao Y, Yang C, Zhao L and Yan C 2021 Al-Intercalated MnO2 cathode with reversible phase transition for aqueous Zn-Ion batteries *Chem. Eng. J.* **422** 130375
[12] Chen W, Li G, Pei A, Li Y, Liao L, Wang H, Wan J, Liang Z, Chen G, Zhang H, Wang J and Cui Y 2018 A manganese-hydrogen battery with potential for grid-scale energy storage *Nat. Energy* **3** 428–35
[13] Chao D, Zhou W, Ye C, Zhang Q, Chen Y, Gu L, Davey K and Qiao S Z 2019 An Electrolytic Zn–$MnO_2$ Battery for High-Voltage and Scalable Energy Storage *Angew. Chemie - Int. Ed.* **58** 7823–8
[14] Li G, Chen W, Zhang H, Gong Y, Shi F, Wang J, Zhang R, Chen G, Jin Y, Wu T, Tang Z and Cui Y 2020 Membrane-Free Zn/$MnO_2$ Flow Battery for Large-Scale Energy Storage *Adv. Energy Mater.* **10** 1–10
[15] Salles R C M, De Oliveira G C G, Díaz S L, Barcia O E and Mattos O R 2011 Electrodeposition of Zn in acid sulphate solutions: PH effects *Electrochim. Acta* **56** 7931–9
[16] Zhong C, Liu B, Ding J, Liu X, Zhong Y, Li Y, Sun C, Han X, Deng Y, Zhao N and Hu W 2020 Decoupling electrolytes towards stable and high-energy rechargeable aqueous zinc–manganese dioxide batteries *Nat. Energy* **5** 440–9
[17] Liu C, Chi X, Han Q and Liu Y 2020 A High Energy Density Aqueous Battery Achieved by Dual Dissolution/Deposition Reactions Separated in Acid-Alkaline Electrolyte *Adv. Energy Mater.* **10** 1903589
[18] Dūrena R, Zukuls A, Vanags M and Šutka A 2022 How to increase the potential of aqueous Zn-$MnO_2$ batteries: The effect of pH gradient electrolyte *Electrochim. Acta* **434**
[19] Liu Z, Qin L, Lu B, Wu X, Liang S and Zhou J 2022 Issues and Opportunities Facing Aqueous $Mn^{2+}$/$MnO_2$-based Batteries *ChemSusChem* **15** e202200348

Journal of Physics: Conference Series **2751** (2024) 012006 doi:10.1088/1742-6596/2751/1/012006

# Topological spin-torque in non-collinear anti-ferromagnetic 3Q state

**Adel Abbout**
King Fahd University of Petroleum and Minerals, Dhahran, Saudi Arabia
E-mail: adel.abbout@kfupm.edu.sa

**Abstract.** We investigate the spin torque in the topological phase of the 3Q antiferromagnetic (AFM) configuration. We first obtain the band structure to identify the topology and nature of the different gaps of the system and then calculate the spin density in the whole system. We demonstrate the presence of a non-vanishing spin-torque component at the edge of the system and analyze its effect on the texture. Moreover, we show that the direction of the torque depends hugely on the way we cut the system to define the edge.

## 1. Introduction

The race behind low-energy consumption devices led to the investigation of a multitude of systems exhibiting topological behaviour [1, 2]. Indeed, the presence of states robust against disorder and deformations with quantized topological properties opens the door to outstanding features [2]. One of the key elements in topological systems is spin-orbit coupling and symmetry breaking. This interaction might lead in some systems to the emergence of a non-vanishing Berry curvature and the presence of an anomalous Hall effect [1, 3]. Recently, the investigation of non-collinear [4, 5] and non-coplanar anti-ferromagnets [6] showed that a topological state can be obtained without the presence of spin-orbit coupling [7, 8]. As a matter of fact, the 3Q state that appears in Mn/Cu(111) interface [9], presents an interesting non-collinear and non-coplanar texture with 4-spin unit cell in a triangular lattice with the spins having the same angle between them. This non-conventional texture is the reslut of a phase transition from a coplanar row-wise AFM configuration [9, 10] to a non-coplanar one due to the presence of 4-spin exchange interaction [9, 11, 12]. Indeed, *ab initio* calculation with a Heisenberg model show that this interaction induces a ground state where the 4-spins of the unit cell are living in 3D while keeping their sum to be zero. In a finite-size sample, the spins at the edge might have a different orientation than those of the bulk because of the absence of the effective field coming from the missing neighbours. Hence, these local spins are expected to feel a larger spin torque from the electronic spin density. This effect is more likely to happen in the case of a state in the topological phase since all the charge carriers propagate at the edge of the sample as shown in Figure 1.

Investigating the effect of edge spin density and its corresponding spin torque on the magnetic texture is very important to specify whether the texture can remain stable or will collapse to another quasi-ground state. An interesting approach to a more realistic investigation would be to run an atomistic simulation and generate the exact texture of the ground state [13]. In parallel, a tight-binding simulation will provide the electronic spin density that can be incorporated in

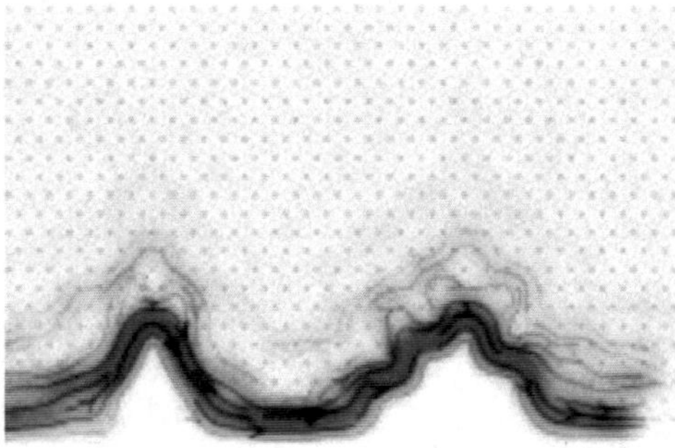

**Figure 1.** Sketch of the edge charge current in the topological phase of a magnetic system. The current is following the deformation, and propagates along the edge.

the effective field of the atomistic simulation. In this study, we will focus on the tight-binding simulation to express the electronic spin densities and their spin torque on the texture in the 3Q state configuration.

## 2. Model

The antiferromagnetic 3Q texture consists of localized spins on a triangular lattice with a spin Hamiltonian $\mathcal{H} = -J_1 \sum_{\langle i,j \rangle} \mathbf{S}_i \mathbf{S}_j - J_2 \sum_{\langle\langle i,j \rangle\rangle} \mathbf{S}_i \mathbf{S}_j$. $J_1$ and $J_2$ are antiferromagnetic exchange coefficients with nearest and next-nearest neighbours interactions respectively [10]. This Hamiltonian is rich in its phase space and one needs to add the 4-spin interaction to stabilize the 3Q state with its 4-spins having the same angle between them ($\theta_0 \approx 109.47°$) as shown in Fig. 2 [9, 12]. The 4 spins making the unit cell are non-collinear and non-coplanar, with a sum $\sum_{i \in \text{cell}} \mathbf{S}_i = 0$ [3, 6, 12]. Unlike a conventional planar antiferromagnetic texture, the chirality $\kappa = \mathbf{S}_k.(\mathbf{S}_i \times \mathbf{S}_j) \neq 0$ which provides an effective spin-orbit coupling for the electrons hopping in the lattice [3, 7]. Consequently, a non-zero-Berry curvature is found and the signature of quantum-anomalous Hall effect is seen in the $\sigma_{xy}$ conductivity of electrons. The Hamiltonian describing electrons in the system tightly coupled to the local spins with sd interaction can be described in the tight-binding approximation as [6, 14]:

$$H = -\sum_{i,j} t_{ij} \hat{c}_i^\dagger \hat{c}_j + \Delta \sum_i \hat{c}_i^\dagger (\mathbf{S}_i.\hat{\boldsymbol{\sigma}}) \hat{c}_i \tag{1}$$

with the first sum is done over the first ($t_{ij} = t_0$) and second ($t_{ij} = t'$) nearest neighbours only. $\Delta$ is the s-d interaction exchange and $\hat{\boldsymbol{\sigma}}$ is the Pauli matrices vector. $\hat{c}_i{}^\dagger$ ($\hat{c}_i$) is the creation (annihilation) operator for electrons on site $i$. (The electron spin index is hidden in the operator $\hat{c}_i \equiv (\hat{c}_{i\uparrow}, \hat{c}_{i\downarrow})$). The lattice is shown in Figure 2 as well as the non-coplanar magnetic structure with an antiferromagnetic unit cell. The band structure for relatively large $\Delta$ (compared to the hopping parameter $t$) exhibits two topological gaps and a trivial one, with bands being two-fold degenerate [3]. As shown in Figure 3, the band for a semi-infinite quasi-1D waveguide shows two edge states at the top and bottom gaps, revealing a signature of a topological behaviour of the system for the corresponding set of energies. These edge states are non-degenerate and provide a quantum of conductance $\frac{e^2}{h}$ in the conductance profile. In the case of an infinite system, the topology appears in the presence of a quantum anomalous Hall effect, manifesting as a quantum conductance in the transverse conductance measurement $\sigma_{xy}$. From the atomistic Landau-Lifshitz-Gilbert simulation point of view, one can construct the structure using the magnetic

Hamiltonian described previously and calculate the effective fields to stabilize iteratively the structure towards the ground state [13]. In this process, the use of periodic boundary condition may speed up the stabilization process. In a real situation, the conditions at the edge are open, and the local spins are missing half of the contribution of the effective field coming from the missing neighbours at the edge. For this reason, one expects the local spins at the edge to deviate from the equilibrium direction of the spins of the bulk. In the presence of quantum transport, for energies in the toplogical gap, the edge states build up an electronic spin density at the edge that couples to the local atomic spins and exert a spin torque on them. The exerted spin torque will challenge the local equilibrium of the spins and can even destroy the non-coplanar phase and thus the topological (electronic and magnetic) properties of the system. For this reason, it is important to study the spin density and the spin torque in such system.

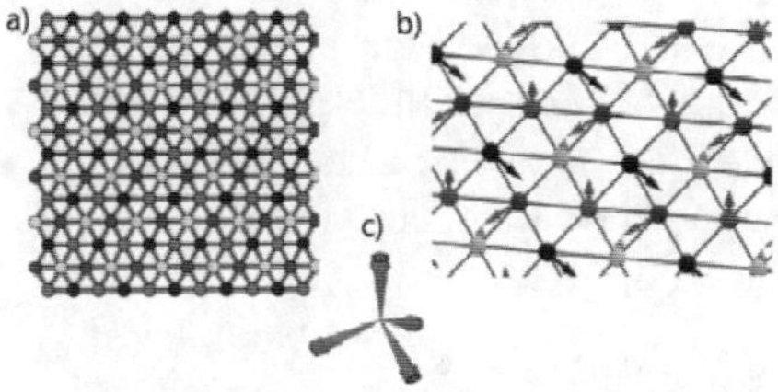


**Figure 2.** Lattice model for the 3Q state. a) The triangular lattice with 4 sites unit cell (the next nearest neighbours hopping are not depicted for clarity of the figure). b) The local spins configuration. All the spins are non-collinear and do not live on the same plane. c) The four different local spins of a unit cell plotted from the same origin. The angles between each two local spins are the same. For the transport properties, the system will be considered infinite in the x-direction while the width is kept fixed.

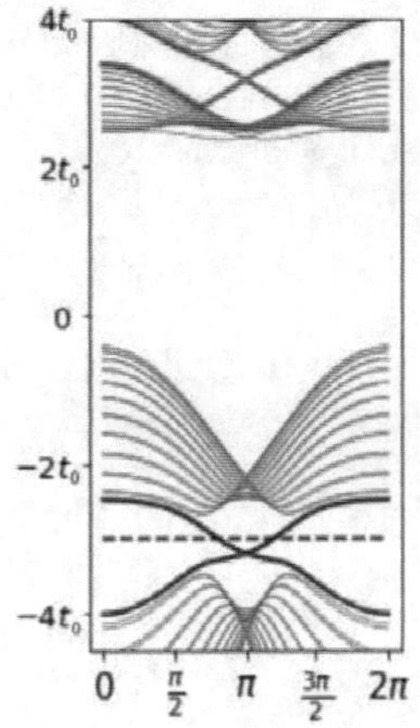


**Figure 3.** The band strucure for non-collinear antiferromanetic 3Q texture. The next nearest hopping is $t' = 0.3t_0$. The blue and red bands represent the two edge states corresponding to the two topological gaps. The dashed line shows the Fermi level $E_F = -3t_0$. The exchange parameter is $\Delta = 3t_0$

To investigate the spin torque, we consider a system with next nearest neighbours $t' = 0.3t_0$ and an exchange strength $\Delta = 3t_0$ . The Fermi energy is set to $E_F = -3t_0$ to allow one single topological mode at the bottom gap. The spherical angles $(\theta, \phi)$ defining the local spins in the unit cell are $(\theta_0, -\frac{\pi}{3})$, $(\theta_0, \frac{\pi}{3})$,$(0,0)$,$(\theta_0, \pi)$, where $\theta_0 = \arccos(-\frac{1}{3})$. The system is infinite in the x-direction and has a width $W = 20a$, $a$ being the lattice constant. We use the Kwant package to calculate the wave function and the spin densities on each site of the system [15].

## 3. Results

Figure 2 shows the studied lattice system with the position of the local moments on the sites and the angles between the spins. The band structure for the quasi 1D system is depicted in figure 3. The middle gap is trivial whereas the top and bottom ones are topological, recognized by the presence of two edge states (shown in red and blue) [3].

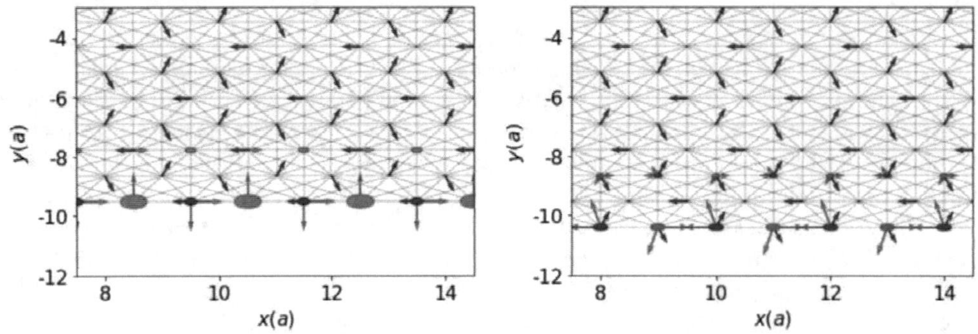


**Figure 4.** The lattice model of the 3Q antiferromagnetic state. The figures show the lower edge of a uniform waveguide infinite in the x-direction. The blue arrows represent the in-plane local moments (spins). The colours for the unit cell sites are green, purple, blue and red for the corresponding spherical angles $(\theta_0, -\frac{\pi}{3})$, $(\theta_0, \frac{\pi}{3})$, $(0, 0)$, $(\theta_0, \pi)$ respectively. The difference between the left and right figure is in the sites lying at the lower edge of the sample. The size of the sites is proportional to the $\langle s_z \rangle$ component of the electron spin density. The in-plane electron spin density $\langle \boldsymbol{s}_{\|} \rangle$ is depicted with a red arrow. The in-plane spin torque $\boldsymbol{\tau}_{\|}$ is depicted with a green vector. (In these figures, "a" is the lattice parameter)

By choosing the Fermi energy in the middle of the lower gap ($E_F = -3t_0$), we select the edge state that propagates at the lower edge of the system. We can see in Figure 4 that the state lives on the edge. Indeed, the size of each site is put to be proportional to the out-of-plane electron spin density and it clearly shows that the state barely penetrates two unit cells and decays exponentially from the lower edge of the system. The local moments at the edge can be different from one sample to another as it can be seen in Figures 4 left and right. In fact, depending on the direction of the cut of the system, we can have any couple of the spherical angles defined previously. The direction of the spin torque depends hugely on the sites lying on the edge since the built-up electronic spin density depends on the magnetization at that site. The in-plane spin density in Figure 4 left is opposite to the local magnetization due to the positive sign of the exchange parameter $\Delta$. This direction of the spin-density is not due solely to the sign of the exchange parameter as it can be deduced from Figure 4 (right). Indeed, a tilt is visible between the spin density (red arrows) and the local moments (blue arrows). Since the electron is hopping between sites of different local moments, one can not predict the direction of the spin-density by hands, without proper calculation, despite that the sd exchange interaction is antiferromagnetic in our model. The spin torque reflects the effect of the spin density on the magnetic structure and how it tends to modify it. The most straightforward and natural way to define it is [16, 17]:

$$\boldsymbol{\tau_i} = \frac{2\Delta}{^-h} \langle \boldsymbol{s_i} \rangle \times \boldsymbol{S_i} \tag{2}$$

where $i$ refers to the site $i$. The spin density at the Fermi level is evaluated using the edge state following the formula:

$$\langle \boldsymbol{s_i} \rangle = \boldsymbol{\psi}_i^{\dagger} \hat{\boldsymbol{\sigma}} \boldsymbol{\psi}_i \tag{3}$$

From equation 2, we deduce that the spin torque in the bulk of the system is exponentially vanishing and only survives near the edge of the system (because so is the wave function). This is confirmed by Figures 4 where the green arrows showing the spin torque are mainly on one or two unit cells from the edge. A non-vanishing torque at the edge means that the local moments at the edge will precess to a new equilibrium state in the presence of damping [13]. This new configuration at the edge will alter the equilibrium of spins in the next layer from the edge and we can continue with the same reasoning for the inner layers until we get deep into the bulk. This change in the magnetic structure can be negligeable and manifests as a small tilt in the local moments without destroying the main transport features of the system. On the other hand, when the magnitude of the spin torque is large enough, the whole magnetic structure starts collapsing and a change of phase will occur with completely different magnetic and transport properties [18]. In order to have a clear picture of the expected magnetic structure, one needs to perform atomistic simulation using Landau-Lifshitz-Gilbert equation to predict the stable ground state for the given density of states [13]. Moreover, we need to make sure to keep open boundary conditions since the main effect we are investigating comes from the presence of the edge.

## 4. Conclusion

The spin density that builds up at the edge of the topological phase of the 3Q antiferomagnetic state does not follow the direction of the local moments and thus exerts a torque on the magnetic structure. The direction of the torque depends on the local moments present at the edge and might vary from one sample to another. The fact that in the topological phase, the edge state is mainly concentrated on a tiny width from the edge, the involved spin densities are expected to be large enough to be accounted in the effective field inducing the equilibrium of the magnetic ground state. Consequently, the stability of the 3Q state depends on the the polarization of the current injected from the leads which can sustain a important spin density on the edge. The non-vanishing exerted torque will most likely modify the anti-ferromagnetic order at the edge and change the local transport properties of the system.

## Aknowledgement

The author gratefully acknowledges the support provided by the Deanship of Research Oversight and Coordination (DROC) at King Fahd University of Petroleum and Minerals (KFUPM) for funding this work through exploratory research grant No. ER221002.

## References

[1] Bonbien V, Zhuo F, Salimath A, Ly O, Abbout A and Manchon A 2021 *Journal of Physics D: Applied Physics* **55** 103002
[2] L Smejkal Y Mokrousov B Y and MacDonald A H 2018 *Nature Physics* **14** 242
[3] Ndiaye P B, Abbout A, Goli V M L D P and Manchon A 2019 *Phys. Rev. B* **100**(14) 144440
[4] S Nakatsuji N K and Higo T 2015 *Nature* **527** 212–215
[5] Nayak A K, Fischer J E, Sun Y, Yan B, Karel J, Komarek A C, Shekhar C, Kumar N, Schnelle W, Kübler J, Felser C and Parkin S S P 2016 *Sci. Adv.* **2** 21501870
[6] Martin I and Batista C D 2008 *Phys. Rev. Lett.* **101**(15) 156402
[7] Chen H, Niu Q and MacDonald A H 2014 *Phys. Rev. Lett.* **112**(1) 017205
[8] J Kübler C F 2014 *Europhys. Lett.* **108**(6) 67001
[9] Kurz P, Bihlmayer G, Hirai K and Blügel S 2001 *Phys. Rev. Lett.* **86**(6) 1106–1109
[10] Jolicoeur T, Dagotto E, Gagliano E and Bacci S 1990 *Phys. Rev. B* **42**(7) 4800–4803
[11] Blügel S 2005 C1 reduced dimensions 1 : Magnetic moment and magnetic structure
[12] Kurz P, Bihlmayer G, Hirai K and Blügel S 2002 *Phase Transitions* **75** 101–112
[13] Evans R F L, Fan W J, Chureemart P, Ostler T A, Ellis M O A and Chantrell R W 2014 *Journal of Physics: Condensed Matter* **26** 103202
[14] Abbout A, Weston J, Waintal X and Manchon A 2018 *Phys. Rev. Lett.* **121**(25) 257203
[15] Groth C W, Wimmer M, Akhmerov A R and Waintal X 2014 *New Journal of Physics* **16** 063065
[16] Akosa C A, Ndiaye P B and Manchon A 2017 *Phys. Rev. B* **95**(5) 054434
[17] Ndiaye P B, Akosa C A, Fischer M H, Vaezi A, Kim E A and Manchon A 2017 *Phys. Rev. B* **96**(1) 014408
[18] Goli V M L D P and Manchon A 2021 *Phys. Rev. B* **103**(10) 104425

# Silicon/Silver nanocomposite-based surface-enhanced Raman spectroscopy (SERS) biosensor for ultrasensitive direct detection of deafness mutations in real system

Arwa Al Safadi[a*], Krithikadevi Ramachandran[b*], Abdelaziz Tlili[a], Kais Daoudi[b,c]

[a] Department of Applied Biology, College of Sciences, University of Sharjah, Sharjah 27272, United Arab Emirates.

[b] Centre for Advanced Materials Research, Research Institute of Sciences and Engineering, University of Sharjah, P. O. Box 27272, Sharjah, United Arab Emirates

[c] Department of Applied Physics and Astronomy, College of Sciences, University of Sharjah, Sharjah 27272, United Arab Emirates.

*Corresponding authors: u20200496@sharjah.ac.ae &

pt_kramachandran@sharjah.ac.ae

**Abstract**

Surface-enhanced Raman scattering (SERS) is a promising emerging, highly sensitive technique for detecting biological samples. The current research aims to bring an effortless biosensor-based approach to detect and discriminate DNA mutations with their fingerprints for each deafness mutation within the GJB2 (gap junction beta 2) gene. Therefore, we developed silver (Ag) modified silicon-based SERS sensing platform to directly screen GJB2 gene mutations for UAE patients with hereditary hearing loss (HL). The prepared AgNPs@Si biosensor was utilized for establishing a silicon-based SERS biosensing protocol for further analyses toward label-free DNA deafness mutations screening. Furthermore, the comparative studies demonstrated the viability of the developed sensor to distinguish between the normal/mutant samples in the real-time system by investigating the DNA assignments and the peaks intensity. Noticeably, under 785 nm laser excitation distinguishable Raman peaks for cytosine bands (C) are enriched with normal DNA (DMF-34) at 780 $cm^{-1}$, 1450 $cm^{-1}$ and 1590 $cm^{-1}$ whereas with the single nucleotide deletion of cytosine band for the mutant DNA correspondingly predicted with Quenched intensities at the parallel Raman peak positions. The sensitivity studies portrayed the limit of detection to be 1 fg/μL. Further their reproducibility studies evaluation conveyed with low values of % RSD as 10. By the end of this study, we expected to update the mutational profile of the GJB2 gene in the UAE population using the advanced simple SERS-based technique for further supplemental applications in terms of genetic counseling and prenatal diagnosis.

Keywords: DNA, SERS, biosensor, AgNPs/Si, EM enhancement, DNA mutation

## 1. Introduction

Sensorineural deafness is one of the most popularr defect, out of 1000 children there is one child suffering from hearing loss. About 80% of hereditary deafness is contributed by Non syndromic hearing loss (NSHL). Mostly autosomal recessive type mutation is the highest frequency mode of inheritance

[1]. GJB2 (gap junction protein beta 2) is the most prevalent GJB2 gene among GCC (Gulf Cooperation Council) countries and has been reported that, GJB2 gene accountable for more than 50 % out of all cases with ARNSHL (Autosomal Recessive Non-Syndromic Hearing Loss) in most populations and particularly within UAE (United Arab Emirates) due to high number of the endogamy. This form of high heterogeneity exhibiting Non syndromic hearing loss (NSHL) would increase the challenge to diagnose and screen, despite the results efficiency for the classical methods like Sanger sequencing, DNA microarray, PCR (Polymerase Chain Reaction) and its modifications. These techniques also include high cost and time consumption [2-4]. From this insight, seeking and developing a rapid, susceptible, and affordable technique to screen such mutations will be of service to improve the diagnosis of deafness mutations in early stages.

SERS (Surface Enhanced Raman Spectroscopy) biosensors-based approach put forward a fascinating scope for biological samples as a rapid and sensitive testing method by means of plasmonic nanoparticles based on electromagnetic enhancement (EM). Silver nanoparticles tend to attract and to create an engaging area with great electrical field that can enhance the signal read out of analyte molecules at their proximity. Label free detection of DNA as well mutations are possibly achieved with advanced SERS technology [5, 6].

Herein, we aim to fabricate Silicon/silver nanocomposite (SSN) platform for direct screening of hearing loss mutations using Surface enhanced Raman scattering (SERS) platform. The sensor characteristics were detailed as well their sensing mechanisms also projected. The fabricated SSN sensor detected DNA mutations with utmost sensitivity providing the base for SERS to be developed as diagnostic platform in the medical field.

## 2. Experimental Details

### 2.1 DNA samples preparation

The overall scheme of work (Figure. 1) starts with extraction of DNA from the saliva samples of the affected persons with details as referred to our previous works [7]. Two PCR (Polymerase Chain Reaction) tests were conducted to amplify the concentration of DNA samples. It was preceded with (GoTaq® G2 Flexi DNA Polymerase) Promega, REF number M7806, targeting the second exon of GJB2 gene using primers (Cx26 2F and Cx26 2R) followed by Gel electrophoresis confirmation.

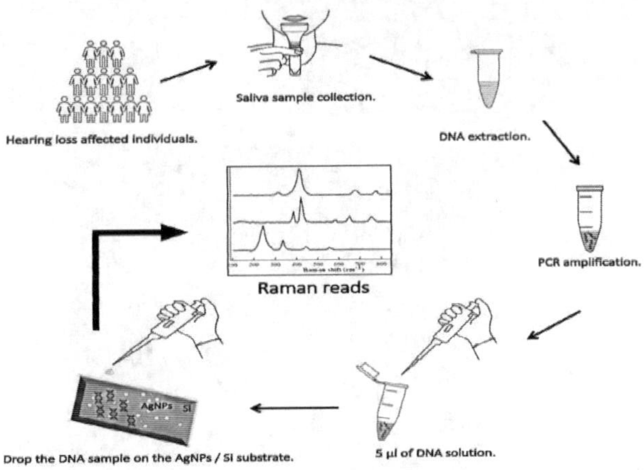


Figure 1. Scheme of the DNA extraction and SERS DNA analysis

Next step, purification of DNA was done according to the manufacturer's protocol (Wizard SV Gel and PCR clean-up system, Catalogue number selected: A9281), followed by DNA concentration measurement by Nanodrop One Microvolume UV–Vis Spectrophotometer (Nanodrop One, Thermo Scientific). Finally, all the samples volume were consolidated by dilution with distilled water to end up with 50 μl with a concentration of 10 ng / μl. The extracted, purified DNA samples were stored at -18°C for further use. DNA sample (DMF16) with pathogenic deafness, autosomal recessive hereditary hearing loss which carry mutation with heterozygous state, deletion in the coding region 235delC position 235 (GGGCC) as well (DMF34) normal sequence (GGGCCC) within GJB2 (gap junction protein beta 2) were taken for further analysis with SERS sensing to detect/ discriminate mutations. Moreover, all experimental procedures and informed consents used in this study were approved by the University of Sharjah Research Ethics Committee (No. REC-15-11-P004) and performed in accordance with the relevant guidelines and regulations.

### 2.2 DNA solution preparation for SERS sensing with SSN sensor

SERS sensor (SSN) was fabricated with respect to our earlier reports by facile wet chemical strategy [6]. The aqueous DNA stock solutions (DMF 16 and DMF 34) stored in -18°C was taken and thawed. After thawing, SERS studies, 5 μl from each DNA samples were dropped on the top of the SSN sensor substrate for direct detection DNA. All the Raman experiments are done for five times for the assurance of concordance results.

## 3. Results and Discussion

### 3.1 SERS sensor characterization

The SSN sensor was fabricated and characterized for its morphology since it lays on the base for SERS improvement.

(a) (b)

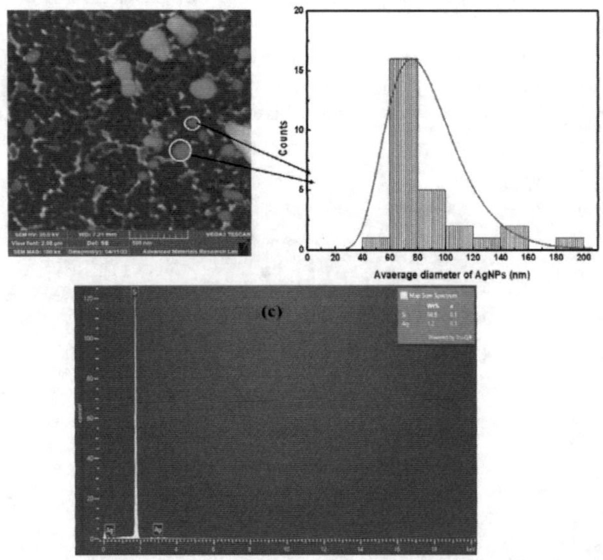


Figure 2. (a) SEM analysis on the morphology of the fabricated SSN sensor (b) Size distribution analysis with Image J software (c) Elemental analysis of the Si/AgNPs nanocomposite SSN sensor.

The sensor exhibited spherical shape with on sight of growth of dumb bells like morphology by the linkage of the adjacent spherical AgNPS by the adopted synthetic strategy (Figure. 2 a). The size distribution calculations (Figure. 2 b) with Image J software conveyed that, the average particle size being 75 nm ± 10 nm. Further their elemental analysis confirmation (Figure. 2 c) portrayed the purity of the fabrication process with only the elements of Si and Ag respectively.

**3.2 SERS sensor (SSN) for Mutant DNA analysis (DMF 16)**

As previously mentioned, DMF- 16 carries mutation with single base pair deletion (GGGCC) against normal healthy DNA sample DMF- 34 (GGGCCC). Under 785 nm laser excitation, the SSN SERS bio sensor detected the DMF-16 DNA sequence with their fingerprint characteristics.

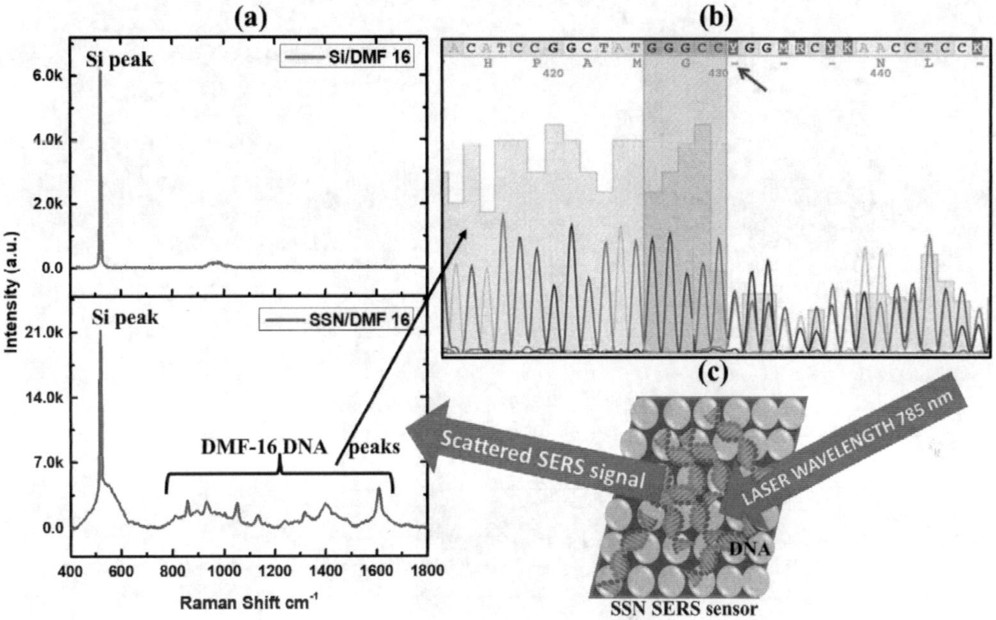


Figure 3: (a) DMF-16 DNA SERS sensing with SSN sensor (b) DMF -16 DNA sequence obtained after PCR with Sager sequencing (c) Mechanism of SERS DNA sensing

The DNA bands of purines and pyrimidines were well explored with utmost sensitivity (Figure. 3 a; lower panel). For reference the DNA analyte solution (DMF-16) was also tested with bare Si substrate without plasmonic nanomaterial AgNPs and found to notify only the presence of Si peak at 520 $cm^{-1}$ (Figure. 3 a; upper panel). The DNA sequence (Figure. 3 b) obtained from Sanger sequencing after PCR studies matched well with SERS Raman peaks (Figure. 3 a). The mechanism of SERS sensing with SSN sensor (Figure. 3 c) is contributed by the EM effect amplified with the Plasmonic AgNPs (Figure. 4 c). Upon laser exposure, the AgNPs excitation was achieved to create collective oscillation of electron clouds activation its unique property of localized surface plasmon resonance (LSPR). The spherical as well dumb bell conjugation of morphology with less interparticle distance enables high density of electron clouds creating the hotspots in the minimal volume contacting the DNA analyte. The hotspots in turn generates the scattered signal carrying the DNA fingerprint information [5, 8]. The field strength in the creation of confined hotspots for SERS sensing was also studied with Finite difference time domain (FDTD) simulation studies (Figure. 4 c). Stronger EM field strength generation was observed with AgNPs of size 80 nm. The $E_{max}$ value obtained in one plane (6.32) portrayed that the plasmonic SERS sensor behaviour is in line with experimental EF values (7.9X $10^6$). Raman mapping (Figure. 4 a) was done with an area of 25 X 25 $cm^2$ to ensure the reproducibility in signal sensing (625 scans). The bright green squares indicate the higher sensitive Raman signals whereas the slight dark squares project the area with weak Raman signal. The weaker signal is quite possible with time since the biomolecules may undergo conformational changes / denaturation upon time. Their statistical validation (Figure. 4 b) with RSD values (% RSD =10) lies in good agreement for sensor studies [6].

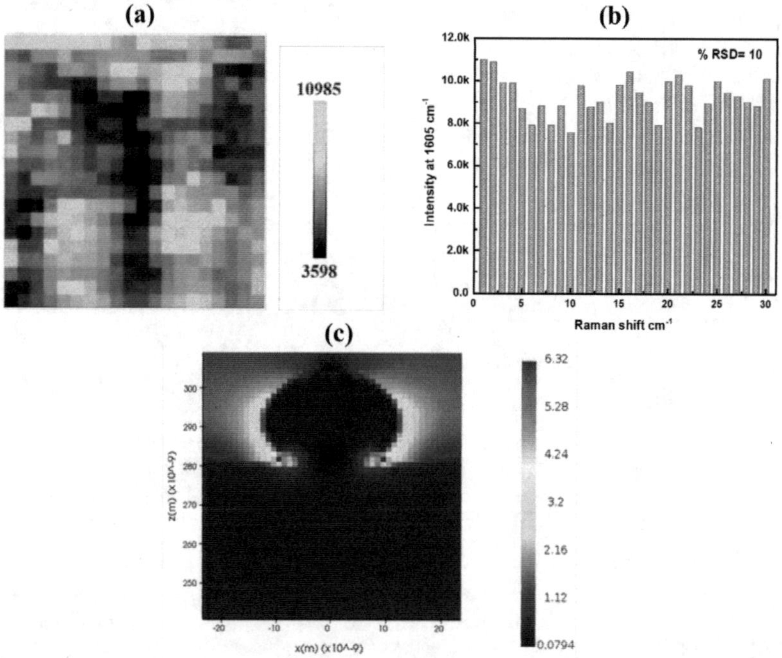


Figure 4: (a) Mapping spectrum of DMF-16 for peak at 1605 $cm^{-1}$ (b) RSD analysis of DMF 16 for peak at 1605 $cm^{-1}$ (c) EM field simulations of fabricated SSN sensor with FDTD.

Detectability studies were evaluated for the fabricated sensor with DMF-16 mutant DNA. The DNA concentration was from 10 ng/μL to 0.001ng/μL were prepared by serial dilutions and subjected to SERS analysis. The DNA features begin are prominent with higher concentration of DMF-16 whereas, the peak intensity and the characteristic peaks were found to be diminished with lower concentrations (Figure. 5 a). The sensor detectability from the analysis was found to be 1 fg/ μL. Regression analysis with linear fit provided that it followed the logarithmic linear behaviour with $R^2$ value being 0.98 (Figure. 5 b).

### 3.3 Relative SERS sensing of DNA mutations with normal healthy DNA

In comparison between two DNA sequences, the mutant sample (DMF-16) suffers from heterozygous frameshift mutation caused by base pair deletion in the coding region exon 2 of GJB2 gene (Figure. 6). With SSN sensor sensing, we observed some of the shared peaks within the range of 900 $cm^{-1}$ – 1300 $cm^{-1}$ and mostly that refer to the DNA backbones vibration, particularly, 920 $cm^{-1}$, 1300 $cm^{-1}$, and 1600 $cm^{-1}$.

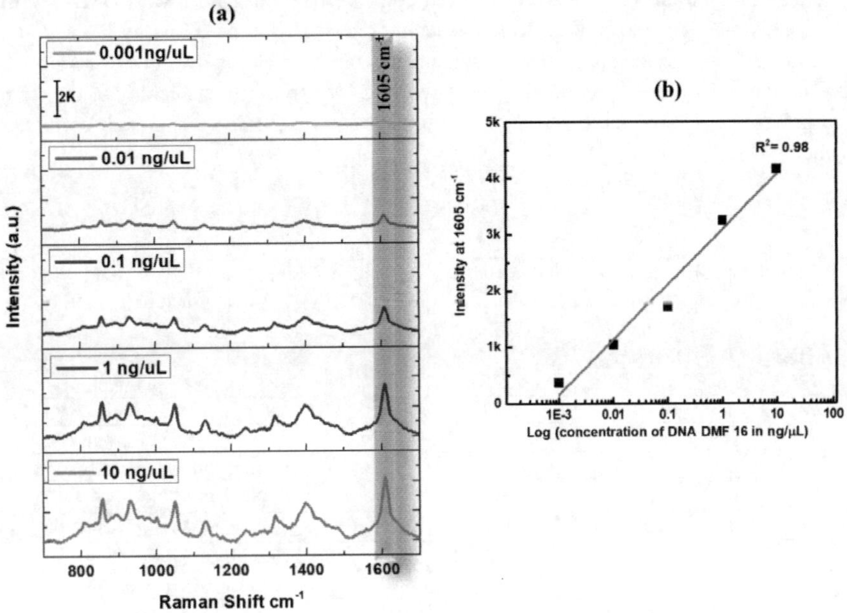


Figure 5: (a) Detectability studies of DMF-16 with SSN sensor (b) Statistical validation analysis

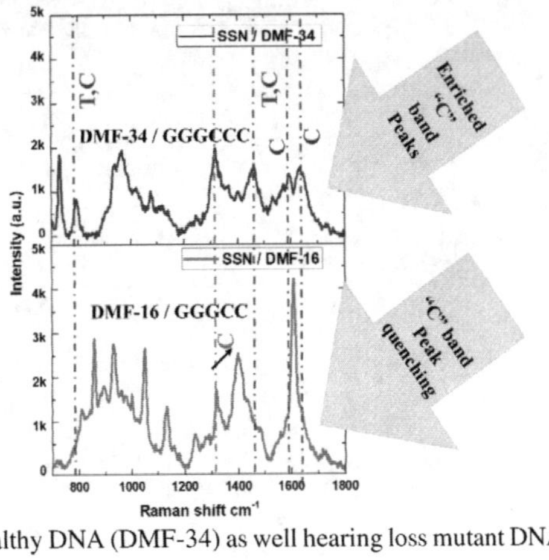


Figure 6: Comparison between healthy DNA (DMF-34) as well hearing loss mutant DNA (DMF-16) SERS sensing with SSN sensor

For the wild type DNA (DMF- 34/ GGGCCC) we perceived several distinct peaks, (GGGCCC) within which the prime peaks account for 720 $cm^{-1}$, 780 $cm^{-1}$, 950 $cm^{-1}$, 1080 $cm^{-1}$, 1450 $cm^{-1}$, 1590 $cm^{-1}$ and 1640 $cm^{-1}$ respectively. The peak assignments are enclosed in Table 1. On the other hand, for the mutant type DNA (DMF-16 / GGGCC) had two featured peaks at Raman shift of 1403 $cm^{-1}$ and 1605 $cm^{-1}$. With reference to the sequence distinctly, the frequency of (C) base pair in the normal sequence (DMF-34) showed as blue color vertical dash lines are highly intense than the mutant sample (DMF-16). The single base pair deletion in DMF -16 was notified with cytosine band quenching at 780 $cm^{-1}$, 1450 $cm^{-1}$ and 1590 $cm^{-1}$ respectively. The red color vertical dash lines with DMF-16 imparts that it contains only one cytosine band at 1304 $cm^{-1}$ which normal to appear for any DNA under study with their sequencing. According to the DNA assignment, Raman peaks at 780 $cm^{-1}$, 1450 $cm^{-1}$ and 1590 $cm^{-1}$ representing (C) which is predominant with DMF- 34 whereas the quenching of the Raman bands at the respective peaks were authenticated with mutant sample DMF-16 also well correlated with PCR sequencing (Figure. 3 b).

| SERS Raman modes ($cm^{-1}$) (DMF-34) | SERS Raman modes ($cm^{-1}$) (DMF-16) | DNA Bands | Detailed Assignments of DNA peaks with Reference to their structure [5, 6, 8] |
|---|---|---|---|
| 720 | - | A | In plane Ring breathing |
| 780 | - | T, C | In plane Ring breathing |
| - | 880 | G | In/out (5-ring def, 6-ring def, wag N9−H, N1−H) |
| 920 | 920 | A | In plane 5th ring deformation |
| 950 | - | A | In plane 6th ring deformation |
| - | 1050 | G | In plane stretching of str N1−C2, C2−N3 |
| - | 1150 | DNA | Backbone phosphate group |
| 1302 | 1304 | C | In-plane ring str C−N |
| - | 1402 | T | In-plane bend N1−H, N3−H |
| 1450 | - | A, T | In-plane str C2−N3, N1−C6, bend C2−H, sciss NH2 (A)<br>In-plane bend CH3 (T) |
| 1590 | - | C | In plane bend NH2 |
| - | 1605 | T, A | In-plane str etching C2=O, C4=O (T)<br>In-plane stretching N3−C4, N1−C6, C5−N7, N7−C8, bend N9−H (A) |
| 1640 | - | A, C, T | In-plane sciss NH2, str C6−N10, C5−C6 (A)<br>Stretching C2=O (C)<br>In-plane stretching C2=O, C4=O (T) |

Table 1: DNA peaks assignment with SSN SERS biosensor

## 4. Conclusion

Congenital hearing loss is one of the prime common sensory disorders over the world and GJB2 gene has been reported to be one of the generality implicated genes within many ethnics group including UAE in terms of recessive non-syndromic hearing loss. Thus, succinctly, in this research we fabricated label-free SERS biosensor (Si/AgNPs- SSN) to detect DNA of healthy person (DMF-34) as well with hearing loss patient DNA mutation (DMF-16) at single site. The SSN sensor detected DNA with utmost sensitivity and the Raman modes of DNA bands were assigned. From the DNA assignment studies clearly, we can state the differences between healthy person and hearing loss patient. However, the study reported is in intermediate level, still it needs some developments to apply for real time diagnosis applications. With the imprinted results we project SERS as a rapid diagnosis tool to screen the hearing loss individuals by the creation of library files from all of the combinations of DNA mutation.

## 5. Acknowledgement

The authors would like to acknowledge the financial support from the University of Sharjah-United Arab Emirates (Grant No. 23021430136).

## References

[1] Najmabadi H and Kahrizi K 2014 *Int. J. Pediatr. Otorhinolaryngol.* 78 2026–36

[2] Tlili A, Al Mutery A, Kamal Eddine Ahmad Mohamed W, Mahfood M and Hadj Kacem H 2017 *Genet. Test. Mol. Biomarkers* 21 686–91

[3] Adadey S M, Wonkam E T, Aboagye E T, Quansah D, Asante-poku A, Quaye O, Amedofu G K, Awandare G A and Wonkam A 2020 *Genes (Basel).* 11

[4] Choi S Y, Kim Y E, Ann D Bin, Kim T H, Choi J H, Lee H R, Hwang S J, Kim U K and Lee S H 2009 *Clin. Exp. Otorhinolaryngol.* 2 44–7

[5] Madzharova F, Heiner Z, Gühlke M, and Kneipp J 2016 J. Phys. Chem. C, 120 15415.

[6] Daoudi K, Ramachandran K, Columbus S, Tlili A, Mahfood M, El Khakani M A and Gaidi M 2021 *Adv. Nat. Sci. Nanosci. Nanotechnol.* 12 0–10

[7] Mahfood M, Chouchen J, Kamal Eddine Ahmad Mohamed W, Al Mutery A, Harati R and Tlili A 2021 *Saudi J. Biol. Sci.* 28 4421–9

[8] Wang H, Jiang X, Wang X, Wei X, Zhu Y, Sun B, Su Y, He S and He Y 2014 *Anal. Chem.* 86 7368–76

# Co-Al layered double oxide activated carbon composite for eliminating lead ions from water

Alaa Abushawish [1,2], Anjaneyulu Chatla [1], Ismail W Almanassra [1], Abdallah Shanableh [1,2], Muataz Ali Atieh [1,3*]

[1] Research Institute of Sciences & Engineering, University of Sharjah, Sharjah, 27272, United Arab Emirates.
[2] Department of Civil and Environmental Engineering, University of Sharjah, Sharjah, 27272, United Arab Emirates.
[3] Chemical and Water Desalination Engineering Program, College of Engineering, University of Sharjah, Sharjah, 27272, United Arab Emirates.
*Correspondence: mhussien@sharjah.ac.ae (M.A.A)

**Abstract.** In the current study, preparation of cobalt-aluminum layered double oxide doped activated carbon (Co-Al LDO/AC) was achieved by the co-precipitation technique and utilized for the remediation of lead ($Pb^{2+}$) from water. Various methods were employed to examine the properties of the composite material, including BET, XRD, FTIR, SEM, and EDS analysis. The material characterization outcomes indicated that the LDO structure was successfully incorporated into the AC matrices with a surface area of 189.4 $m^2/g$. The influence of adsorption parameters including Co-Al LDO/AC dosage, period of contact, initial $Pb^{2+}$ loading, and initial solution pH were investigated. Moreover, the isotherm and kinetic models were investigated to provide a deeper understanding of the elimination mechanism of $Pb^{2+}$ ions. The adsorption results illustrated that pH has a substantial influence on $Pb^{2+}$ removal with a highest removal effectiveness at pH = 6 and a fast adsorption rate within 7 h. The kinetic data were well aligned with the pseudo-second-order model while the isotherm data obeyed the Sips model ($R^2$>0.966). The highest adsorption uptake, estimated by the Sips model was 25.09 mg/g. Considering the modeling and characterization of the spent Co-Al LDO/AC, a chemical interaction process was involved in the elimination process and mainly controlled by ion exchange, electrostatic interactions, and surface complexation mechanisms. Accordingly, the Co-Al LDO/AC could have great potential as a promising hybrid for the purification of toxic $Pb^{2+}$ ions from contaminated water streams.

## 1. Introduction:

In the water environment, various contaminants have been detected and reported to cause pollution and serious human diseases [1]. Heavy metals, organic wastes, radioactive wastes, and pharmaceuticals are among the major pollutants that gained considerable attention due to their severe impact on the human body and the ecological system [2]. Lately, heavy metals attracted numerous attention due to their toxicity, nonbiodegradability, and tendency to persist in the environment [3]. Although some minerals are beneficial for human, animal, and plant health such as iron (Fe) and zinc (Zn); other elements such as lead (Pb), chromium (Cr), cadmium (Cd), and mercury (Hg), are extremely harmful in trace concentrations can lead to determinantal impacts [4].

Lead exists mainly in an ionic form of $Pb^{2+}$ in an aqueous solution which was found to be among the major heavy metals that lead to high intoxication in humans and other organisms. $Pb^{2+}$ is a well-known neurotoxin that damages the central nervous system, affects the nerves' development in both aquatic creatures and humans, and causes renal and hepatic disorders [5]. Based on the World Health Organization (WHO) report, 143,000 deaths worldwide are attributed to lead exposure, where children are among the most affected categories [6]. Although $Pb^{2+}$ can exist naturally in the environment, human activities played a major role in increasing its concentration. Leaded-petroleum spills, leaded pipes for water distribution, and wastewater discharge from lead-based industries such as paints, dyes, electroplating, and battery manufacturing are the major sources of $Pb^{2+}$ exposure in water [7], [8]. Therefore, according to the WHO standards, potable water must not exceed 10 μg/L of $Pb^{2+}$ [9].

To remediate such contamination, several treatment technologies have been developed including electro-coagulation [10], ion replacement [11], chemical precipitation [12], and adsorption [13] techniques. However, adsorption is viewed to be a promising treatment method for $Pb^{2+}$ ions removal due to its simplicity, low cost, efficiency in removing trace concentrations, facile regeneration of its adsorbent material, and ability to recover the pollutant [14]. Numerous adsorbents have been introduced to eliminate $Pb^{2+}$ from contaminated water including natural clay [15], activated carbon (AC) [16], fly ash [17], layered double hydroxide/oxide (LDH/LDO)[18], zeolite [19], biochar [20], and sugarcane bagasse [21]. ACs are recently attracting more research for heavy metal removal because of their advantageous features such as large SA, renewability, extraordinary adsorption efficiency, synthesis using environmentally friendly materials, chemical stability, and resistance to strong acids and bases [22]. However, attention has been directed to improve the AC and enhance its adsorption capacity and its removal rate towards $Pb^{2+}$ ions by modifying its surface. Examples are iron oxide-doped AC [23], polypyrrole-based AC [24], and nanoscale zero-valent iron/AC composite [25]. Despite the successful modifications presented by these studies, there has been a growing emphasis on conducting further research to investigate the viability of using LDH/LDO-based AC composite for $Pb^{2+}$ removal.

Layered double hydroxides/oxides (LDH/LDO) were proposed as unique materials for the elimination of various pollutants including heavy metals, mainly accredited to their special layered structure, high ion exchange capacity, and chemical stability [26]. Zhao et. al. [27] investigated the use of Mg-Al LDH and illustrated a $Pb^{2+}$ adsorption uptake of 75.18 mg/g. While Rahmanian et. al. [28] revealed the highest adsorption uptake of 94.3 mg/g using citric intercalated Zn-Fe LDH. Whereas Liang et. al., [29] showed a remarkable adsorption uptake of 170.09 mg/g using diethylenetriaminepentaacetic acid (DTPA) intercalated Mg-Al LDH. Due to the high capacity of the LDHs for $Pb^{2+}$ adsorption, several research studies were undertaken to evaluate the potential doping of LDH/LDOs onto the surface of different supports for $Pb^{2+}$ removal such as carbon nanofiber loaded with Co-Al LDH and hematite [30], Mg-Al LDH modified palygorskite clay [31], and Mg-Fe LDH/biochar composite [32]. AC as a cheap material, has plenty of functional groups with high surface area (SA) that could be an effective carrier of LDO/LDH structure and contribute to better distribution of these particles and consequently might provide better removal efficiency toward contaminants such as $Pb^{2+}$ ions [33]. To the best of the authors' knowledge, the remediation of $Pb^{2+}$ ions by Co-Al LDO-doped AC has not been explored in the literature yet.

Accordingly, this paper aims to investigate the use of Co-Al LDO-doped AC for the removal of $Pb^{2+}$ ions. The main objectives are (I) synthesizing Co-Al LDO/AC following the co-precipitation method, (II) characterizing the LDO/AC composite for its surface and morphological structure, and (III) investigating the impact of LDO/AC dosage, period of contact, and initial solution pH ($pH_i$) on $Pb^{2+}$ adsorption in addition to the kinetic and isotherm models.

## 2. Material and methods

### *2.1. Materials*

Granular coconut shell AC (Jacobi Carbon, HY-101, China) was purchased. For Co-Al LDO/AC preparation, Aluminum nitrate nonahydrate (Sisco Research Laboratories, India, 98.5%), Cobalt (II) nitrate hexahydrate (Sigma-Aldarich, USA, ≥99%), Sodium hydroxide pellets (Sigma-Aldrich, USA, ≥98%) were used. In the adsorption experiments, preparation of 1000 mg/L $Pb^{2+}$ stock solution was conducted using Lead(ll) nitrate (Merck, Germany, 99%) and deionized (DI) water. To adjust the liquid pH and assess the adsorbent point of zero charge (PZC), HCl (Merck Co., Germany, 37%) and NaCl, (Merck, Germany – Supelco) were used.

### *2.2. Synthesis of Co-Al LDO/AC composite*

Co-Al LDO/AC was fabricated by the co-precipitation approach. In brief, 43.65 g of $Co(NO_3)_2 \cdot 6H_2O$ and 18.76 g of $Al(NO_3)_3 . 9H_2O$ (Co/Al ratio = 3) were dissolved in 300 mL DI water and agitated for 30 min to obtain a homogenous metal solution. Next, 5 g of raw AC was added to the metal solution and stirred for 1 h. Adjustment of the solution at pH 11 was conducted using 1 M NaOH solution and kept for 2 h under agitation at 70 °C. Afterward, the heat was turned off and the Co-Al/AC solution was stirred overnight at room temperature (RT). Separation and DI washing of the precipitant were then carried out using a centrifuging machine (Hitachi CR22N) where the resultant composite material was oven-dried at 100 °C for a day. The dried Co-Al LDH/AC was then subjected to calcination at 500 °C for 4 h and using a rate of heating of 5°C/min. The composite synthesized was labeled as Co-Al LDO/AC.

### *2.3. Characterization of Co-Al LDO/AC composite*

For the raw AC and Co-Al LDO/AC composite, the average pore size, pore size distribution, and BET-SA were determined following Barrett–Joyner–Halenda (BJH) and $N_2$ adsorption-desorption isotherm methods using NOVAtouch LX2 analyzer where the samples were degassed at 300 °C for 4 h. Scanning Electron Microscopy (EDS) (Thermo scientific Apreo C) linked with Energy dispersed X-ray spectroscopy (EDS) were utilized to assess the morphology and the chemical composition of the raw and LDO/AC composite. Fourier-Transform Infrared (FT-IR) Spectroscopy (JASCO 6300) was further utilized to investigate the chemical functional groups of the composite surface (KBr pellet method). The determination of the crystalline structure of the adsorbents was achieved through X-ray diffraction (XRD) analysis (D8 Advance instrument, Burker, Germany). Finally, the PZC was measured following the pH drift method for the raw and LDO/AC composite. The PZC was determined by dispersing 30 mg of the raw AC and Co-Al LDO/AC in 30 mL of 0.1 M NaCl solution. The PZC was obtained by measuring the pH before and after agitation of the materials for 72 h at 150 rpm.

### *2.4. Batch adsorption experiments*

Batch adsorption tests were conducted to determine the adsorption uptake and the elimination efficiency of $Pb^{2+}$ using Co-Al LDH/AC composite. **Table 1** illustrates the conditions set for each adsorption study where all the experiments were conducted using conical flasks and test tubes and at fixed Co-Al LDO/AC dosage (2 g/L). $Pb^{2+}$ adsorption capacity and removal efficiency were determined using eq. 1 and 2.

$$q = \frac{(C_i - C_e)}{S_o} \tag{1}$$

$$R\% = \frac{C_i - C_e}{C_i} \times 100 \tag{2}$$

where $q$ is the $Pb^{2+}$ adsorption capacity of Co-Al LDO/AC (mg/g), $C_i$ is the original $Pb^{2+}$ loading and $C_e$ is the exit loading (mg/L), $S_o$ is the Co-Al LDO/AC dosage (g/L), and $R\%$ is the removal efficiency of $Pb^{2+}$. The residual $Pb^{2+}$ concentration was determined by Inductively Coupled Plasma Optical Emission

Spectroscopy (ICP-OES) (Thermo Scientific iCAP 7000 Series) where calibration standards were prepared between 5 to 50 mg/L of $Pb^{2+}$.

**Table 1:** Adsorption kinetic models using linearized equations.

| Parameter | Range examined | Other adsorption conditions |
|---|---|---|
| Co-Al LDO/AC dosage | 0.7 – 3 g/L | $[Pb^{2+}]_i$ = 50 mg/L, t = 24 h, $pH_i$ =6.1-6.5, RT. |
| Contact time | 0 – 24 h | $[Pb^{2+}]_i$ = 50 mg/L, $S_o$ = 2 g/L, $pH_i$ =6.1-6.5, RT. |
| Initial $Pb^{2+}$ concentration ($[Pb^{2+}]_i$) | 50 – 600 mg/L | $S_o$ = 2 g/L, $pH_i$ =6.1-6.5, t = 24 h, RT. |
| $pH_i$ | 2-6 | $[Pb^{2+}]_i$ = 50 mg/L, $S_o$ = 2 g/L, t = 24 h, RT. |

*2.5. Kinetic and isotherm modeling*

To determine the rate-limiting step of $Pb^{2+}$ removal and the adsorption uptake of Co-Al LDO/AC, the adsorption kinetics were examined. This was conducted by analyzing the experimental data using the linearized intra-particle diffusion (IPD), Elovich, pseudo-first-order (PFO), and pseudo-second-order (PSO) models (**Table 2**). While to evaluate the adsorption isotherm, the experimental data were optimized using nonlinearized Langmuir, Redlich-Peterson (R-P), Freundlich, and Dubinin-Radushkevich (D-R) models (**Table 3**). The presented isotherm data were compared by computing the determination factor ($R^2$) and the sum of square errors (SSE) using an Excel solver. Calculation of SEE was carried out using eq. 3 [14].

$$SSE = \sum_{i=1}^{n}(q_{i(exp)} - q_{i(cal)})^2 \tag{3}$$

where $q_{i(exp)}$ and $q_{i(cal)}$ are the experimental and estimated $Pb^{2+}$ adsorption uptake of Co-Al LDO/AC, respectively.

**Table 2:** Adsorption kinetic models using linearized equations.

| Kinetic models | Linearized equation | x vs. y |
|---|---|---|
| PFO | $log\,(q_e - q_t) = log(q_e) - (\frac{k_1}{2.303})t$ | $t$ vs. $log\,(q_e - q_t)$ |
| PSO | $\frac{t}{q_t} = \frac{1}{k_2\, q_e{}^2} + (\frac{1}{q_e})t$ | $t$ vs. $\frac{t}{q_t}$ |
| Elovich | $q_t = \frac{1}{\beta} ln\,(\alpha\beta) + (\frac{1}{\beta}) ln(t)$ | $ln(t)$ vs. $q_t$ |
| IPD | $q_t = k_{diff}\sqrt{t} + c$ | $\sqrt{t}$ vs. $q_t$ |

**Table 3:** Non-linearized equations of adsorption isotherm models.

| Isotherm Models | Non-linearized model | Parameters |
|---|---|---|
| Langmuir | $q_e = \frac{Q_m K_L C_e}{(1 + K_L C_e)}$ | $Q_m$, maximum $Pb^{2+}$ adsorption uptake (mg/g). $K_L$, Langmuir consistent (L/mg) |
| Freundlich | $q_e = K_F C_e{}^{\frac{1}{n_F}}$ | $K_F$, uptake capacity coefficient (mg/g/(mg/L)$^{1/n}$). $n_s$, adsorption intensity coefficient. |

| | | |
|---|---|---|
| Sips | $q_e = \frac{Q_s K_s {C_e}^{n_s}}{1 + K_s {C_e}^{n_s}}$ | $Q_s$, maximum adsorption uptake (mg/g). $K_s$, the model's constant (L/mg). $n_s$, exponent of Sips model. |
| R-P | $q_e = \frac{K_R C_e}{1 + b_R {C_e}^{n_R}}$ | $K_R$ and $b_R$ are constants (L/g) and ($mg^{-1}$) respectively. *n*, R-P exponent. |

## 3. Results and Discussion

### 3.1. Composite characterization

#### 3.1.1. Structural properties

The SA of raw AC and Co-Al LDO/AC composite was analyzed and estimated using the $N_2$ adsorption-desorption method. According to the categorization of the IUPAC, and as depicted in **Figure 1(a)**, the raw AC exhibited type IV isotherm with an H4 hysteresis loop usually obtained for mesoporous materials [34]. While the Co-Al LDO/AC composite, shown in **Figure 1(b),** demonstrates type II adsorption isotherm with an H3 hysteresis loop representing a microporous structure material [33]. While in the BET-SA result, a reduction in the SA was noticed where the raw AC and Co-Al LDO/AC exhibited 829.6 and 189.4 $m^2/g$, respectively. This implies that the agglomeration of Co and Al led to AC pore blocking and the combustion of the carbon during the calcination process may have contributed to the reduction in the BET-SA. [33].

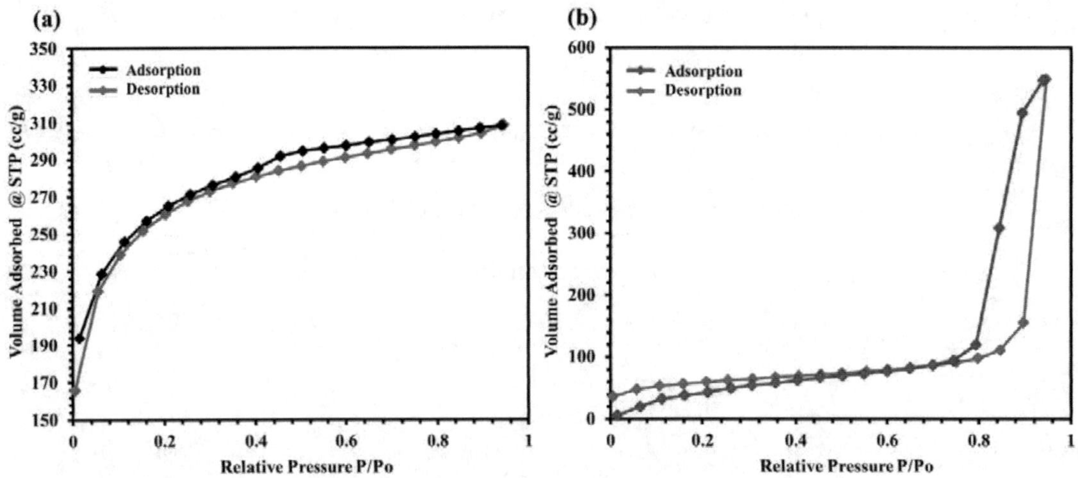


**Figure 1:** $N_2$ adsorption-desorption isotherm data of (a) raw AC and (b) Co-Al LDO/AC composite.

The structural morphology and elemental mapping of the raw AC and Co-Al LDO/AC composite were investigated by SEM-EDS and their related monographs were illustrated in **Figure 2.** As seen in **Figure 2(a),** the raw AC exhibited a porous material and the EDS images (**Figure 2(c)**) confirmed that the carbon was the main component of the adsorbent (97.4 Wt%) with a low percentage of Fe (0.7 Wt.%), Al (0.5 Wt.%), and Si (0.7 Wt.%). **Figure 2(b)** displayed that the material maintained its porosity and roughness with irregular shapes after loading the Co-Al LDO. Moreover, the density of shiny particulates looks more in the modified material confirming the successful doping of the Co-Al oxides. The AC pores look covered by the metal oxides suggesting a reduction in the SA confirming the BET analysis. The EDS mapping, demonstrated in **Figure 2(d)**, clearly illustrated the homogeneous distribution of the doped elements with significant improvement in oxygen element onto the surface of the AC implying the successful doping of the Co and Al elements onto the AC matrix.

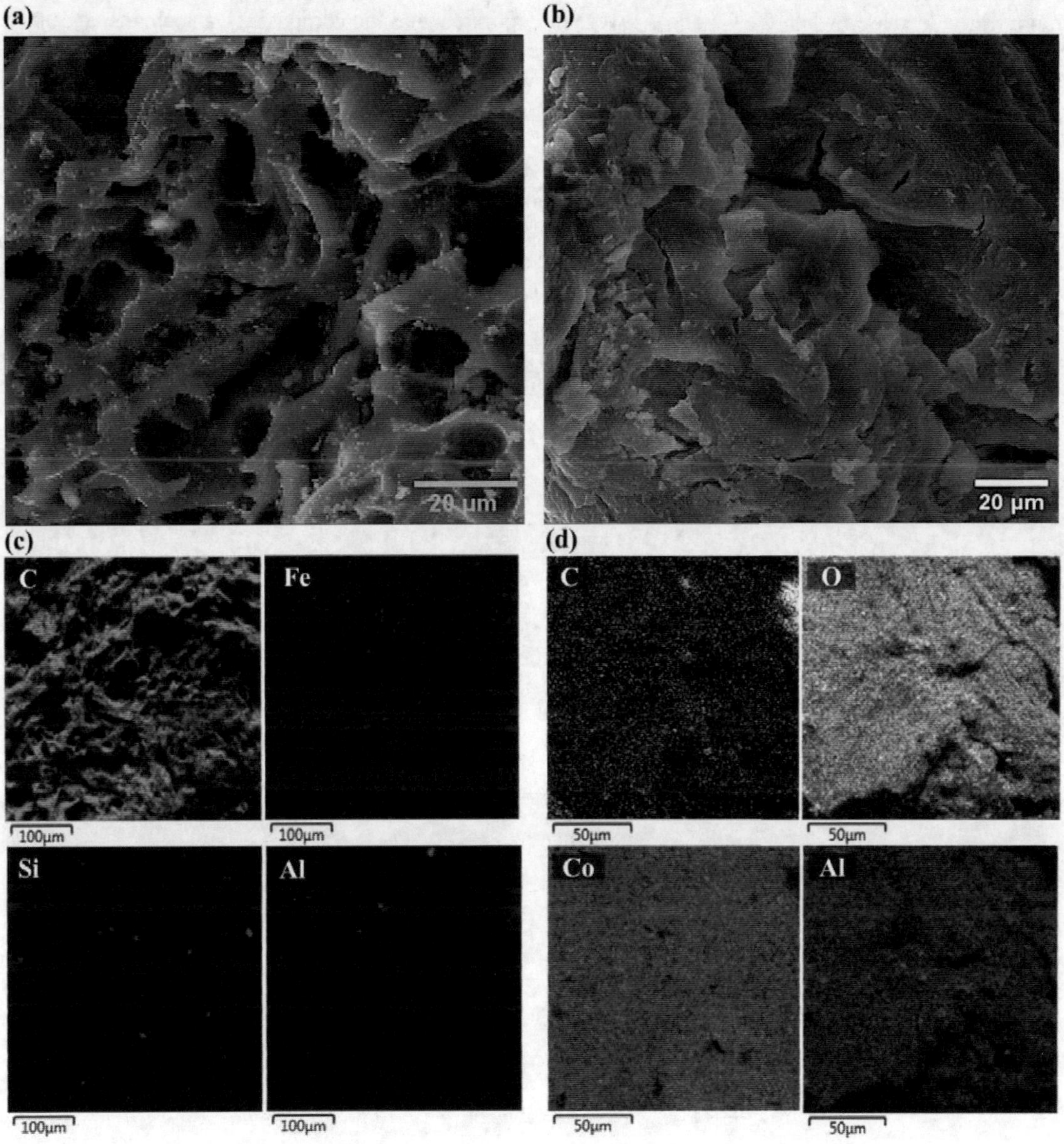


**Figure 2:** SEM-EDS images of (a)(c) raw AC and (b)(d) Co-Al LDO/AC composite.

*3.1.2. Surface functional groups*

FT-IR assessment was attained to explore the functional groups of the raw and Co-Al LDO/AC composite's surface as depicted in **Figure 3(a)**. Based on the IR spectrum, peaks were noticed at 3446 $cm^{-1}$ and 1636 $cm^{-1}$ which related to the stretching vibrations of the – OH functional group [35]. These peaks were indicative of the presence of the absorbed water on the AC's surface. The peak at 2922 $cm^{-1}$ was ascribed to the C–H stretching vibration and the peak at 2359 $cm^{-1}$ was ascribed to the stretching vibration of the atmospheric $CO_2$ [36]. Moreover, peaks observed after 1200 $cm^{-1}$ were attributed to the C=C vibrations and C–O stretching [14]. For the Co-Al LDO/AC composite, the sharpness of the 3446 $cm^{-1}$ band was dramatically decreased when the Co and Al were incorporated, and the composite was calcined at 500 °C. In addition, a characteristic peak at 950 $cm^{-1}$ was observed due to the C–O–C group. While distinct peaks were observed too at 673 $cm^{-1}$ and 568 $cm^{-1}$ which were recognized to the vibrational modes of M-O-H (Co-O-H and Al-O-H) and M-O-M (Co-O-Al) lattice vibrations, respectively [30]. This was an indication of the successful doping of Co-Al LDH onto the AC matrix.

The XRD analysis of the raw and modified AC were depicted in **Figure 3(b)**. The raw AC presented two distinct peaks at 24.9° and 43.6°, which represent the (002) and (100) carbon planes [37]. While the peaks emerged at 31.3°, 36.8°, and 59.5° representing (012), (104), and (0012) phases of rhombohedral crystalline $Co_2Al_2(\text{-}OH)_{16}{\cdot}4H_2O$, respectively [30]. These particular peaks have also been documented in earlier studies on Co-Al LDH material [30], [38]. These findings collectively indicate successful doping of Co-Al LDH onto the AC matrix, as the data obtained from EDS, FTIR, and XRD analyses are all in agreement..

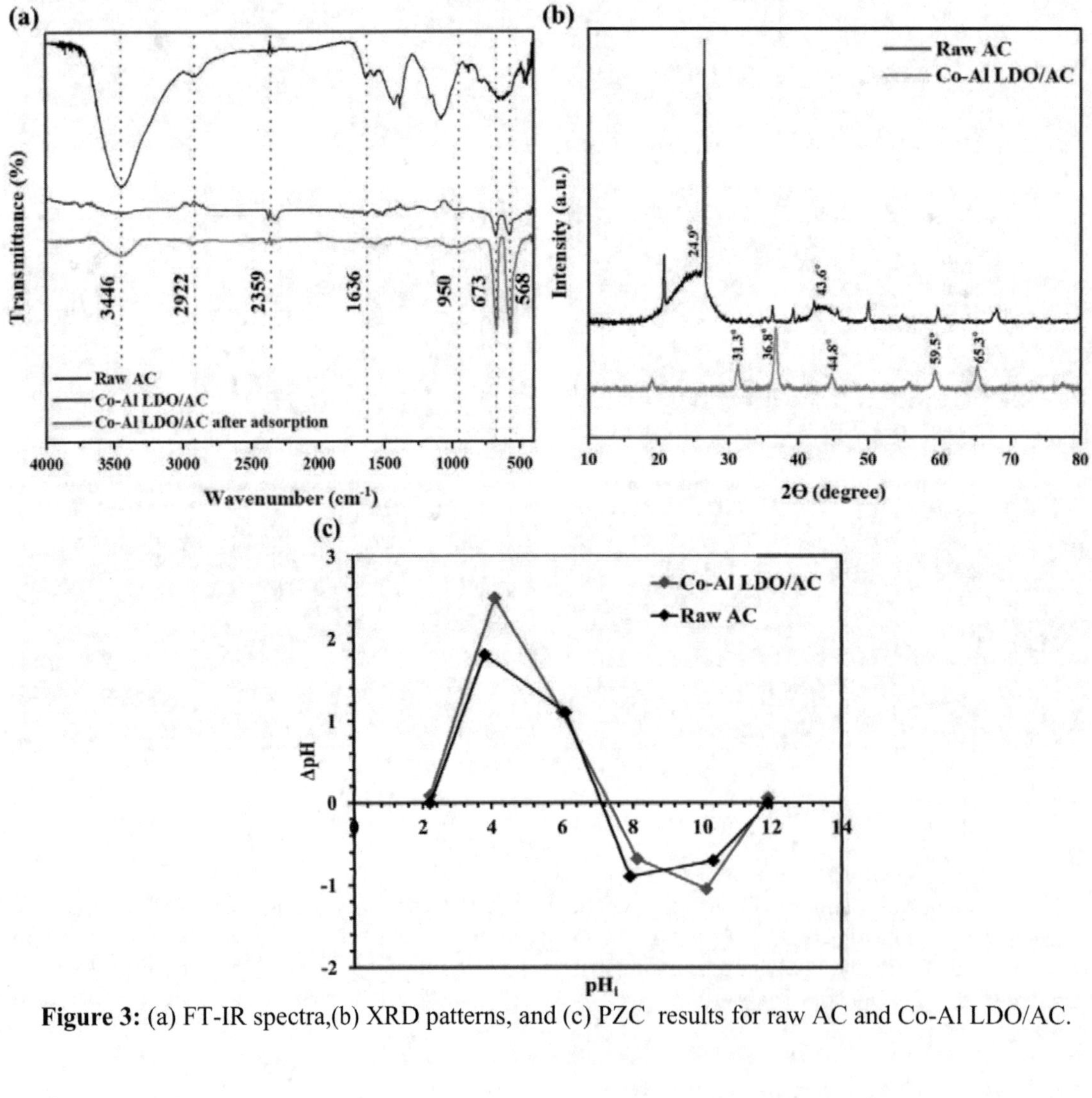


**Figure 3:** (a) FT-IR spectra,(b) XRD patterns, and (c) PZC results for raw AC and Co-Al LDO/AC.

#### *3.1.3. PZC Determination*

The PZC for the raw AC and Co-Al LDO/AC composite were determined via the pH drift method and the results were illustrated in **Figure 3(c).** The raw AC demonstrated a PZC value at pH 7.1, After loading Co-Al LDO onto the raw AC, the PZC slightly increased to 7.5. These outcomes imply that the metal oxides have increased the positivity of the AC. Generally, the material holds a positive charge at a pH lower than the PZC and a negative charge at a pH higher than the PZC. Although, increasing the positivity of the material might not be favorable for the removal of positively charged contaminants such as the $Pb^{2+}$ ions, the affinity of Co and Al oxides toward $Pb^{2+}$ ions is expected to significantly increase the removal efficiency as will be investigated in the coming sections.

### *3.2. Batch adsorption analyses*

#### *3.2.1. Adsorbent dosage impact*

The influence of Co-Al LDO/AC dosage on $Pb^{2+}$ ions adsorption efficiency was examined, and outcomes were displayed in **Figure 4(a)**. It was clearly noted that by elevating the composite dosage from 0.7 to 3 g/L, the $Pb^{2+}$ elimination efficiency was significantly enhanced from 61.4% to 97.3%. A higher dosage of the adsorbent implies more vacant active adsorption sites, therefore; more $Pb^{2+}$ ions are exposed to the Co-Al LDO/AC surface leading to higher removal efficiency. However, since there was no dramatic increase in the $Pb^{2+}$ removal efficiency from dosage 2 g/L (84.9%) to 3 g/L (97.3%), 2 g/L Co-Al LDO/AC dosage was recommended for the remaining experiments. This trend was in agreement with a recently reported study that investigated the use of CoAl-LDH@biochar composite for various heavy metals removal [39].

#### *3.2.2. Contact time effect*

The influence of contact time on the elimination of $Pb^{2+}$ ions by Co-Al LDO/AC was demonstrated in **Figure 4(b)**. The outcomes showed that more than 40 % of the $Pb^{2+}$ were removed after 40 min of contact. This removal is mainly attributed to the external adsorption of $Pb^{2+}$ ions on the surface of the adsorbent. Next, a slower rate of removal was noticed where the removal efficiency increased to 72% after 440 min. This is mainly ascribed to the diffusion of Pb2+ into the internal pores of the adsorbent. After that, the adsorption process started to reach equilibrium, and no more changes in the removal efficiency were obtained after 7 h of contact. These outcomes suggest that 7 h is the saturation time of our adsorption process.

#### *3.2.3. Impact of* $pH_i$

Since the $pH_i$ has a substantial influence on the uptake capacity of a pollutant, the influence of this parameter was evaluated at a pH range of 2 to 6. As depicted in **Figure 4(c)**, it was clearly shown that by improving the $pH_i$ from 2.2 to 5.9, the $Pb^{2+}$ elimination efficiency has dramatically increased from 9.1% to 99.2% with an upsurge in the adsorption uptake from 2.15 mg/g to 23.52 mg/g. After pH 6, some studies reported an improvement in the elimination efficiency of $Pb^{2+}$ ions which was ascribed to the formation of $Pb^{2+}$ hydroxide precipitation [23], [40]. At pH 2.2, the adsorbent holds a highly positive charge explaining the low adsorption capacity at this pH. However, increasing the pH to 5.9 suggests less electrostatic interactions and hence improved adsorption capacity. Higher pH values were not considered, a white precipitate was noticed during raising the pH more than 7 before adding the adsorbent, this is due to the precipitation of $Pb^{2+}$ hydroxides. As precipitation is occurring, the adsorption process cannot be considered as a removal mechanism and therefore pH higher than 7 was not investigated in our study.

#### 3.2.4. Adsorption isotherm

Langmuir, Freundlich, Sips, and Redlich-Peterson (R-P) nonlinearized models were utilized to examine the equilibrium experimental data of $Pb^{2+}$ adsorption. The fitted data were depicted in **Figure 4(d)** and their modeling factors and numerical coefficients were presented in **Table 4**. The Sips model was the best model describing the experimental data with the greatest $R^2$ (> 0.966) and the lowest SSE (0.076), where the highest obtained adsorption uptake was 25.09 mg/g, which is a considerable capacity compared with other materials (see Table 5). Freundlich ($R^2$ > 0.962, SSE = 0.084) and R-P models ($R^2$ > 0.963, SSE = 0.083) were also aligned with the experimental data whereas Langmuir has the lowest $R^2$ (> 0.902) and highest SSE (0.218). This suggests that the binding sites of the composite surface were more likely to have heterogenous and unequal energy distribution than homogenous [30], [41].

**Table 4:** Determined factors of isotherm study for $Pb^{2+}$ adsorption on Co-Al LDO/AC composite.

| Isotherm models | Parameters | Value | $R^2$ | SSE |
|---|---|---|---|---|
| Langmuir | $Q_m$ | 20.42 | 0.902 | 0.218 |
| | $K_L$ | 0.68 | | |
| Freundlich | n | 21.09 | 0.962 | 0.084 |
| | $K_F$ | 16.37 | | |
| Sips | $Q_s$ | 25.09 | 0.966 | 0.076 |
| | $K_s$ | 1.61 | | |
| | $n_s$ | 0.21 | | |
| R-P | $K_r$ | 53.09 | 0.963 | 0.083 |
| | n | 0.96 | | |
| | $b_r$ | 3.08 | | |

**Table 5:** Comparing various AC and LDH adsorbents' adsorption capacities for $Pb^{2+}$ uptake.

| Adsorbents | Adsorbent dosage (g/L) | Initial Pb concentration (mg/L) | pH | $q_{max}$ (mg/g) | Ref. |
|---|---|---|---|---|---|
| **Mg-Al LDH modified palygorskite clay** | 1 | 100 | 5 | 49 | [31] |
| **Fe-AC** | 1.75 | 5-100 | 5.6 | 11.9 | [23] |
| **$H_2O_2$ modified AC** | 2 | 6-200 | 5.3-6.1 | 17.2 | [42] |
| **Olive branches AC** | 20 | 50-1000 | 5 | 41.3 | [43] |
| **NZVI/AC** | 0.8 | 10-300 | 6 | 59.4 | [25] |
| **Albizia lebbeck seed shells AC** | 2 | 10-50 | 5 | 4.4 | [44] |
| **Melia azedarach seed shells AC** | 2 | 10-50 | 5 | 3.4 | [44] |
| **Co-Al LDO/AC** | 2 | 50-600 | 6.1-6.5 | 25.1 | Current work |

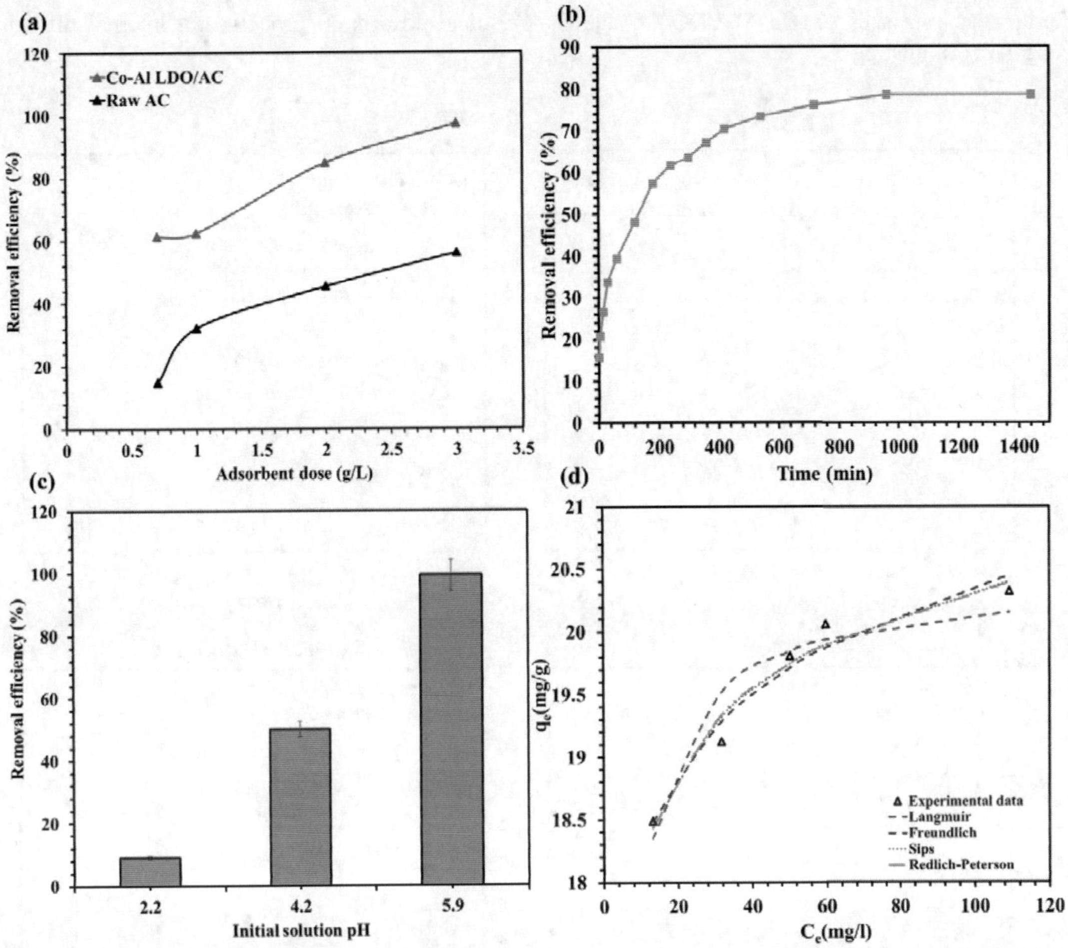


**Figure 4**: (a) Co-Al LDO/AC dosage effect on $Pb^{2+}$ removal efficiency. $[Pb^{2+}]_i = 50$ mg/L, t = 24 h, pHi =6.1-6.5, RT. (b) Contact time effect on $Pb^{2+}$ removal efficiency. $[Pb^{2+}]_i = 50$ mg/L, $S_o = 2$ g/L, $pH_i$ =6.1-6.5, RT. (c) $pH_i$ effect on $Pb^{2+}$ adsorption capacity. $[Pb^{2+}]_i = 50$ mg/L, $S_o = 2$ g/L, t = 24 h, RT. (d) $Pb^{2+}$ adsorption isotherm and the fitted models. $S_o = 2$ g/L, $pH_i$ =6.1-6.5, t = 24 h, and at RT.

*3.2.5. Adsorption kinetic:*

The determination of $Pb^{2+}$ adsorption kinetic has a major influence on assessing the adsorption rate and the mechanism of Co-Al LDO/AC in $Pb^{2+}$ uptake. Hence, the experimental data of 7 h adsorption were fitted into four linearized kinetic models, i.e., PFO, PSO, Elovich, and IPD model, as depicted in **Figure 5,** and their factors were presented in **Table 6**. Based on the obtained results, the experimental data exhibited strong conformity with PFO and PSO models with an $R^2$ of 0.986 and 0.987, respectively. Nonetheless, the theoretical adsorption capacity of the PFO model (13.8 mg/g) was not compatible with the experimentally obtained equilibrium adsorption capacity (19.8 mg/g). While the PSO model presented a better fitting with a high determination coefficient ($R^2 > 0.987$) and a theoretical adsorption uptake of 18.1 mg/g, indicating that the experimental data was best presented by the PSO model. Therefore, it can be anticipated that both chemisorption and physisorption processes between $Pb^{2+}$ and Co-Al LDO/AC play a role in the adsorption of $Pb^{2+}$ ions. Elovich and inter-particle diffusion models were examined too where both models complied with the experimental data, with an $R^2 > 0.928$ and 0.985, respectively. This could suggest the presence of heterogeneous surfaces on Co-Al LDO/AC with a chemical adsorption process and internal diffusion mechanism dominating.

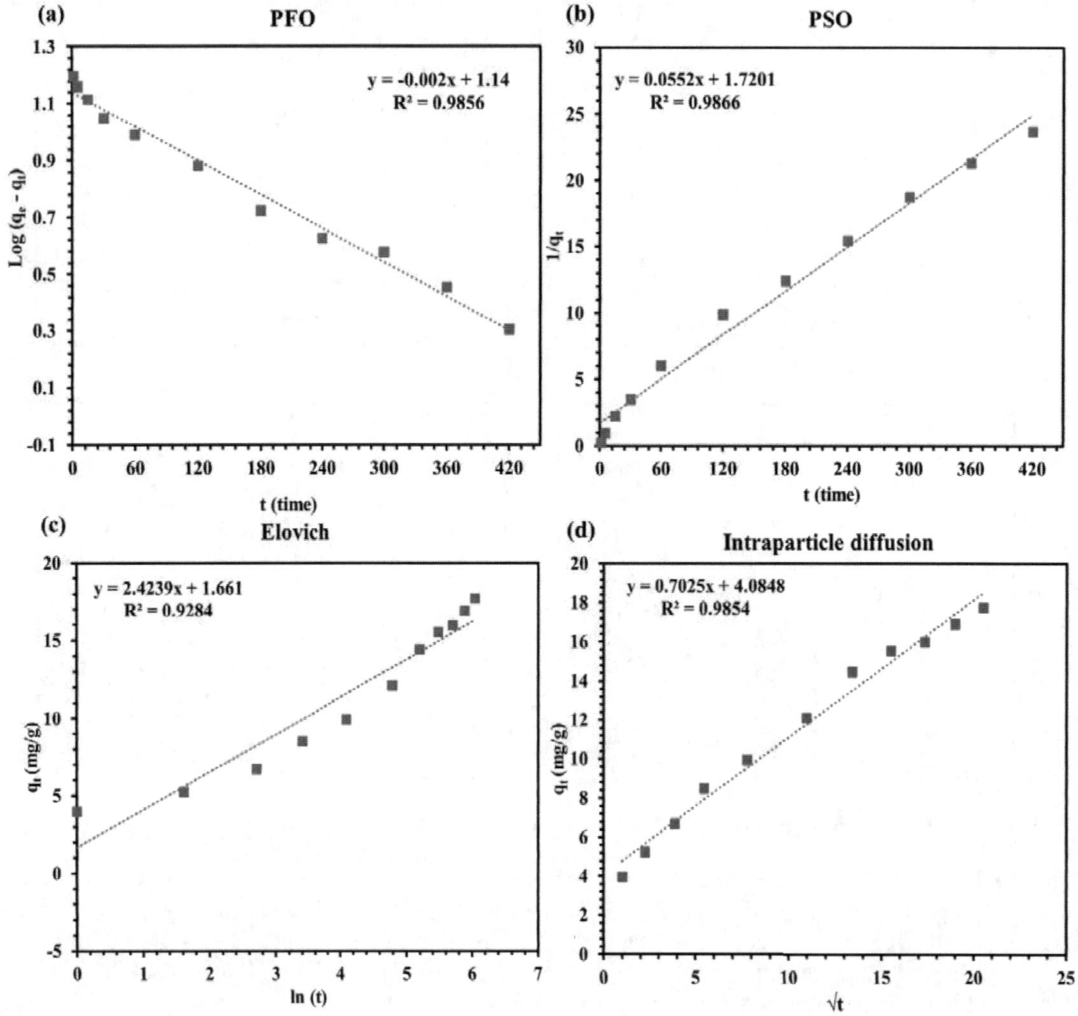


**Figure 5:** Adsorption kinetics of $Pb^{2+}$ adsorption onto Co-Al LDO/AC composite using linearized models of (a) PFO, (b) PSO, (c) Elovich, and (d) IPD.

**Table 6:** Parameters and determination coefficient of the four kinetic models for $Pb^{2+}$ adsorption onto Co-Al LDO/AC.

| Kinetic models | Parameters | Value |
|---|---|---|
| PFO | $q_e$ (mg/g) | 13.8 |
| | $k_1$ ($min^{-1}$) | 4.6 x $10^{-3}$ |
| | $R^2$ | 0.986 |
| PSO | $q_e$ (mg/g) | 18.1 |
| | $k_2$ (g/mg/min) | 0.59 |

Journal of Physics: Conference Series **2751** (2024) 012008 doi:10.1088/1742-6596/2751/1/012008

| | | |
|---|---|---|
| | $R^2$ | 0.987 |
| Elovich | $\beta$ (g/mg) | 0.41 |
| | $\alpha$ (mg/g/min) | 4.8 |
| | $R^2$ | 0.928 |
| IPD | c (mg/g) | 4.1 |
| | $K_{diff}$ (mg/g/min$^{-0.5}$) | 0.70 |
| | $R^2$ | 0.985 |

### *3.3. Adsorption mechanism*

In order to examine the mechanism of $Pb^{2+}$ uptake onto Co-Al LDO/AC composite, the surface functional groups of the Co-Al LDO/AC were examined after $Pb^{2+}$ adsorption, and the FTIR spectra were elucidated in **Figure 3(a).** As depicted in the figure, after the $Pb^{2+}$ elimination process, the broad band at the 3446 cm$^{-1}$ region was noticed which corresponds to the stretching vibration of OH groups as well as interlayer and surface $H_2O$ molecules. The stretching frequency of this was more intense after $Pb^{2+}$ adsorption indicating the presence of hydroxyl groups that were intercalated into the Co-Al LDO/AC interlayers [45]. While the intensity of the characteristic peaks at 673 cm$^{-1}$ and 568 cm$^{-1}$ was noticeably intensified which may be ascribed to the vibrational modes of M-O-H (Pb-O-H) and M-O-M (Al-O-Pb) [32]. This suggests that successful adsorption of the $Pb^{2+}$ ions might occur through either physical adsorption into the AC porous material or an ion exchange mechanism involving $Pb^{2+}$ and $H^+$ [18]. Additionally, it is suggested that there is a chemical binding between $Pb^{2+}$ ions and some of the $OH^-$ groups on the LDO surface, resulting in an inner-sphere complexes formation or through electrostatic interactions [27]. These processes can be expressed by eq. 4 and eq. 5 [46].

Chemical binding adsorption:

$$Sur - OH + Pb^{2+} \rightarrow Sur - O - Pb^{+} + H^{+} \tag{4}$$

Electrostatic binding adsorption:

$$Sur - O^{-} + Pb^{2+} \rightarrow Sur - O^{-} \ldots Pb^{2+} \tag{5}$$

where "Sur" indicates the LDO surface and "..." represents the effect of electrostatic binding.

## 4. Conclusion

In the current study, the synthesis of Co-Al LDO/AC composite was achieved following the co-precipitation method for the elimination of $Pb^{2+}$ from an aqueous solution. The surface and structural characteristics of the hybrid were thoroughly investigated and the results revealed a successful incorporation of Co and Al onto the AC matrix with a SA of 189.4 m$^2$/g. The batch adsorption experiments demonstrated that by comparing Co-Al LDO/AC with the raw AC, $Pb^{2+}$ removal efficiency was enhanced by approximately 50%. The optimum dosage of Co-Al LDO/AC was 2 g/L. The outcomes further revealed that equilibrium was achieved within 7 h and an optimum pH of 5.9. The kinetic experimental data were well-fitted with the PSO model ($R^2 > 0.987$) while the isotherm data were best explained by the Sips model ($R^2 > 0.966$). The maximum Sips adsorption capacity was 25.09 mg/g. These modeling and characterization outcomes proposed that the removal of $Pb^{2+}$ could be controlled by ion exchange, electrostatic interaction, and surface complexationadsorption mechanisms. Our outcomes suggested that Co-Al LDO/AC could be a potential material for the remediation of $Pb^{2+}$ ions from water.

References

[1] A. Abushawish, I. Almanassra, and L. Jaber, "Desalination Pretreatment Technologies : Current Status and Future Developments," no. April, 2023, doi: 10.3390/w15081572.

[2] S. Zhang *et al.*, "Applications of water-stable metal-organic frameworks in the removal of water pollutants: A review," *Environ. Pollut.*, vol. 291, no. September, p. 118076, 2021, doi: 10.1016/j.envpol.2021.118076.

[3] S. Mitra *et al.*, "Impact of heavy metals on the environment and human health: Novel therapeutic insights to counter the toxicity," *J. King Saud Univ. - Sci.*, vol. 34, no. 3, p. 101865, 2022, doi: 10.1016/j.jksus.2022.101865.

[4] S. Sen Raychaudhuri, P. Pramanick, P. Talukder, and A. Basak, *Polyamines, metallothioneins, and phytochelatins—Natural defense of plants to mitigate heavy metals*, 1st ed., vol. 69. Elsevier B.V., 2021.

[5] M. Wołowiec, M. Komorowska-kaufman, A. Pruss, G. Rzepa, and T. Bajda, "Removal of Heavy Metals and Metalloids from Water Using Drinking Water Treatment Wołowiec, M., Komorowska-Kaufman, M., Pruss, A., Rzepa, G., & Bajda, T. (2019).," *Minerals*, vol. 9, no. Table 1, pp. 1–17, 2019.

[6] PAHO - WHO, "Lead Contamination," *Pan Amarican Health Organization - World Health Organization*, 2016. https://www3.paho.org/hq/index.php?option=com_content&view=article&id=8206:2013-lead-contamination&Itemid=39800&lang=en (accessed Apr. 23, 2022).

[7] E. Obeng-Gyasi, "Sources of lead exposure in various countries," *Rev. Environ. Health*, vol. 34, no. 1, pp. 25–34, 2019, doi: 10.1515/reveh-2018-0037.

[8] N. Qafisheh, M. Manar, A. G. Dana, K. Ayah, E. Olla, and K. Bisma, "Water pollution in the Arabian gulf," *Glob. Environ.*, vol. 13, no. 2, pp. 261–284, 2020, doi: 10.3197/ge.2020.130202.

[9] WHO, "Drinking-water," *World Health Organization*, 2022. https://www.who.int/news-room/fact-sheets/detail/drinking-water#:~:text=In 2020%2C 74%25 of the,needed%2C and free from contamination. (accessed Apr. 23, 2022).

[10] A. Prasetyaningrum, D. Ariyanti, W. Widayat, and B. Jos, "Copper and lead ions removal by electrocoagulation: Process performance and implications for energy consumption," *Int. J. Renew. Energy Dev.*, vol. 10, no. 3, pp. 415–424, 2021, doi: 10.14710/ijred.2021.31665.

[11] A. Lalmi, K. E. Bouhidel, B. Sahraoui, and C. el H. Anfif, "Removal of lead from polluted waters using ion exchange resin with Ca(NO3)2 for elution," *Hydrometallurgy*, vol. 178, no. 2017, pp. 287–293, 2018, doi: 10.1016/j.hydromet.2018.05.009.

[12] P. Zhang, S. Ouyang, P. Li, Z. Sun, N. Ding, and Y. Huang, "Ultrahigh removal performance of lead from wastewater by tricalcium aluminate via precipitation combining flocculation with amorphous aluminum," *J. Clean. Prod.*, vol. 246, p. 118728, 2020, doi: 10.1016/j.jclepro.2019.118728.

[13] X. Yang, L. Chen, D. Ren, S. Wang, and Z. Ren, "Adsorption of Pb(II) from water by treatment with an O-hydroxyphenyl thiourea-modified chitosan," *Int. J. Biol. Macromol.*, vol. 220, no. May, pp. 280–290, 2022, doi: 10.1016/j.ijbiomac.2022.08.090.

[14] A. Abushawish *et al.*, "High-efficiency removal of hexavalent chromium from contaminated water using nitrogen-doped activated carbon: kinetics and isotherm study," *Mater. Chem. Phys.*, vol. 291, no. September, p. 126758, 2022, doi: 10.1016/j.matchemphys.2022.126758.

[15] A. Kushwaha, R. Rani, and J. K. Patra, "Adsorption kinetics and molecular interactions of lead [Pb(II)] with natural clay and humic acid," *Int. J. Environ. Sci. Technol.*, vol. 17, no. 3, pp. 1325–1336, 2020, doi: 10.1007/s13762-019-02411-6.

[16] S. A. Sajjadi *et al.*, "A novel route for preparation of chemically activated carbon from pistachio wood for highly efficient Pb(II) sorption," *J. Environ. Manage.*, vol. 236, no. September 2018, pp.

34–44, 2019, doi: 10.1016/j.jenvman.2019.01.087.

[17] K. Tomasz, K. Anna, and C. Ryszard, "Effective adsorption of lead ions using fly ash obtained in the novel circulating fluidized bed combustion technology," *Microchem. J.*, vol. 145, no. August 2018, pp. 1011–1025, 2019, doi: 10.1016/j.microc.2018.12.005.

[18] X. Feng, R. Long, L. Wang, C. Liu, Z. Bai, and X. Liu, "A review on heavy metal ions adsorption from water by layered double hydroxide and its composites," *Sep. Purif. Technol.*, vol. 284, no. November 2021, p. 120099, 2022, doi: 10.1016/j.seppur.2021.120099.

[19] N. Bu *et al.*, "Synthesis of NaY zeolite from coal gangue and its characterization for lead removal from aqueous solution," *Adv. Powder Technol.*, vol. 31, no. 7, pp. 2699–2710, 2020, doi: 10.1016/j.apt.2020.04.035.

[20] J. Qu *et al.*, "High-efficiency decontamination of Pb(II) and tetracycline in contaminated water using ball-milled magnetic bone derived biochar," *J. Clean. Prod.*, vol. 385, no. September 2022, p. 135683, 2023, doi: 10.1016/j.jclepro.2022.135683.

[21] P. Praipipat, P. Ngamsurach, and A. Sanghuayprai, "Modification of sugarcane bagasse with iron(III) oxide-hydroxide to improve its adsorption property for removing lead(II) ions," *Sci. Rep.*, vol. 13, no. 1, pp. 1–23, 2023, doi: 10.1038/s41598-023-28654-5.

[22] J. Wang, R. Wang, J. Ma, and Y. Sun, "Study on the Application of Shell-Activated Carbon for the Adsorption of Dyes and Antibiotics," *Water (Switzerland)*, vol. 14, no. 22, 2022, doi: 10.3390/w14223752.

[23] L. Jaber *et al.*, "Adsorptive Removal of Lead and Chromate Ions from Water by Using Iron-Doped Granular Activated Carbon Obtained from Coconut Shells," *Sustainability*, vol. 14, pp. 1–24, 2022, doi: https://doi.org/10.3390/su141710877.

[24] A. A. Alghamdi, A. B. Al-Odayni, W. S. Saeed, A. Al-Kahtani, F. A. Alharthi, and T. Aouak, "Efficient adsorption of lead (II) from aqueous phase solutions using polypyrrole-based activated carbon," *Materials (Basel).*, vol. 12, no. 12, 2019, doi: 10.3390/ma12122020.

[25] X. Liu, D. Lai, and Y. Wang, "Performance of Pb(II) removal by an activated carbon supported nanoscale zero-valent iron composite at ultralow iron content," *J. Hazard. Mater.*, vol. 361, no. April 2018, pp. 37–48, 2019, doi: 10.1016/j.jhazmat.2018.08.082.

[26] S. Iftekhar, M. E. Küçük, V. Srivastava, E. Repo, and M. Sillanpää, "Application of zinc-aluminium layered double hydroxides for adsorptive removal of phosphate and sulfate: Equilibrium, kinetic and thermodynamic," *Chemosphere*, vol. 209, pp. 470–479, 2018, doi: 10.1016/j.chemosphere.2018.06.115.

[27] D. Zhao, G. Sheng, J. Hu, C. Chen, and X. Wang, "The adsorption of Pb(II) on Mg2Al layered double hydroxide," *Chem. Eng. J.*, vol. 171, no. 1, pp. 167–174, 2011, doi: 10.1016/j.cej.2011.03.082.

[28] O. Rahmanian, S. Amini, and M. Dinari, "Preparation of zinc/iron layered double hydroxide intercalated by citrate anion for capturing Lead (II) from aqueous solution," *J. Mol. Liq.*, vol. 256, pp. 9–15, 2018, doi: 10.1016/j.molliq.2018.02.018.

[29] X. Liang *et al.*, "Sorption of lead ion by layered double hydroxide intercalated with diethylenetriaminepentaacetic acid," *Colloids Surfaces A Physicochem. Eng. Asp.*, vol. 366, no. 1–3, pp. 50–57, 2010, doi: 10.1016/j.colsurfa.2010.05.012.

[30] M. B. Poudel, G. P. Awasthi, and H. J. Kim, "Novel insight into the adsorption of Cr(VI) and Pb(II) ions by MOF derived Co-Al layered double hydroxide @hematite nanorods on 3D porous carbon nanofiber network," *Chem. Eng. J.*, vol. 417, no. March, p. 129312, 2021, doi: 10.1016/j.cej.2021.129312.

[31] F. Yang, S. Sun, X. Chen, Y. Chang, F. Zha, and Z. Lei, "Mg-Al layered double hydroxides modified clay adsorbents for efficient removal of Pb2+, Cu2+ and Ni2+ from water," *Appl. Clay Sci.*, vol. 123, pp. 134–140, 2016, doi: 10.1016/j.clay.2016.01.026.

[32] Y. Jia *et al.*, "A novel magnetic biochar/MgFe-layered double hydroxides composite removing Pb

2+ from aqueous solution: Isotherms, kinetics and thermodynamics," *Colloids Surfaces A Physicochem. Eng. Asp.*, vol. 567, no. January, pp. 278–287, 2019, doi: 10.1016/j.colsurfa.2019.01.064.

[33] A. K. A. Khalil, F. Dweiri, I. W. Almanassra, A. Chatla, and M. A. Atieh, "Mg-Al Layered Double Hydroxide Doped Activated Carbon Composites for Phosphate Removal from Synthetic Water: Adsorption and Thermodynamics Studies," *Sustainability*, vol. 14, no. 12, pp. 1–23, 2022, doi: 10.3390/su14126991.

[34] X. Wu *et al.*, "Fabrication of the composite sepiolite @ polyethyleneimine / sodium alginate and its excellent adsorption performance for heavy metal ions," *Appl. Clay Sci.*, vol. 228, no. June, p. 106647, 2022, doi: 10.1016/j.clay.2022.106647.

[35] J. S. Calisto *et al.*, "Adsorption kinetic and thermodynamic studies of the 2, 4 – dichlorophenoxyacetate (2,4-D) by the [Co–Al–Cl] layered double hydroxide," *Heliyon*, vol. 5, no. 12, p. e02553, 2019, doi: 10.1016/j.heliyon.2019.e02553.

[36] M. M. B. Quiroga, M. I. Juncà, and N. A. Merino, "Study of the effect of the LDH cations precursors in the removal of arsenic in aqueous solution," *Water Sci. Technol.*, vol. 83, no. 10, pp. 2518–2525, 2021, doi: 10.2166/wst.2021.153.

[37] J. Bedia, M. Peñas-Garzón, A. Gómez-Avilés, J. J. Rodriguez, and C. Belver, "Review on Activated Carbons by Chemical Activation with FeCl3," *J. Carbon Res.*, vol. 6, no. 2, p. 21, 2020, doi: 10.3390/c6020021.

[38] Z. Tian, Q. Li, J. Hou, L. Pei, Y. Li, and S. Ai, "Platinum nanocrystals supported on CoAl mixed metal oxide nanosheets derived from layered double hydroxides as catalysts for selective hydrogenation of cinnamaldehyde," *J. Catal.*, vol. 331, pp. 193–202, 2015, doi: 10.1016/j.jcat.2015.08.020.

[39] X. Zhang, X. Zhang, C. An, and S. Wang, "Electrochemistry-enhanced peroxymonosulfate activation by CoAl-LDH@biochar for simultaneous treatment of heavy metals and PAHs," *Sep. Purif. Technol.*, vol. 311, no. January, p. 123341, 2023, doi: 10.1016/j.seppur.2023.123341.

[40] J. Wu, T. Wang, Y. Zhang, and W. P. Pan, "The distribution of Pb(II)/Cd(II) adsorption mechanisms on biochars from aqueous solution: Considering the increased oxygen functional groups by HCl treatment," *Bioresour. Technol.*, vol. 291, no. July, p. 121859, 2019, doi: 10.1016/j.biortech.2019.121859.

[41] J. A. Ramos-Guivar *et al.*, "Improved removal capacity and equilibrium time of maghemite nanoparticles growth in zeolite type 5a for pb(Ii) adsorption," *Nanomaterials*, vol. 10, no. 9, pp. 1–19, 2020, doi: 10.3390/nano10091668.

[42] X. Song, H. Liu, L. Cheng, and Y. Qu, "Surface modification of coconut-based activated carbon by liquid-phase oxidation and its effects on lead ion adsorption," *Desalination*, vol. 255, no. 1–3, pp. 78–83, 2010, doi: 10.1016/j.desal.2010.01.011.

[43] A. M. Alkherraz, A. K. Ali, and K. M. Elsherif, "Removal of Pb (II), Zn (II), Cu (II) and Cd (II) from aqueous solutions by adsorption onto olive branches activated carbon: Equilibrium and thermodynamic studies," *J. Chem. Int.*, vol. 6, no. 1, pp. 11–20, 2020.

[44] M. Ullah *et al.*, "The effective removal of heavy metals from water by activated carbon adsorbents of Albizia lebbeck and Melia azedarach seed shells," *Soil Water Res.*, vol. 15, no. 1, pp. 30–37, 2020, doi: 10.17221/212/2018-SWR.

[45] Y. H. Jiang, A. Y. Li, H. Deng, C. H. Ye, and Y. Li, "Phosphate adsorption from wastewater using ZnAl-LDO-loaded modified banana straw biochar," *Environ. Sci. Pollut. Res.*, vol. 26, no. 18, pp. 18343–18353, 2019, doi: 10.1007/s11356-019-05183-1.

[46] X. Liang, W. Hou, and J. Xu, "Sorption of Pb(II) on Mg-Fe layered double hydroxide," *Chinese J. Chem.*, vol. 27, no. 10, pp. 1981–1988, 2009, doi: 10.1002/cjoc.200990333.

# Effect of Heat Treatments on Microstructure and Mechanical Properties of Fe-Mn-Ni-Al-Gd Shape Memory Alloy

**Faisal Mustafa [1,2*], Mehmet Egilmez [1,2], Wael Abuzaid [2,3*], Sami El-Khatib [1,2]**

[1] Department of Physics, American University of Sharjah, Sharjah 26666, United Arab Emirates
[2] Materials Science and Engineering Program, American University of Sharjah, Sharjah 26666, United Arab Emirates
[3] Department of Mechanical Engineering, American University of Sharjah, Sharjah 26666, United Arab Emirates
*Corresponding authors' e-mail: b00090953@aus.edu, wabuzaid@aus.edu

**Abstract**. There are significant scientific and industrial efforts to develop and optimize Iron-based shape memory alloys (SMA) such as FeMnNiAl for cost-sensitive applications. This alloy system shows shape memory and superelastic properties across a large temperature range. However, many studies have pointed out the need for rather complex thermo-mechanical treatments for the optimization of the SMA properties. In addition, works considering the effects of alloying on the development of microstructures that are more conducive to pseudo-elasticity in this system remain limited. Hence, systematic studies aiming at the investigation of the microstructural evolution of the FeMnNiAl(Gd) system are of great interest. In this study, solution heat treatment is done to tune the microstructure for optimum mechanical properties. The effect of phase distribution on mechanical properties is investigated at different heat treatments. Whereas cyclic heat treatment induced abnormal grain growth (AGG) in all samples, so large grains were obtained. The phase variation and elemental composition are analyzed by X-ray diffraction and Energy Dispersive Spectroscopy, respectively. The microstructure and phase distribution are observed using Scanning Electron Microscope and then related to the microhardness results. The microstructure has a good correlation with mechanical properties where the fine distribution of phases results in a higher hardness number.

## 1. Introduction

Traditional alloying methods based on single principle elements in materials science are progressing towards unconventional multi-principle elements alloying (MPEAs) approaches during the last two decades. Many electronic [1], magnetic [2], mechanical [3], and electrochemical functionalities [4] have emerged from this new group of materials [5][6][7]. These alloys gain more popularity due to properties like high cryogenic strength, shape memory and superelasticity, high fracture toughness, and ductility, etc. Many shape memory alloy (SMA) systems, such as Ni-Ti, Ni-Mn-Ga, Cu-Al-Ni, and Ni-Co-Mn-In alloys, [8][9][1,3,10][11] have been explored since Au-Cd alloys' shape memory behavior was discovered.

Currently, the Ni-Ti system dominates the SMA market however, there is a collaborative worldwide effort on iron-based shape memory alloys because they are cost-effective and offer larger compositional and temperate tuneability of the functional mechanical properties along with some electronic and magnetic functionalities. Reversible martensitic transition in Fe-based shape memory alloys can be grouped into four systems: (i) Fe-noble metal-based alloys, such as Fe–Pt [12] and Fe-Pd [13], (ii) The

Fe-Ni-Co-based, (iii) Fe-Ni-Co-based alloys, such as Fe-Ni-Co-Al-Ta-B [14], Fe-Ni-Co-Al-Nb-B [15], and Fe-Ni-Co-Al-Ti-B [16][17][18][19][20] and (iv) BCC Fe-Mn-based alloys, in particular, can show extraordinary superelasticity.

The Fe-Mn-Al-Ni shape memory alloy exhibits a wide temperature range of superelasticity owing to the negligible entropy change during the martensitic transformation [21][22][23][24][25][26]. Certain parameters influence superelastic behavior, including grain size in proportion to sample cross-sectional area [21][27][28], precipitate size and their volume percentage [29][30][31], and crystallographic orientation [32][33][34]. In this alloy system [35], Abnormal Grain Growth (AGG) generated by cyclic heat treatment [36][37] further improves the usefulness of alloys. Subgrain structure in the BCC phase and the sub-boundary energy driving an AGG have been discovered following cyclic heat treatment between the single-phase BCC and the dual phase BCC+FCC regions. By cyclic heat treatment [38], relatively large-sized crystals can be produced which is conducive to superelasticity in this system, avoids prestress losses and tunes the transformation anisotropy [39].

The originally published Fe-34Mn-15Al-7.5Ni (at. %) alloy [21] was tweaked to solve a number of practical issues, such as; the crack development during final quenching, the difficulties of producing significantly bigger single-crystal samples via AGG, and the insufficient superelasticity.
Rare earth (RE) elements such as Nd, Gd, Ce, etc., have been shown to improve the microstructure and mechanical characteristics of a wide spectrum of alloy systems [40][41][42][41]. Because each RE element has a unique character, combining additions of various RE elements to examine their interactions is particularly vital. In this context, Xie et al. [42] looked into how RE (Y and Nd) elements affected the AZ81 alloy's microstructures and mechanical characteristics. The findings revealed that the right amount of RE elements could drastically purify the grains, resulting in the formation of the $Al_2Y$ and $Al_2Nd$ phases.

In this study, the effects of heat treatment on the microstructure and abnormal grain growth of the FeMnAlNiGd alloy are investigated in order to finetune optimum microstructures and their relationship with mechanical properties such as hardness. The dual-phase structure is thoroughly studied both in terms of crystallography and elemental composition. The sample with optimum microstructure and hardness can be a potential candidate to replace existing SMAs.

## 2. Materials and Methods

$Fe_{40.02}Mn_{32.65}Al_{5.90}Ni_{7.69}Gd_{13.74}$ alloy in wt.% was fabricated from high-purity metals by induction melting process under an inert atmosphere. The cast ingot was then homogenized at 1200 °C for 1440 min in an inert Ar atmosphere. Following homogenization, 4 mm thick slices were cut from the ingot using a water jet. Rectangular specimens with a 4×4 mm² cross-section and 8 mm length were machined from the 4 mm strips. Specimens were subjected to various heat treatments including solution heat treatment (SHT) and cyclic heat treatments. To prevent oxidation during such high-temperature treatments, all specimens were encapsulated in quartz tubes, evacuated, and backfilled with Ar.

| Sample designation | Homo | 1000-20 | 1000-60 | 1225-60 | 1300-60 | 1300-120 | Cyclic-80 | Cyclic-120 |
|---|---|---|---|---|---|---|---|---|
| S. Temp. (°C) | - | 1000 | 1000 | 1225 | 1300 | 1300 | 1250 | 1250 |
| Hold Time (min) | - | 20 | 60 | 60 | 60 | 120 | 80 | 120 |
| Quenching | - | Water | Water | Water | Water | Water | Water | Water |

**Table 1.** Sample classification with their respective heat treatment parameters.

The conducted solution heat treatment conditions are summarized in Table 1. As noted above, all treatments were done following homogenization. The full details of the treatment conditions (i.e., temperature setpoints, heating rates, holding times, and quenching temperature) are explained schematically in Fig. 1 for the solution treatment and in Fig. 2 for the cyclic heat treatments. In all cases, the heating rates were fixed to 3°C/min for the entire heat treatment. Before testing, a low-temperature aging treatment was conducted at 200°C for 240 min.

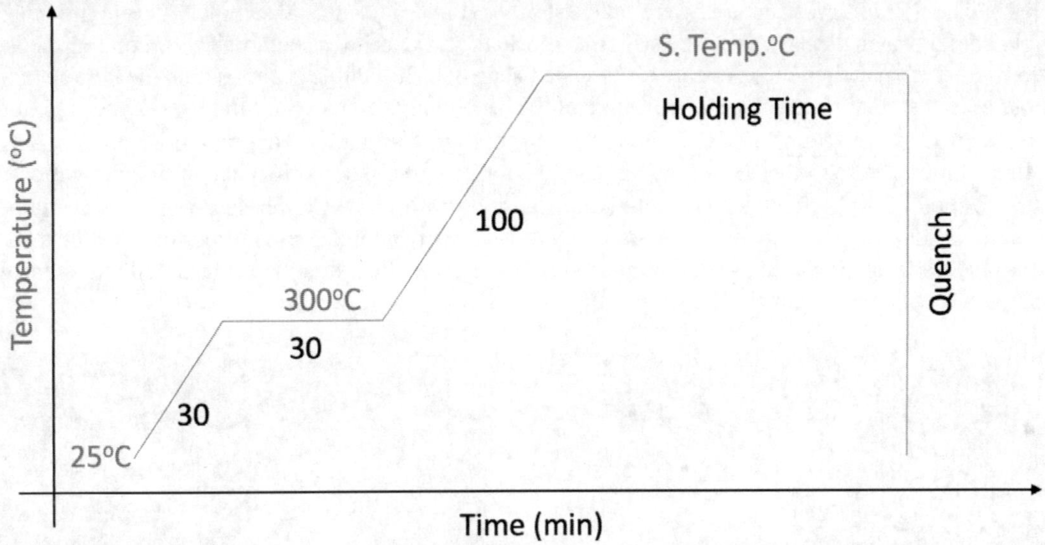


**Figure 1.** Schematic diagram of the solution heat treatment process

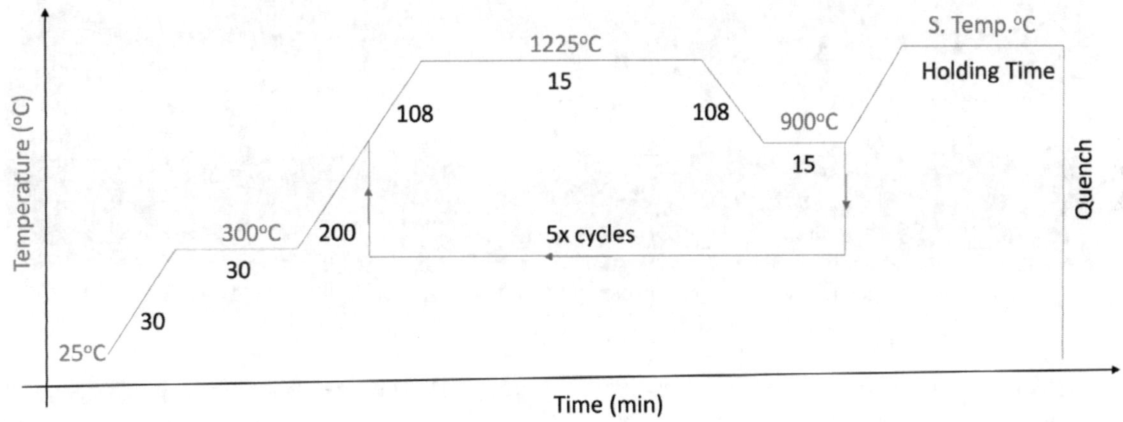


**Figure 2.** Schematic diagram of the cyclic heat treatment process

The heat-treated samples were ground and polished using a metallurgical-grade grinding setup for the examination. Crystal structure, phase purity, and chemical homogeneity of the studied material were examined by Panalytical X′pert [3] powder X-ray diffraction (XRD) and TESCAN VEGA-3 Scanning Electron Microscope (SEM) which was supported by energy-dispersive X-ray spectroscopy (EDX, Oxford Instruments). The target compositions were confirmed by EDX analysis within a 1% error margin. Micro Vickers hardness tests were performed on the polished surfaces for the hardness analysis to deduce a correlation between microstructure and mechanical properties. A force of 1.96 N was applied for a dwell time of 10 sec.

## 3. Results and Discussions

SEM analysis shows a dual-phase structure (α-BCC and γ-FCC phases) in all the samples. Both phases have a unique crystallographic and elemental composition which is further confirmed by XRD and EDS analysis. The SEM microstructures were taken at 500x (Fig. 3) for all the samples. The homogenized sample contains a dual-phase structure which is more like a dendritic structure. The second phase is in the form of sharp platelets and is evenly distributed. The 1000-20 sample is almost like the homogenized sample in terms of microstructure, but the size of the second phase increased. In general, the size of the cluster/second phase grains (γ-FCC) increased with higher treatment temperatures and/or longer treatment times. On the other hand, the 1300-120 sample exhibits dendritic like structures which are mostly formed of α-phase along with relatively flat and uniform γ-FCC phases. With the cyclic heat treatment, abnormal grain growth is favored, resulting in more coarse grains of both phases instead of matrix phase segregation at the grain boundaries. During cyclic heat treatment, the smaller and equally distributed second phase can also be seen within the matrix.

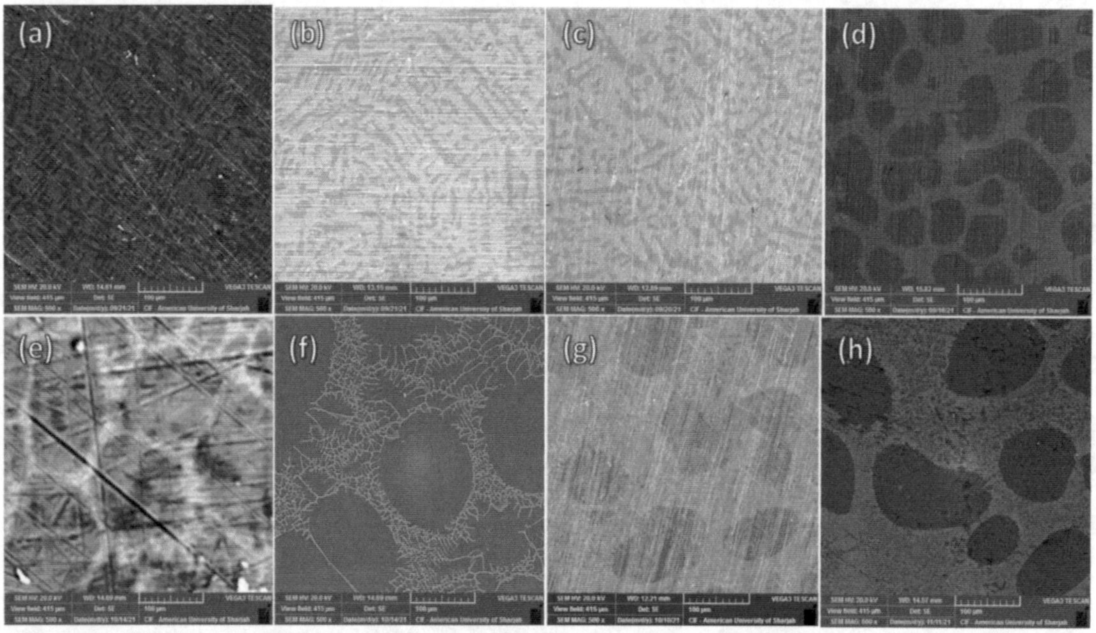


**Figure 3.** SEM images at 500x of (a) homogenized, (b) 1000-20, (c) 1000-60, (d) 1225-60, (e) 1300-60, (f) 1300-120, (g) cyclic-80, and (h) cyclic-120 sample.

Representative EDS results from two specimens (homogenized and the cyclic-80 samples) are shown in Figs. 4-5. In the homogenized conditions, the α-phase (BCC) consists of all the constituents initially present in the alloy, whereas the second phase (show) has only Fe, Mn, Al, and Ni with the absence of Gd. The composition of the other constituent elements also differs between the two phases, which will be discussed later in this section.

The composition in each of the observed phases for a representative case is presented in Fig. 6. Spectrum 1 was obtained from the Gd-free phase (FCC) while Spectrum 2 was collected from a BCC region. For the reported case in Fig. 6, the FCC phase had a composition of $Fe_{53.2}Mn_{36.6}Ni_{4.2}Al_{5.9}Gd_{0.0}$ wt.% while the BCC region had a composition of $Fe_{24.1}Mn_{29.6}Ni_{12.8}Al_{7.8}Gd_{25.7}$. Note that both the Fe and Mn have intrinsic BCC structures adverse to Ni and Al that have FCC structures.

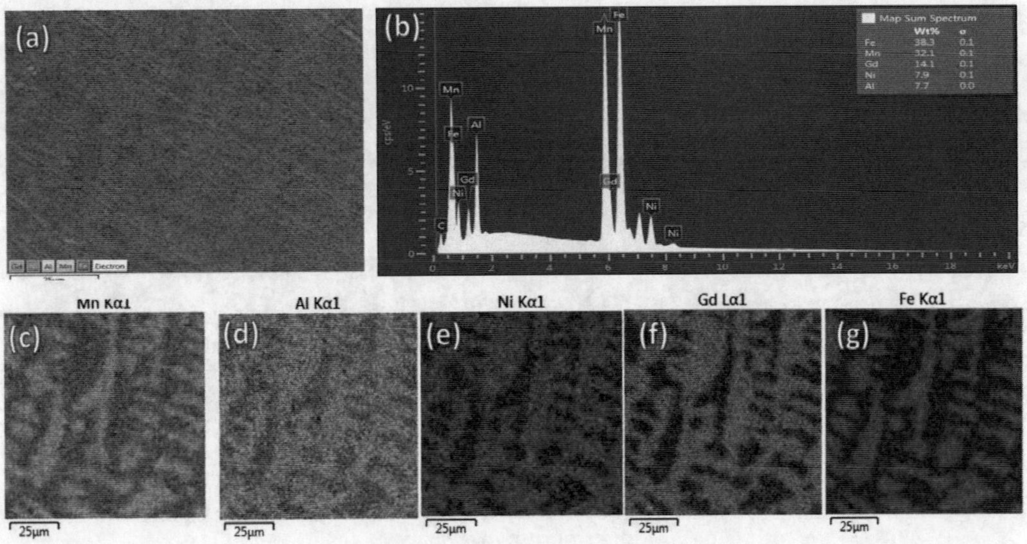


**Figure 4.** EDS area mapping of Homogenized sample, (a) microstructure of selected area, (b) elemental peak intensity versus wavelength and elemental distribution in wt.%, and elemental distribution of (c) Mn, (d) Al, (e) Ni, (f) Gd, (g) Fe.

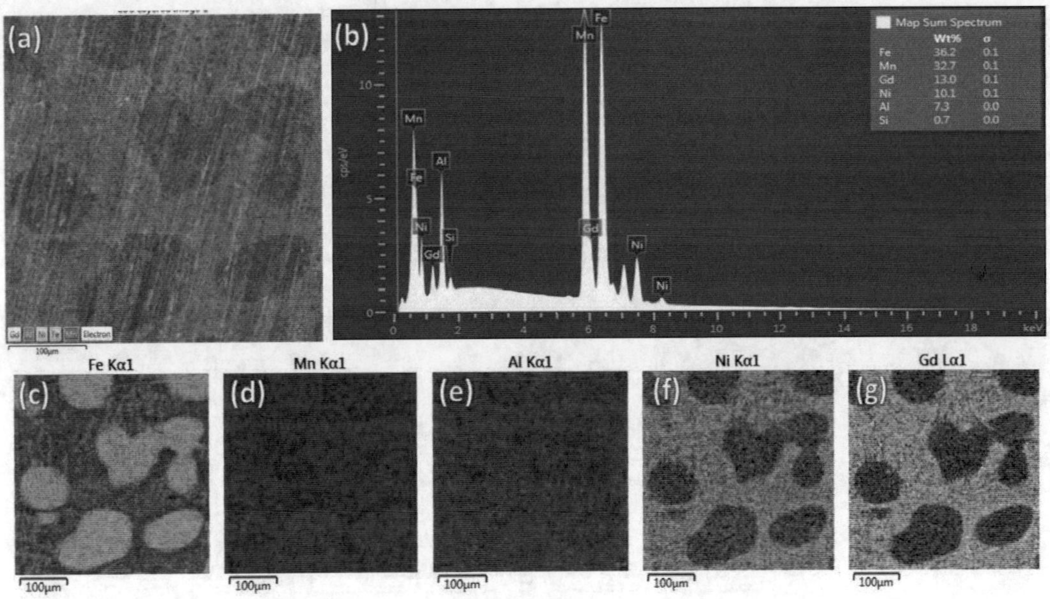


**Figure 5.** EDS area mapping of Cyclic-80 sample, (a) microstructure of the selected area, (b) elemental peak intensity versus wavelength and elemental distribution in wt.%, and elemental distribution of (c) Fe, (d) Mn, (e) Al, (f) Ni, (g) Gd.

Hence, we believe Gd segregation could be a result of competition between two structural phases. More specifically, BCC elements are consuming Gd (HCP) phase and form the BCC structure in the considered alloy. It is also observed that the BCC phase is comparatively Fe deficient (24.1 wt.%) while the FCC phase is Fe dominant phase (53.2 wt.%), whereas Ni is behaving the other way around.

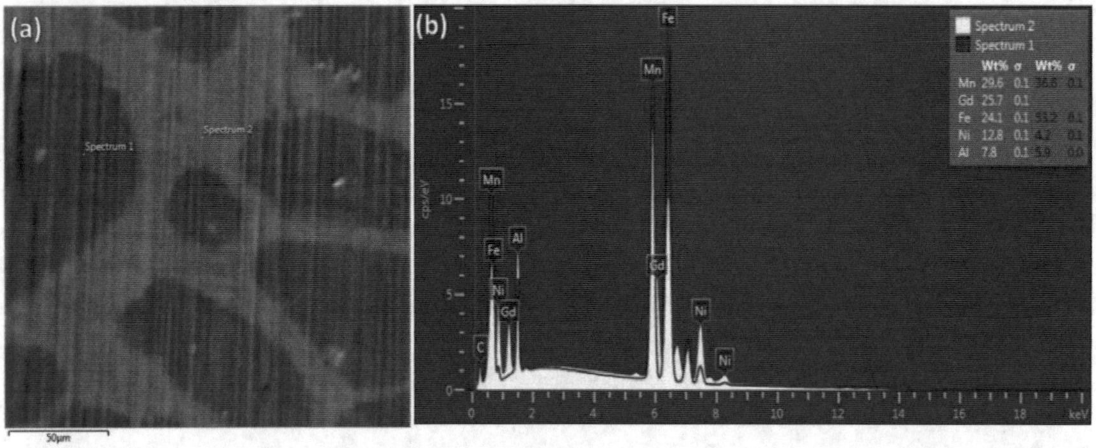


**Figure 6.** EDS point analysis of 1225-60 sample, (a) microstructure of sample showing selected regions for Spectrum 1 and Spectrum 2, (b) elemental peak intensity versus wavelength and elemental distribution in wt.%.

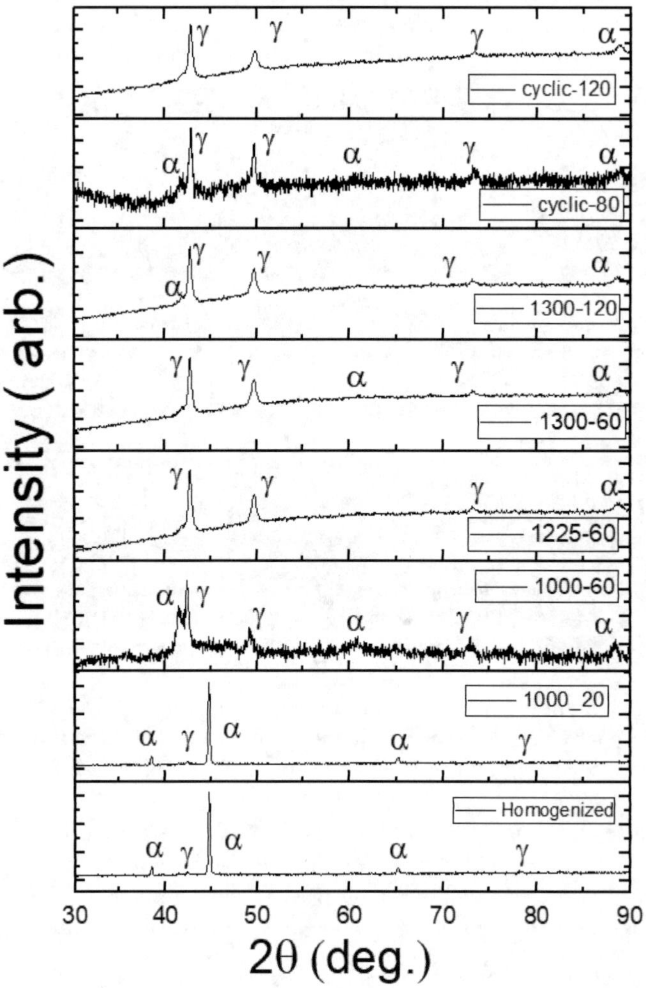


**Figure 7.** XRD analysis of solution heat-treated samples

In line with the SEM analysis, XRD peaks show a dual phase structure confirming the presence of both FCC and BCC phases as shown in Fig. 7. FCC (γ) peaks appeared at 2θ values of around 43°, 50° and 73° whereas for BCC (α) peaks are at 38°, 45 °, 65 ° and 88° [43][6]. With the addition of Gd, BCC (α) phase became the dominant phase in the structure without heat treatment. However, the stability of BCC (α) changes for different SHT processes. The presence of FCC phase compared to BCC is suppressed in the homogenized sample but with the increase in temperature, the clusters have grown, resulting in the higher peak intensity of γ-phase. Comparable α and γ coexistence is observed in the cyclic-80 sample. This α-phase stability is associated with the increment in the Gd-rich phase in SEM microstructures.

Micro Vickers hardness tests were performed to analyze the mechanical properties in relation to the various microstructures introduced through the different SHT processes. The hardness results are shown in Fig. 8, indicating a clear relation between the microstructural evolution and hardness variation. In general, reduced hardness magnitudes are obtained with higher SHT temperatures. This is attributed to the coarser grain structure obtained under such conditions. The minimum hardness of 250.12 HV was achieved in the 1300-120 sample while the maximum hardness of 392.03HV was obtained for the 1000-20 sample. The cyclic sample behaves differently because of abnormal grain growth in both phases and because of finer second-phase distribution in the matrix phase. This finer distribution is more prominent in the cyclic-120 sample which resulted in a higher hardness value of 310.15 HV than 285.35 HV in the cyclic-80 sample. The lower hardness value of cyclic-80 is because the time to form a uniformly distributed phase inside the matrix is less due to the lesser holding time.

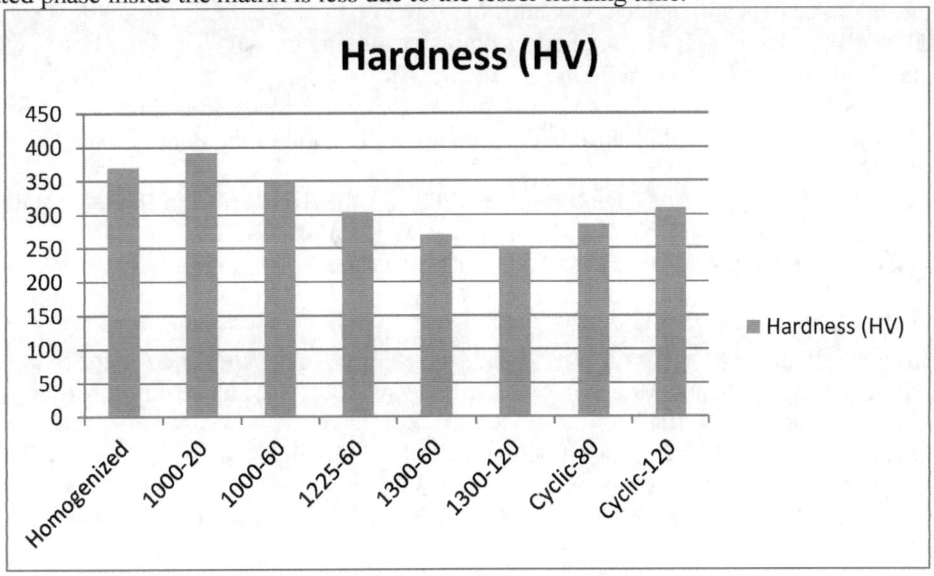


**Figure 8.** Histogram of Micro Vickers Hardness of heat-treated samples

## 4. Conclusions

The microstructural evolution of FeMnAlNiGd alloy is studied at different solution heat treatment cycles. FeMnAlNiGd alloy exhibits a dual-phase structure having BCC and FCC phases forming the matrix and the secondary phase, respectively. Both phases have unique chemical compositions with a prominent difference in the presence of Gd, which exists only in the matrix phase. The secondary phase (γ) is Fe and Mn-rich phase whereas more Ni and Al are found with Gd in the matrix phase (α). Large grains were more easily formed, and AGG was induced in all samples via cyclic heat treatment. Microstructure evolution at different SHT parameters correlates well with mechanical properties such as hardness. The fine distribution of phases results in a higher Vickers number. 1000-20 sample is showing the finest microstructure amongst all the samples, yet suitable for high strength applications.

But the cyclic-120 sample is showing optimum properties due to AGG and significant hardness values, which is potentially suitable for shape memory and superelastic behavior.

## 5. Future Perspective

FeMnNiAl is a shape memory material that shows superelastic properties so heat treatment can be further optimized for the FeMnNiAlGd system to get shape memory effects. Also, the magnetic properties of (other than the 1225-60 sample) can be investigated in detail with electric transport properties. Other mechanical properties including compression and tribology can be further investigated for the self-lubricating steels.

**Acknowledgments**

M.E. and S.E.-K. acknowledge support from the American University of Sharjah through faculty research Grants No. FRG21-M-S34, and No. FRG22-E-S93.

## 6. References

[1] Egilmez M, Batal O, Abuzaid W, Salman Z, Alkhader M, Akamine H, Nishida M and Robinson J W A 2021 Superconductivity in Ti67Zr19Nb11.5Sn2.5 shape memory alloy *Phys. Rev. Mater.* **5** 1–9

[2] Egilmez M and Abuzaid W 2021 Magnetic, electrical and mechanical properties of Fe40Mn40Co10Cr10 high entropy alloy *Sci. Rep.* **11** 1–12

[3] Abuzaid W and Sehitoglu H 2018 Plastic strain partitioning in dual phase Al13CoCrFeNi high entropy alloy *Mater. Sci. Eng. A* **720** 238–47

[4] Mohamed O, Hassan M, Egilmez M, Abuzaid W, Ibrahim T and Khamis M 2022 Corrosion behavior of CoCrNi / mild steel medium entropy alloy thin films *Mater. Today Commun.* **30** 103015

[5] Abuzaid W, Egilmez M and Chumlyakov Y I 2021 TWIP–TRIP effect in single crystalline VFeCoCrNi multi-principle element alloy *Scr. Mater.* **194** 113637

[6] El Batal O, Abuzaid W, Egilmez M, Alkhader M, Patriarca L and Casati R 2022 Deformation Rate and Temperature Sensitivity in TWIP/TRIP VCrFeCoNi Multi-Principal Element Alloy *Metals (Basel).* **12** 1510

[7] Mustafa F, Egilmez M, Abuzaid W, El-Khatib S, Nawaz T, Ahmad S and Alagoz S 2023 Strange Metallicity and Magnetic Order in the CoNi (Cr/V) Medium-Entropy Alloy System *Materials (Basel).* **16** 1044

[8] Ölander A 1932 AN ELECTROCHEMICAL INVESTIGATION OF SOLID CADMIUM-GOLD ALLOYS *J. Am. Chem. Soc.* **54** 3819–33

[9] Otsuka K 1998 Mechanism of shape memory effect and superelasticity *Shape Mem. Mater.* 27–48

[10] Kakeshita T and Ullakko K 2002 Giant magnetostriction in ferromagnetic shape-memory alloys *MRS Bull.* **27** 105–9

[11] Kainuma R, Imano Y, Ito W, Sutou Y, Morito H, Okamoto S, Kitakami O, Oikawa K, Fujita A and Kanomata T 2006 Magnetic-field-induced shape recovery by reverse phase transformation *Nature* **439** 957–60

[12] Dunne D P and Wayman C M 1973 The effect of austenite ordering on the martensite transformation in Fe-Pt alloys near the composition Fe 3 Pt: I. Morphology and transformation characteristics *Metall. Trans.* **4** 137–45

[13] Sohmura T and FE F 1980 Thermoelastic FCC-FCT martensitic transformation in Fe-Pd alloy

[14] Tanaka Y, Himuro Y, Kainuma R, Sutou Y, Omori T and Ishida K 2010 Ferrous polycrystalline shape-memory alloy showing huge superelasticity *Science (80-. ).* **327** 1488–90

[15] Omori T, Abe S, Tanaka Y, Lee D Y, Ishida K and Kainuma R 2013 Thermoelastic martensitic transformation and superelasticity in Fe–Ni–Co–Al–Nb–B polycrystalline alloy *Scr. Mater.* **69** 812–5
[16] Lee D, Omori T and Kainuma R 2014 Ductility enhancement and superelasticity in Fe–Ni–Co–Al–Ti–B polycrystalline alloy *J. Alloys Compd.* **617** 120–3
[17] Abuzaid W and Sehitoglu H 2018 Superelasticity and functional fatigue of single crystalline FeNiCoAlTi iron-based shape memory alloy *Mater. Des.* **160** 642–51
[18] Sato A, Chishima E, Soma K and Mori T 1982 Shape memory effect in $\gamma \rightleftarrows \epsilon$ transformation in Fe-30Mn-1Si alloy single crystals *Acta Metall.* **30** 1177–83
[19] Sawaguchi T, Maruyama T, Otsuka H, Kushibe A, Inoue Y and Tsuzaki K 2016 Design concept and applications of Fe–Mn–Si-based alloys—from shape-memory to seismic response control *Mater. Trans.* **57** 283–93
[20] La Roca P, Baruj A and Sade M 2017 Shape-memory effect and pseudoelasticity in Fe–Mn-based alloys *Shape Mem. Superelasticity* **3** 37–48
[21] Omori T, Ando K, Okano M, Xu X, Tanaka Y, Ohnuma I, Kainuma R and Ishida K 2011 Superelastic effect in polycrystalline ferrous alloys *Science (80-. ).* **333** 68–71
[22] Tseng L W, Ma J, Wang S J, Karaman I, Kaya M, Luo Z P and Chumlyakov Y I 2015 Superelastic response of a single crystalline FeMnAlNi shape memory alloy under tension and compression *Acta Mater.* **89** 374–83
[23] Xia J, Xu X, Miyake A, Kimura Y, Omori T, Tokunaga M and Kainuma R 2017 Stress-and magnetic field-induced martensitic transformation at cryogenic temperatures in Fe–Mn–Al–Ni shape memory alloys *Shape Mem. Superelasticity* **3** 467–75
[24] Patriarca L, Abuzaid W, Carlucci G, Belelli F and Casati R 2021 Pseudoelasticity in FeMnNiAl shape memory alloy lattice structures produced by Laser Powder Bed Fusion *Mater. Lett.* **302** 130349
[25] Abuzaid W and Sehitoglu H 2019 Shape memory effect in FeMnNiAl iron-based shape memory alloy *Scr. Mater.* **169** 57–60
[26] Abuzaid W, Wu Y, Sidharth R, Brenne F, Alkan S, Vollmer M, Krooß P, Niendorf T and Sehitoglu H 2019 FeMnNiAl Iron-Based Shape Memory Alloy: Promises and Challenges *Shape Mem. Superelasticity* **5** 263–77
[27] Omori T, Okano M and Kainuma R 2013 Effect of grain size on superelasticity in Fe-Mn-Al-Ni shape memory alloy wire *APL Mater.* **1** 32103
[28] Tseng L W, Ma J, Vollmer M, Krooß P, Niendorf T and Karaman I 2016 Effect of grain size on the superelastic response of a FeMnAlNi polycrystalline shape memory alloy *Scr. Mater.* **125** 68–72
[29] Tseng L W, Ma J, Hornbuckle B C, Karaman I, Thompson G B, Luo Z P and Chumlyakov Y I 2015 The effect of precipitates on the superelastic response of [1 0 0] oriented FeMnAlNi single crystals under compression *Acta Mater.* **97** 234–44
[30] La Roca P, Baruj A, Sobrero C E, Malarria J A and Sade M 2017 Nanoprecipitation effects on phase stability of Fe-Mn-Al-Ni alloys *J. Alloys Compd.* **708** 422–7
[31] Ozcan H, Ma J, Wang S J, Karaman I, Chumlyakov Y, Brown J and Noebe R D 2017 Effects of cyclic heat treatment and aging on superelasticity in oligocrystalline Fe-Mn-Al-Ni shape memory alloy wires *Scr. Mater.* **134** 66–70
[32] Tseng L W, Ma J, Wang S J, Karaman I and Chumlyakov Y I 2016 Effects of crystallographic orientation on the superelastic response of FeMnAlNi single crystals *Scr. Mater.* **116** 147–51
[33] Poklonov V V, Chumlyakov Y I, Kireeva I V and Kirillov V A 2018 Superelastic response in< 1 2 2>-oriented single crystals of FeMnAlNi shape memory alloy in tension and compression *Mater. Lett.* **233** 195–8
[34] Tseng L W, Ma J, Chumlyakov Y I and Karaman I 2019 Orientation dependence of superelasticity in FeMnAlNi single crystals under compression *Scr. Mater.* **166** 48–52

[35] Omori T, Iwaizako H and Kainuma R 2016 Abnormal grain growth induced by cyclic heat treatment in Fe-Mn-Al-Ni superelastic alloy *Mater. Des.* **101** 263–9

[36] Omori T, Kusama T, Kawata S, Ohnuma I, Sutou Y, Araki Y, Ishida K and Kainuma R 2013 Abnormal grain growth induced by cyclic heat treatment *Science (80-. ).* **341** 1500–2

[37] Kusama T, Omori T, Saito T, Kise S, Tanaka T, Araki Y and Kainuma R 2017 Ultra-large single crystals by abnormal grain growth *Nat. Commun.* **8** 1–9

[38] Vollmer M, Arold T, Kriegel M J, Klemm V, Degener S, Freudenberger J and Niendorf T 2019 Promoting abnormal grain growth in Fe-based shape memory alloys through compositional adjustments *Nat. Commun.* **10** 1–10

[39] Vollmer M, Bauer A, Frenck J-M, Krooß P, Wetzel A, Middendorf B, Fehling E and Niendorf T 2021 Novel prestressing applications in civil engineering structures enabled by FeMnAlNi shape memory alloys *Eng. Struct.* **241** 112430

[40] Zhou H T, Zhang Z D, Liu C M and Wang Q W 2007 Effect of Nd and Y on the microstructure and mechanical properties of ZK60 alloy *Mater. Sci. Eng. A* **445** 1–6

[41] Peng Q, Hou X, Wang L, Wu Y, Cao Z and Wang L 2009 Microstructure and mechanical properties of high performance Mg–Gd based alloys *Mater. Des.* **30** 292–6

[42] JIANG N, Lei C, Linggang M, Canfeng F, Hai H A O and ZHANG X 2016 Effect of neodymium, gadolinium addition on microstructure and mechanical properties of AZ80 magnesium alloy *J. Rare Earths* **34** 632–7

[43] Mishra R K, Sahay P P and Shahi R R 2019 Alloying, magnetic and corrosion behavior of AlCrFeMnNiTi high entropy alloy *J. Mater. Sci.* **54** 4433–43

# Thermal Properties of Backfill Material for Underground Heat Exchange Applications

**Bashir M. Suleiman*,[1], Daniil Moraitis[1], Mohamed I. Abdel-Fattah[1], Atta G. Attaelmanan[2], Hamdan Hamdan[1] and Marwan Alzaylaie[3].**

[1]Applied Physics and Astronomy Department, College of Sciences, University of Sharjah, Sharjah, UAE.

[2] College of Arts, Sciences and Information Technology, University of Khorfakkan, Sharjah, UAE

[3]. Dubai Development Authority, Senior Manager-Geotechnical, Government of Dubai, UAE.

*Corresponding author. Tel.: +97165050374 Fax: +97165050351.

E-mail address: bashir@sharjah.ac.ae (B. M. Suleiman).

**Abstract:**

We are investigating the effect of the thermal properties of backfill material on the efficiency of underground thermal energy systems. We have used aluminum to increase the thermal properties of clay-bentonite and to investigate its potential as a backfill material in underground heat exchange applications. The measurements of thermal properties were made at room temperature using the Transient Plane Source (TPS) technique. The measurements were based on an infinite homogeneous medium using different composites of aluminum and clays in amounts of aluminum concentration ranging from 6 to 25 percent of the sample's weight. The results confirmed an overall relative increase of 170% and 135% in the thermal conductivity and diffusivity, respectively as the Al content increased by 25%. The estimated average of the thermal transmittance (U-value ) of these samples was in the range from 0.40 to 1.1 W/m²K, which was directly proportional to the relative increase in the measured properties. Our findings can be applied to underground thermal energy systems, ground source heat pumps, and solid dielectric underground transmission lines.

## 1. Introduction

Backfilling is vital in the building process, following the ground pouring and fitting utility lines. After each layer is added, the backfill material must be placed in the excavation layers and then compacted [1]. The use of clay-bentonite as backfill material in building construction has been widely explored because of its outstanding characteristics, including low hydraulic conductivity, high swelling capacity, and high compressibility [2]. However, its low thermal conductivity is a major drawback that restricts its use in other applications. In this paper, we have examined the effect of aluminum on enhancing the thermal properties of clay-bentonite and its potential as a backfill material for underground thermal energy systems. We have used the Transient Plane Source (TPS) method [3] for measuring the thermal properties of the clay-bentonite samples with varying aluminum concentrations[4]. Our results show a substantial improvement in the heat conductivity and diffusivity values of the samples as the aluminum content increased. These findings suggest that the addition of aluminum can effectively improve the heat conductivity of the samples and make it a

viable option for use in various applications including underground energy-storing structures, geothermal pumps, and horizontal ground heat transfer devices [5]. The emphasis of this paper is dedicated to the effect of Al on the heat properties (both conductivity and diffusivity) to improve the heat conduction performance in the soils, particularly, of bentonite clay composites that could be used as backfill materials. Mainly, for underground horizontal heat-exchanging applications. Our selection of aluminum is in line with the typical application of modified aluminum clays as a stabilizer in environmental contexts to improve the sedimentation of particles holding phosphorus [6].
It's worth mentioning that it has been reported recently on improving the thermal properties of nanocomposite by a solution intercalation method [7-8] the thermal conductivity of the nanocomposite increased with increasing aluminum content and reached a maximum value of 0.83 W/mK at 20 wt% aluminum loading, which was 3.3 times higher than that of pure clay-bentonite. The thermal stability of the nanocomposite also improved with increasing aluminum content, as indicated by higher decomposition temperature and lower weight loss. However, one of the principal concerns when incorporating nano clays (e.g., bentonite) into a polymer matrix is the difficulty in uniformly dispersing them in the matrix. These studies suggest that adding aluminum nanoparticles to clay-bentonite can effectively improve its thermal properties, as well as its mechanical and flame-retardant properties. However, some challenges remain, such as achieving uniform dispersion of aluminum nanoparticles in the clay matrix, optimizing the processing parameters, and evaluating the environmental impact of the nanocomposites.

## 2. Experimental

### *2.1 TPS technique*

The TPS technique utilizes a resistive heating element. It is a device that converts electrical energy into heat through the process of Joule heating. It is a sensor made of a thin sheet of metal covered on both sides with thin layers of insulating material. The experiment is executed by simultaneously recording the voltage/resistance variations over the TPS element, as its temperature is slightly raised by a constant electrical current pulse. The resistance as a function of time over the element is expressed as

$$R_t = R_r[\ 1 + \alpha\,\Delta T(t)\ ] \qquad (1)$$

where $R_r$ ($\approx 4\Omega$ at ambient temperature) of the element prior to the start of the transient recording, $\alpha$ is the temperature coefficient of resistance ($\approx 4.0\text{x}10^{-3}\ K^{-1}$ at ambient temperature), and $\Delta T(t)$ is the temperature rise as a function of time of the element.
The method is based on the analysis of the three dimensions heat flux within the sample. The sample can be considered as an infinite homogeneous medium if the transient recording time is stopped before the thermal wave reaches the sample boundaries and causes edge effects. The measurement of temperature variation ΔT(t) within the heater is influenced by various aspects such as the total power output in the TPS element, the design constraints of the sensor, the thermal conductivity properties of the sample, The thermal conductivity, and diffusivity of a disk-shaped samples can be calculated from ΔT(t), which is defined by the following equation.

$$\Delta T\ (\tau) = P_o\left(\pi^{3/2} a\, \kappa\right)^{-1} D(\tau) \qquad (2)$$

$P_o$ is the output power, $\kappa$ is the thermal conductivity coefficient of the sample, and $a$ is the sensor's radius. $D(\tau)$ is the mathematical formula that represents the time-varying temperature rise and describes the heat transfer pattern of a disk-shaped sensor, assuming that the disk is composed of multiple concentric ring sources [3]. To simplify calculations, the average temperature change of the sensor is expressed using a dimensionless variable τ, where $\tau = (\eta t/a^2)$ or $\tau = (t/\theta)^{1/2}$. Here, t represents the time elapsed since the beginning of transient heating, $\theta = a^2/\eta$ is the characteristic time, and $\eta$ is the diffusivity of the sample.

By comparing the recorded temperature changes ΔT(t) obtained using Equation (1) with the calculated temperature changes $\Delta T(\tau)$ obtained from Equation (2), all the necessary information can be obtained.

### *2.2 Samples preparation*

Five clay samples were tested for thermal conductivity with different amounts of aluminum mixing concentrations (0%, 6%, 8%, 13%, and 20%) of the sample weight. The sample powder was prepared by mixing a ratio of 10 g of Aluminum ) with 20 g of clay in 1 L of water then settling for 24 then decanting the free water and oven-drying the rest. The size of each disk-shaped sample (pellets) is 30 mm in diameter and 10 mm in thickness. The detailed mixing process is given elsewhere[6]

To verify the chemical makeup and mineral composition of these samples XR analysis was performed using an acquisition time of 150 s, with 1.2 mm, an X-ray tube voltage of 50 kV, and a current of 0.641 mA. Then the mineralogical and chemical compositions of these samples are calculated from the XRF analysis. The results for a typical sample are depicted in Figure 1 as shown below. For thermal properties measurements, each sample consisted of two polished circular plates to assure thermal contact among the samples and the sensor.

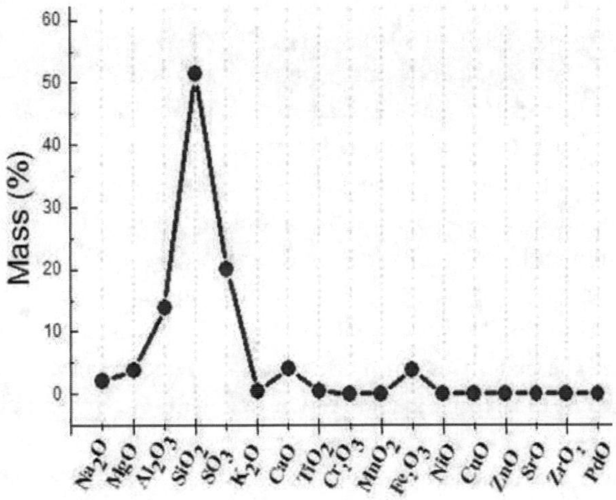


Fig. 1. The compositions of mass percent of a typical sample of modified clay.

## 3. Results and Discussion

Table 1 shows the obtained measurements as the Al concentration increased in clay-bentonite samples. The results indicated a significant increase of 170% and 135% in the heat conductivity and heat diffusivity values, respectively as the aluminum content increased by 25% in the sample weight.

Table 1. The average value of five trials of the heat properties for each clay sample modified with aluminum.

| Clay sample Alum | 0% | 6.25% | 8.33% | 12.5% | 25% | Total % |
|---|---|---|---|---|---|---|

| content(%) | | | | | | Increase |
|---|---|---|---|---|---|---|
| Apparent Density in kg/m$^3$ | 1700 | 1763 | 1783 | 1825 | 1950 | 15 |
| Heat Conductivity in W/m-k | 0.25 | 0.50 | 0.54 | 0.55 | 0.68 | 170 |
| Heat Diffusivity x $10^{-6}$ m$^2$/s | 0.09 | 0.14 | 0.26 | 0.16 | 0.33 | 130 |

Thermal conductivity is a measure of how easily heat flows through a material, and it has a significant impact on the thermal response of the backfill material on the efficiency of underground heat exchange applications such as underground heat pumps and underground electrical transmission lines(cables). Higher thermal conductivity in the soil allows for greater heat transfer between the ground and the cables or horizontal pipes, which can result in more efficient operation and lower energy costs. Conversely, lower thermal conductivity in the soil can reduce the efficiency of the system, leading to excessive heating of the cables and to higher energy consumption of the pipes. Our measurements were performed using different amounts of aluminum concentration ranging from 6 to 25 percent of the sample weight on a uniform basis. The results indicated a significant increase in heat conductivity.

Thermal diffusivity is an essential physical parameter that indicates how a medium responds to a thermal disturbance. It measures how fast heat spreads through a material and can also affect the thermal reaction of backfill materials on either the performance of transmission cables or horizontal underground loop heat exchanging pipes. Thus, measurements of this parameter can uncover the dynamics behind different heat transfer mechanisms and the impact of microstructure on the thermal reaction of backfill materials. In other words, soils with higher thermal diffusivity allow heat to spread faster, which can help to maintain a more consistent temperature profile in the ground and improve the overall efficiency of the system. Our measurements of this parameter show an overall increase in thermal diffusion. This is consistent with desired properties that will reduce the rate of excessive heating of the cables and/or maintain a desired stable temperature profile, more details about the U-values for cables are given in Ref. [9]. To sum up, the heat properties of the soil, such as its heat conductivity and heat diffusivity, which is an important parameter for thermal dissipation dynamics that has a significant impact on the thermal response of underground heat exchange applications [10 & 11]. This will lead to a more consistent temperature profile and better functioning of the geophysical ecosystem. In addition to its use for underground power transmission cables. In the case of nanocomposites[13] the stability and reusability of the composite due to the corrosion and aggregation of aluminum nanoparticles [14 ]

## 4. Conclusions

We have used the TPS technique for instantaneous measurements of thermal conductivity, and thermal diffusivity of Al. added clay-bentonite backfill materials. The addition of aluminum to clay-bentonite backfill materials can significantly increase both thermal conductivity and thermal diffusivity. The study found that adding aluminum to clay-bentonite increased its thermal conductivity by 170% and its thermal diffusivity by 135%. This can lead to more efficient operation and lower energy costs in underground heat exchange applications such as underground heat pumps and underground electrical transmission lines. Besides, it will maintain a more consistent temperature profile on the ground and improve the overall efficiency of the geophysical ecosystem.

## References

[1] https://safetyculture.com/topics/backfilling/ and Backfilling in Construction and its Different Stages & Types - Civil Gyan Dec. 2022

[2] Takai A., Inui T, and Katsumi T.2016 Soils and Foundations, 56(2), 277. https://theconstructor.org/geotechnical/backfilling-types-procedure-foundation/1648/

[3] Suleiman B, Izhar-Ul-Haq, Karawacki S, and Gustafsson S 1992), J. Phys. D Appl. Phys. 25, 813. https://doi.org/10.1088/0022-3727/25/5/011

[4] Suleiman B, Daoudi K., Attaelmanan A., Al Zaylaie M., Gustavssonc M. (2020). Thermal Science and Engineering Progress, Volume 19. https://doi.org/10.1016/j.tsep.2020.100589

[5] Mahmoud, M.et. al 2021 Thermofluids 10 (2021) 100070. https://doi.org/10.1016/j.ijft.2021.100070

[6] Al Zaylaie M 2007 *MScThesis* University of Sharjah

[7] António B. Mapossa, Afonso Henrique da Silva Júnior,Carlos Rafael Silva de Oliveira and Washington Mhike, A Review Polymers 2023, 15(16),3443; https://doi.org/10.3390/polym15163443

[8] Essaleh, M., Bouferra, R., Chihab, Y. et al. Chemistry Africa (2023). https://doi.org/10.1007/s42250-023-00693-5

[9] Bergaya F and Lagaly V 2013, Handbook of Clay Science, 2nd edition https://www.sciencedirect.com/book/9780080993645/handbook-of-clay-science

[10] Zheng X, Zhang L., Ren Q., and Qian H 2013, Energy and Buildings 56, 146–149 https://doi.org/10.1016/j.enbuild.2012.09.020

[11] Selker J and Or D 2021, Soil Hydrology and Biophysics Soil Hydrology and Biophysics – Simple Book Publishing (Oregon state. education)

[12] Al-Kaisi M and Lowery B 2017, Soil Health and Intensification of Agroecosystems https://doi.org/10.1016/B978-0-12-805317-1.00001-4

[13] Khandelwal, N., Singh, N., Tiwari, E. and Darbha, G.K., 2019. RSC advances, 9(20), pp.11160-11169. https://doi.org/10.1039/C9RA00742C

[14] Sah, P.K., Kumar, S.S. & Sreedeep, S. (2023) Int J Thermophys 44, 57. https://doi.org/10.1007/s10765-023-03163-9

Journal of Physics: Conference Series **2751** (2024) 012011 doi:10.1088/1742-6596/2751/1/012011

# Effect of Gd, V, and Y Alloying on the Corrosion Properties of Fe-Mn-Ni-Al Fe-based Shape Memory Alloy

**Faisal Mustafa [1,2*], Shahbaz Ahmad [1,2], Arhum Hassan [1,2], Mehmet Egilmez [1,2*], Sami El-Khatib [1,2], Taleb Ibrahim [2,3]**

[1] Department of Physics, American University of Sharjah, Sharjah 26666, United Arab Emirates
[2] Materials Science and Engineering Program, American University of Sharjah, Sharjah 26666, United Arab Emirates
[3] Department of Chemical Engineering, American University of Sharjah, Sharjah 26666, United Arab Emirates
*Corresponding authors' e-mail: b00090953@aus.edu, megilmez@aus.edu

**Abstract**. Shape Memory Alloys (SMA) are widely utilized in numerous industrial applications nowadays. Significant attention has been given to Fe-based SMA in civil engineering applications due to their high strength and reasonable cost. Researchers are striving to improve the corrosion resistance of these alloys so they can be utilized in marine environments. This article focuses on the effect of Gd, V, and Y alloying on FeMnNiAl based SMA. The phases and morphological effects were characterized using X-Ray diffractometer (XRD), Scanning Electron Microscopy (SEM), and Energy Dispersive Spectroscopy (EDS). To examine the corrosive nature of the Fe-based SMA system, a potentiodynamic analysis and electrochemical impedance spectroscopy (EIS) tests were performed in 3.5% NaCl solution. Results show that the addition of Gd rare earth metal has improved the overall corrosion efficiency of Fe-based SMAs by 48%.

## 1. Introduction

Multi-principle element alloying techniques have been utilized during the last two decades due to their advantage in producing mechanical, electrical, and magnetic functionalities [1–4]. One of the mechanical functionalities of great interest is the shape memory effect (SME) and superelasticity (SE). Shape memory materials (SMMs) are a class of materials that possess the ability to regain their initial shape after undergoing a considerable and seemingly plastic deformation whenever a specific stimulus is applied. In SMMs, properties such as superelasticity (in alloys) and visco-elasticity (in polymers) can be observed under certain conditions. Shape memory alloys (SMAs) go through a reversible martensitic phase transition from the high-symmetry austenite to low-symmetry martensite phase under the influence of temperature or stress field, resulting in the SME and superelasticity, respectively. Since the discovery of the shape memory effect in the 1970s by Wayman [5], The functional and mechanical properties (such as pre-stressing, damping capacity, and exceptional workability) of Fe-based shape memory alloys (SMAs) have demonstrated their potential as a viable option for various industrial applications [6][7][8]. In particular, SMAs are often explored for use in civil engineering applications since they are a more cost-effective option than their non-ferrous counterparts.

Fe-Mn-Ni-Al is a new and promising SMA candidate which was first reported by Omori et al. in 2011 [9]. Numerous researchers worldwide have conducted extensive studies on Fe-Mn-based shape memory alloys (SMAs) with the aim of enhancing their mechanical and corrosion characteristics

[10][11][12][13][14][15][16][17][18]. This group of materials exhibits excellent superelastic characteristics over a broad temperature spectrum. This Fe-based Shape Memory Alloy (SMA) experiences a thermoelastic transformation across a broad temperature spectrum, involving a transition from the Body Centered Cubic (BCC) austenitic phase to a Face Centered Cubic (FCC) martensitic phase. This transformation is primarily attributed to the formation of nano-sized precipitates within the matrix [19][20][21][22]. Moreover, studies showed a good corrosion resistance in this alloy which is on par with that of stainless steels owing to the addition of Nickel and Chromium elements [23]. This makes Fe-based SMA with Cr and Ni addition a better candidate in coastal/offshore engineering construction applications which have chloride environments [24]. In particular, Lin et al. showed enhancement of corrosion resistance in Fe-Mn-based SMA when Cr and Ni are introduced to the structure [25]. Moreover, Jaehoon et al. reported that the corrosion resistance of the SMA was improved with the addition of Mn and Cr in Fe-based SMA due to the formation of a stable passivation oxide layer in alkaline environments for different pH [26]. The enhancement of corrosion resistance by alloying is reported not only for Fe-based SMA but also by other SMA alloys. For example, Cu-Al-Ni-based SMA showed a reduction in corrosion rate due to the presence of 1.5 wt.% of Ti [27].

Recently, the addition of rare earth elements has shown an improvement in the corrosion resistance of SMA alloys. Xin et al. studied the influence of Nd on the corrosion behavior of Cu-Al-Ni high-temperature SMA [28]. The authors reported a reduction in the corrosion current density of the SMA alloy owing to the presence of acicular martensite with the addition of Nd. Nawal and Abdul studied the alloying effect of titanium on the corrosion resistance of Cu-based SMA [29]. The presence of Ti in the passivation layer enhanced the alloy resistance from the aggressive action of $Cl^-$ ions in the NaCl solution compared to the alloy without Ti Addition.

The current work focuses on the influence of adding various metals including rare earth metals (Gd, V, and Y) on the corrosion behavior of Fe-Mn-Ni-Al-based SMA.

## 2. Materials and Methods

All Fe-based SMA materials were induction melted as per the composition given in Table 1. After solidification, the samples were homogenized at 1200 °C for 24 hours in an inert Ar atmosphere. These homogenized samples were machined using wire Electric Discharge Machining (EDM) according to the desired dimension of 1cm x 1cm and having a thickness of 1mm. The samples were ground and polished for corrosion and microstructural study.

**Table 1.** Alloying composition for the Fe-based SMA (at. %).

| Sample ID | Fe | Mn | Ni | Al | Add-on |
|---|---|---|---|---|---|
| Fe-SMA | 43.5 | 34 | 7.5 | 15 | - |
| Fe-Gd-SMA | 40.7 | 32.4 | 7.3 | 6.8 | 12.9 |
| Fe-V-SMA | 42.4 | 33.6 | 9.7 | 8.5 | 5.7 |
| Fe-Y-SMA | 41.1 | 33.7 | 8.2 | 7.5 | 9.5 |

Phase and crystallographic analyses of all samples were performed using X-Ray Diffractometer (Malvern Panalytical's X'Pert³, Malvern, UK). Morphological and elemental composition were investigated with Scanning Electron Microscope (TESCAN VEGA 3) coupled with Energy-Dispersive X-ray Spectroscopy (Tescan, Brno, Czechia). Drop Shape Analyzer (DSA-100, KRUSS, Matthews, NC, USA) was utilized to carry out the water contact angle measurements.

The accelerated potentiodynamic corrosion tests and Electrochemical Impedance Spectroscopy (EIS) were performed in three electrode setup connected to CHI-660E Potentiostat (CH Instruments, USA). The corresponding three electrode setup was composed of Platinum coil (counter electrode), Ag/AgCl filled with 3M KCl solution (reference electrode) and the studied samples with a $1cm^2$ exposed area as a working electrode. All three electrodes were immersed in 3.5 wt.% NaCl electrolyte.

## 3. Results and Discussions

### *3.1. X-Ray Diffractometry*

XRD patterns of all the FeMnNiAl family of alloys under observation showed a dual phase structure consisting of a BCC and FCC phase, as shown in Fig. 1. Prominent FCC (γ) peaks appeared at 2θ values of 43° and 49° whereas for BCC (α) peaks were observed at 45° and 65° [30][31]. BCC was a dominant phase in the structure regardless of alloy addition. It is because, inherently, FeMnNiAl is in a dual phase, i.e., a secondary phase inside the matrix phase. So, the addition of further alloys results in preferential alloying with one phase, which was confirmed via EDS area mapping. This affinity towards a preferred state can be illustrated with the help of a 5-element phase diagram which is beyond the scope of the current work. But this dual-phase structure has its impact on the corrosion properties of the materials. Usually, it is easy to predict the corrosion behavior of an alloy with a single phase under certain circumstances. However, with a double phase, one should consider not only the intrinsic behavior of each phase but also understand the effect of one phase on another. The dual phases can have different chemical and electronegative potentials, which can result in a galvanic cell formation and might also result in severe corrosion of one phase than the other. The morphology and distribution (phase and elemental) of these phases are further elaborated using SEM and EDS.

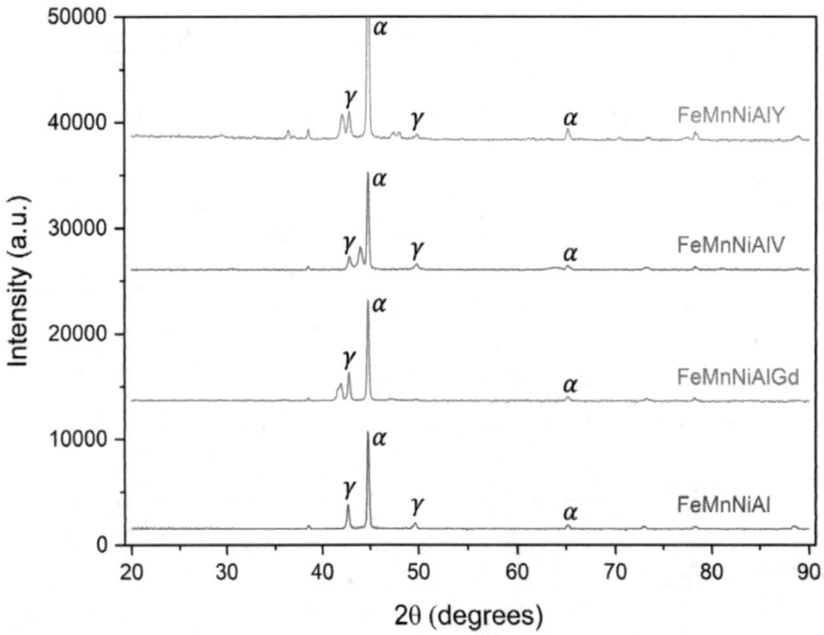


**Figure 1.** XRD patterns of the FeMnNiAl SMA with the addition of different alloying elements.

### *3.2. Scanning Electron Microscopy*

Scanning Electron Microscopy (SEM) images showed the dual phase structures having a matrix (α) and a secondary phase (γ) (Fig. 2). Figure 2(a) shows the dual phase structure of the FeMnNiAl system where the secondary phase has a dendritic (lamellar) structure. The phases have distinct chemical compositions due to which two unique electronegativity potentials exist and will result in corrosion on the phase boundary between the two phases, shown in Figure 2(e). In the Fe-Gd system, the phases are still dual but have slightly bigger secondary phase grains inside the structure, as seen in Figure 2(b).

Whereas, after corrosion (Fig. 2(f)) the matrix phase seems more oxidized (corroded) than the secondary phase and it is further validated by EDS results.

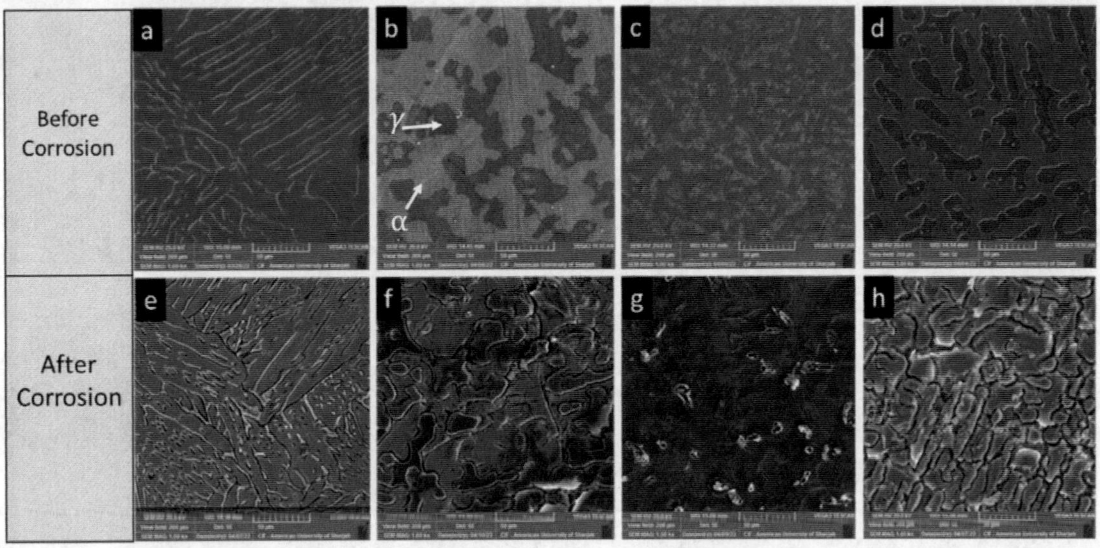


**Figure 2.** SEM images of (a, e) base alloy (FeMnNiAl), alloyed with (b, f) Gd, (c, g) V, and (d, h) Y. SEM for images (a-d) were collected before corrosion tests and (e-h) after corrosion tests.

The SEM image of the Fe-V sample before corrosion is shown in Figure 2(c); this sample has a dual-phase structure with a fine distribution of the secondary phase in the matrix. The contrast variation shows a slight change in phase distribution and hence instead of galvanic corrosion, pitting is observed after the corrosion attack. The SEM of the Fe-V sample after corrosion is shown in Figure 2(g) having pits that are clearly visible on the surface. Fe-Y sample with large grains of secondary phase (Fig. 2(d)) showed severe uniform corrosion under seawater environment (Fig. 2(h)). It is hard to identify the phases after corrosion. Therefore, EDS was performed to show the phase distribution with an elemental variation using elemental area mapping.

By analyzing all the SEM images, it was observed that the Fe-Gd sample was the least corroded sample as indicated by the surface oxides after corrosion. The base sample Fe-SMA showed better corrosion resistance than Fe-V and Fe-Y alloys as the corrosion product is mostly on the phase boundaries. Fe-V has the worst corrosion resistance as revealed by the formation of pits in the structure, which are dangerous from an application point of view. Fe-Y is also not very corrosion resistant but has a relatively better corrosion resistance than Fe-V.

### *3.3. Energy Dispersive X-ray Spectroscopy*

Energy Dispersive X-ray Spectroscopy (EDS) analysis was performed to show the elemental composition of the overall alloy (EDS area mapping) and the elemental composition of each phase (point EDS). The overall elemental composition through area mapping of each alloy is given in Table 1, while the elemental phase compositions through point scan on each phase for each alloy before and after corrosion is given in Table 2 and Table 3, respectively.

**Table 2.** Phase compositions were obtained using point scans for the different phases in each alloy system before corrosion (wt. %).

| Sample ID | α Fe | γ Fe | α Mn | γ Mn | α Ni | γ Ni | α Al | γ Al | α Add. | γ Add. |
|---|---|---|---|---|---|---|---|---|---|---|

| | | | | | | | | | | |
|---|---|---|---|---|---|---|---|---|---|---|
| Fe-SMA | 46.4 | 36.2 | 37.4 | 27.1 | 7.7 | 19.5 | 8.5 | 17.2 | - | - |
| Fe-Gd-SMA | 32.8 | 50.0 | 26.5 | 40.0 | 9.7 | 4.7 | 8.0 | 5.3 | 23.0 | - |
| Fe-V-SMA | 40.3 | 41.8 | 30.9 | 36.1 | 11.6 | 9.9 | 10.9 | 7.7 | 6.3 | 4.5 |
| Fe-Y-SMA | 36.9 | 49.9 | 30.8 | 39.3 | 10.0 | 4.9 | 8.2 | 5.9 | 14.0 | - |

**Table 3.** Phase compositions were obtained using point scan for the different phases in each alloy system after corrosion (wt. %).

| Sample ID | α Fe | γ Fe | α Mn | γ Mn | α Ni | γ Ni | α Al | γ Al | α Add. | γ Add. | α O | γ O |
|---|---|---|---|---|---|---|---|---|---|---|---|---|
| Fe-SMA | 44.9 | 33.5 | 36.6 | 23.8 | 7.5 | 22.6 | 7.7 | 20 | - | - | 3.2 | - |
| Fe-Gd-SMA | 13.1 | 41.8 | 4.9 | 32.5 | 53.9 | 10.9 | 2.2 | 4.3 | 3.8 | - | 21.4 | 10.4 |
| Fe-V-SMA | 31.9 | 41.2 | 21.8 | 37.4 | 12.9 | 6.6 | 8.9 | 4.3 | 7.6 | 4.3 | 16.2 | 5.7 |
| Fe-Y-SMA | 10.8 | 39.2 | 4.6 | 27.5 | 56.8 | 16.6 | 2.5 | 3.8 | 0.7 | - | 24.6 | 12.9 |

Fe-SMA (FeMnNiAl) showed a dual-phase structure where both phases have distinct elemental compositions. The matrix phase (α) is enriched in Fe and Mn while the secondary phase (γ) has Ni and Al in abundance. Point EDS analysis of each phase shows the individual elemental composition of each phase in the Fe-SMA (Table 2). These phase compositions result in corrosion at the phase boundaries due to differences in chemical potential and the formation of a galvanic cell. The EDS mapping of each phase after corrosion (Table 3) showed that the matrix phase is corroding as a sacrificial anode where the oxide formation of Fe, Mn, Ni, and Al can be observed easily, and the oxygen content is only observed in the matrix phase.

The Fe-Gd sample also has compositional variation across both phases. In this alloy, unlike Fe-SMA, Fe and Mn form the secondary phase while Ni, Al, and Gd have higher concentrations in the matrix phase. The exact phase compositions are shown in Table 2, where Gd is forming the matrix phase along with Ni and Al and is not present in the secondary phase. At the same time, all other elements are present in the secondary phase with higher weight percentages of Fe and Mn. Table 3 represents the elemental variation along with oxide content in each phase after corrosion, showing that the matrix phase has more oxide content and hence it was deduced that more corrosion occurred in this phase. A similar observation was made from the SEM images (Fig. 2). Oxides of Fe and Mn are the main corrosion products in the matrix phase and at the same time, the oxides of secondary phase are also of similar elements but with lesser concentration.

Unlike Fe-Gd-SMA, Fe-V shows composition-based phase duality, but V is present in both phases. The concentration of V is more in the matrix phase along with Ni and Al than in the secondary phase. The elemental variation along the two phases is very close yet different from each other. All the elements are present in both the phases with a slight variation of Fe and Mn abundance in the secondary phase and vice versa. Due to this, the chemical potential difference between the phases is not prominent and such kinds of phases will result in either uniform or pitting corrosion. The EDS analysis of both phases before and after corrosion depicts that the oxides of Fe and Mn are abundant in the secondary phase, while oxides of Al and Ni are formed in the matrix phase. But overall, more oxide products appeared in the matrix region.

The EDS result of Fe-Y alloy having a secondary phase containing Fe and Mn in abundance is shown in Table 2. Ni and Al are in a higher concentration in the matrix phase (α) along with Y. The presence of yttrium in this dual-phase system of FeMnNiAl SMA is similar to that of Gd as it is only present in the matrix phase along with major concentrations of Ni and Al. The individual phase composition verifies the presence of Y in the matrix phase but not in the γ-phase. Table 3 shows the EDS analysis of the Fe-Y sample after corrosion showing oxidation on the surface, majorly in the matrix region. Fe, Mn, and Ni were heavily oxidized along with Y in the matrix phase while Ni remains stable while lesser oxide formation is found on the γ-phase, having no Y, concluding that Y is accelerating the oxide

formation. Hence severe oxidation is observed in this alloy after corrosion in the seawater; both phases form oxides that are not passive, and the corrosion properties of this system are not as good as the other studied compositions.
The water contact angle (WCA) method was utilized to conduct a surface wettability analysis on all samples. The hydrophobic nature of Fe-SMA, Fe-Gd-SMA, Fe-V-SMA, and Fe-Y-SMA polished electrode surfaces is demonstrated in Figure 3, where the water droplet in contact with each material exhibits a contact angle of 110.2°, 110°, 110.5°, and 113.2°, respectively. The hydrophobic surface impedes the adhesion of aqueous corrosive agents, thereby endowing the surface with favorable anticorrosive characteristics [32].

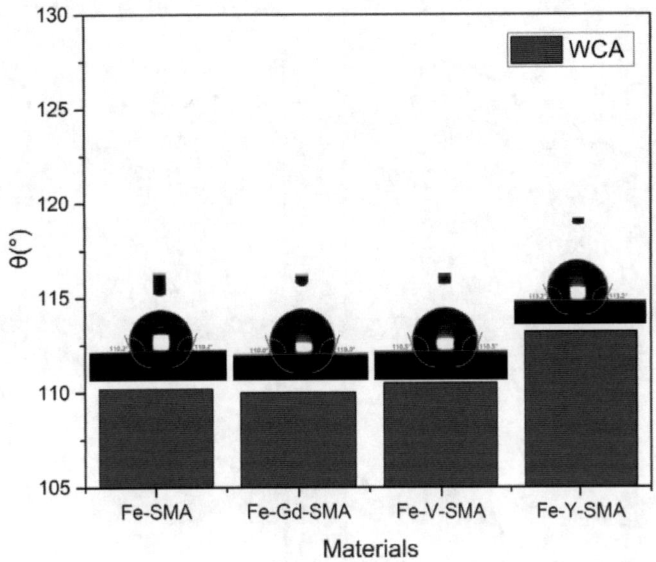


**Figure 3**. Water contact angle measurements of SMA samples

### *3.4. Corrosion Analysis*
The anti-corrosive properties of Fe-based SMAs were investigated to further verify the argument based on microstructural analysis. Figure 4 (a-d) illustrates the Potentiodynamic polarization curves for all SMA samples. The relationship between corrosion potential ($V_{corr}$) and corrosion current ($I_{corr}$) were determined by fitting a slope through Tafel region in polarization curve where E vs log j is linear over a significant range. From Ohm's law $I_{corr}$=E\$R_p$, where "E" is the difference between cathode and anode potential, "$R_p$", is the polarization resistance during potentiodynamic measurement, and is an arithmetic sum of the solution resistance of the corrosive medium ($R_s$) and the interfacial charge transfer resistance ($R_{ct}$) of the immersed working metal electrode. Table 4 shows that $R_{ct,}$ measured from the EIS Nyquist plots in (Figure 5(a)) by fitting an equivalent Randle's circuit (Figure 5(b)) exhibits a higher value for Fe-Gd-SMA (7458 Ω.cm$^2$) followed by Fe-SMA (6964 Ω.cm$^2$) sample while Fe-Y-SMA (1267 Ω.cm$^2$) and Fe-V-SMA (1047 Ω.cm$^2$) showed comparatively small resistance values.
The corrosion rate (CR) and protection efficiency (P.E) of all SMA samples were determined from Figure 4 (a-d) using the following equations [33][34,35].

$$CR = \frac{j_{corr}M}{DZF} \quad (1)$$

Where $j_{corr}$ (mA/cm$^2$) is the corrosion current density, M (g/mol) is the equivalent molecular weights of elements in the alloy, D is the density of corroding metal (g/cm$^3$), Z is the metal valency, and F is the Faraday constant (96485 As/mol).

$$P.E\% = \frac{j_{corr} - j'_{corr}}{j_{corr}} \quad (2)$$

Herein, $j_{corr}$ and $j'_{corr}$ denote the corrosion current densities prior to and subsequent to the incorporation of alloying elements (Gd, V, and Y), respectively.

The anticorrosive property of material is evaluated by its protection efficiency (P.E), which is determined by the relative decrease in the current density (j) of the sample in comparison to the Fe-SMA reference. The Fe-Gd-SMA specimen displays superior anti-corrosive properties relative to other specimens, as inferred from equations 1 and 2, with corresponding CR and P.E values of 8.9 μm/year and 48 % relative to reference Fe-SMA substrate. However, a substantial increase in CR of 14.9 and 14.8 μm/year was observed for Fe-V-SMA and Fe-Y-SMA showing poor corrosion properties with higher corrosion rates.

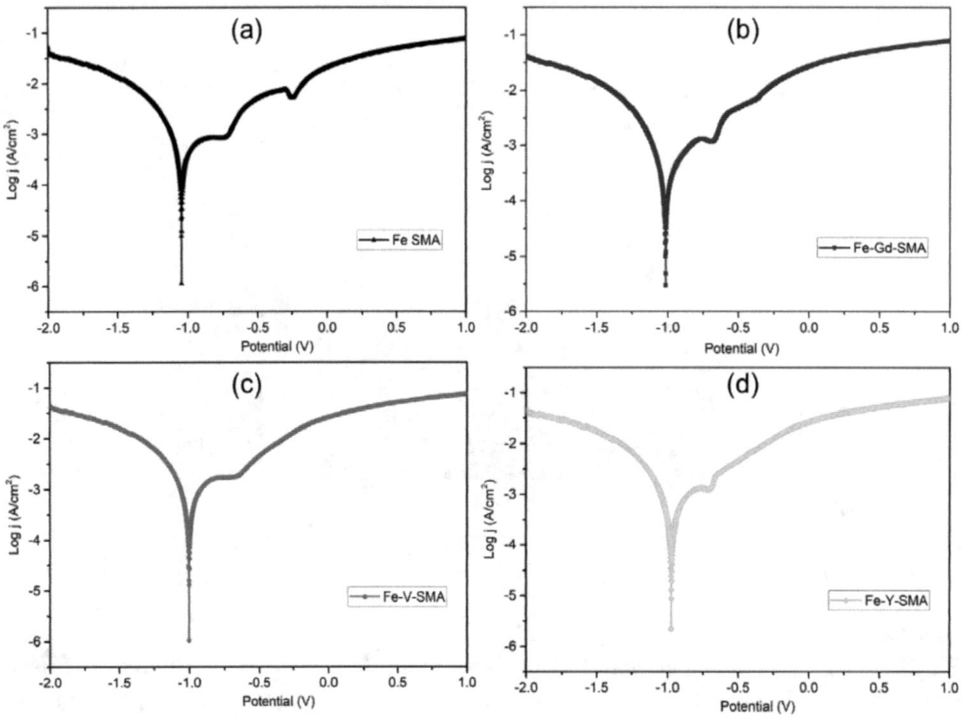


**Figure 4.** Electrochemical potentiodynamic polarization plots: (a) Fe-SMA, (b) Fe-Gd-SMA, (c) Fe-V-SMA, (d) Fe-Y-SMA.

**Table 4.** Summary of electrochemical corrosion parameters extracted from Tafel and EIS plots.

| Samples | $E_{corr}$ (V) | $j_{corr}$ (μA/cm²) | CR μm/Year | $R_{ct}$ (Ω.cm²) | P.E (%) |
|---|---|---|---|---|---|
| Fe-SMA | -1.042 | 540 | 12.1 | 6964 | Reference |
| Fe-Gd-SMA | -1.016 | 283 | 8.9 | 7458 | +48 |
| Fe-V-SMA | -1.004 | 639 | 14.9 | 1047 | -18 |
| Fe-Y-SMA | -0.968 | 598 | 14.8 | 1267 | -11 |

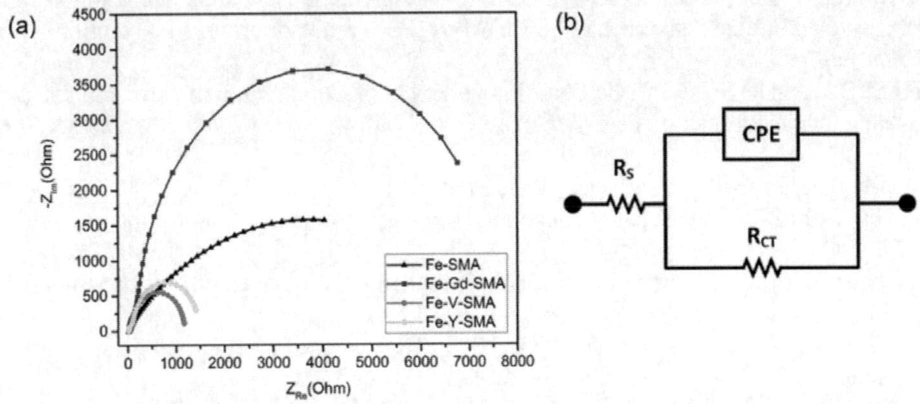


**Figure 5.** (a). EIS Nyquist plots of Fe-based SMAs, (b) Equivalent Randle's circuit.

## 4. Conclusions

In this work, alloying of transition metals and rare earth elements (Gadolinium, Vanadium and Yttrium) in FeMnNiAl shape memory alloy was carried out through induction melting. The structural and compositional characteristics of the alloys were studied by XRD and SEM. Electrochemical investigations were performed using Tafel polarization and Electrochemical Impedance Spectroscopy in 3.5 wt% NaCl solution to replicate the seawater environment. The SEM images indicated the formation of two phases in each of these alloys, verified using XRD. XRD analysis shows that these phases correspond to two different crystal structures, i.e., BCC and FCC. Tafel polarization plots indicated that the addition of Gd in FeMnNiAl SMA enhanced the anticorrosion properties, such as the slowest corrosion rate of 8.9 μm/Year, which was much better as compared to its other reference alloys. The presence of Gd helped in forming a self-passivating oxide layer during accelerated corrosion. The SEM/EDS analysis also showed low corrosion products. On the other hand, the addition of V and Y creates a potential difference between the phases. This results in a high charge transfer rate which makes the alloy less suitable for corrosion applications. The protection efficiency of the alloy made by the addition of Gd is approximately 50%, while the addition of V and Y resulted in negative efficiencies. The study concludes that the addition of Gd in the FeMnNiAl system has shown promising results for anticorrosive applications in the marine environment.

## References

[1] Egilmez M and Abuzaid W 2021 Magnetic , electrical and mechanical properties of - entropy alloy *Sci. Rep.* 1–12
[2] Abuzaid W, Egilmez M and Chumlyakov Y I 2021 TWIP – TRIP effect in single crystalline VFeCoCrNi multi-principle element alloy *Scr. Mater.* **194** 113637
[3] Mustafa F, Egilmez M, Abuzaid W, El-khatib S, Nawaz T and Ahmad S 2023 Strange Metallicity and Magnetic Order in the CoNi ( Cr / V ) Medium-Entropy Alloy System 1–14
[4] George E P, Raabe D and Ritchie R O 2019 High-entropy alloys *Nat. Rev. Mater.* **4** 515–34
[5] Wayman C M 1971 On memory effects related to martensitic transformations and observations in β-brass and Fe3Pt *Scr. Metall.* **5** 489–92
[6] Sato A, Chishima E, Soma K and Mori T 1982 Shape memory effect in $\gamma \rightleftarrows \epsilon$ transformation in Fe-30Mn-1Si alloy single crystals *Acta Metall.* **30** 1177–83
[7] Sato A, Chishima E, Yamaji Y and Mori T 1984 Orientation and composition dependencies of shape memory effect in Fe-Mn-Si alloys *Acta Metall.* **32** 539–47
[8] Janke L, Czaderski C, Motavalli M and Ruth J 2005 Applications of shape memory alloys in

civil engineering structures—Overview, limits and new ideas *Mater. Struct.* **38** 578–92
[9] Fernández-Domene R M, Blasco-Tamarit E, García-García D M and García-Antón J 2013 Passive and transpassive behaviour of Alloy 31 in a heavy brine LiBr solution *Electrochim. Acta* **95** 1–11
[10] HU B, BAI P, DONG Z and CHENG J 2009 Effect of Cu addition on corrosion resistance and shape memory effect of Fe-14Mn-5Si-9Cr-5Ni alloy *Trans. Nonferrous Met. Soc. China* **19** 149–53
[11] Söderberg O, Liu X W, Yakovenko P G, Ullakko K and Lindroos V K 1999 Corrosion behaviour of Fe–Mn–Si based shape memory steels trained by cold rolling *Mater. Sci. Eng. A* **273** 543–8
[12] Ariapour A, Yakubtsov I and Perovic D D 1999 Effect of nitrogen on shape memory effect of a Fe–Mn-based alloy *Mater. Sci. Eng. A* **262** 39–49
[13] Wan J, Huang X, Chen S and TY H X Z 2002 Effect of nitrogen addition on shape memory characteristics of Fe-Mn-Si-Cr alloy *Mater. Trans.* **43** 920–5
[14] Charfi A, Bouraoui T, Feki M, Bradai C and Normand B 2009 Surface treatment and corrosion behaviour of Fe–32Mn–6Si shape memory alloy *Comptes Rendus Chim.* **12** 270–5
[15] Patriarca L, Abuzaid W, Carlucci G, Belelli F and Casati R 2021 Pseudoelasticity in FeMnNiAl shape memory alloy lattice structures produced by Laser Powder Bed Fusion *Mater. Lett.* **302** 130349
[16] Sidharth R, Mohammed A S K, Abuzaid W and Sehitoglu H 2021 Unraveling frequency effects in shape memory alloys: NiTi and FeMnAlNi *Shape Mem. Superelasticity* **7** 235–49
[17] Abuzaid W, Wu Y, Sidharth R, Brenne F, Alkan S, Vollmer M, Krooß P, Niendorf T and Sehitoglu H 2019 FeMnNiAl Iron-Based Shape Memory Alloy: Promises and Challenges *Shape Mem. Superelasticity* **5** 263–77
[18] Abuzaid W and Sehitoglu H 2019 Shape memory effect in FeMnNiAl iron-based shape memory alloy *Scr. Mater.* **169** 57–60
[19] Vollmer M, Kriegel M J, Walnsch A, Klemm V, Leineweber A and Niendorf T 2019 On the microstructural and functional stability of Fe-Mn-Al-Ni at ambient and elevated temperatures *Scr. Mater.* **162** 442–6
[20] La Roca P, Baruj A, Sobrero C E, Malarria J A and Sade M 2017 Nanoprecipitation effects on phase stability of Fe-Mn-Al-Ni alloys *J. Alloys Compd.* **708** 422–7
[21] Tseng L W, Ma J, Hornbuckle B C, Karaman I, Thompson G B, Luo Z P and Chumlyakov Y I 2015 The effect of precipitates on the superelastic response of [1 0 0] oriented FeMnAlNi single crystals under compression *Acta Mater.* **97** 234–44
[22] Omori T, Nagasako M, Okano M, Endo K and Kainuma R 2012 Microstructure and martensitic transformation in the Fe-Mn-Al-Ni shape memory alloy with B2-type coherent fine particles *Appl. Phys. Lett.* **101** 231907
[23] Cladera A, Weber B, Leinenbach C, Czaderski C, Shahverdi M and Motavalli M 2014 Iron-based shape memory alloys for civil engineering structures: An overview *Constr. Build. Mater.* **63** 281–93
[24] Wang X X and Zhao L C 1992 The effect of thermal-mechanical training on the formation of stress-induced ϵ martensite in an Fe□Mn□Si□Ni□Co alloy *Scr. Metall. Mater.* **26** 1451–6
[25] Lin H C, Lin K M, Lin C S and Ouyang T M 2002 The corrosion behavior of Fe-based shape memory alloys *Corros. Sci.* **44** 2013–26
[26] Joo J, Kang M, Shin D, Seo E, Kim D, Yeon Y, Hong K, Lee W and Lee J 2020 Corrosion resistance of shape recoverable Fe-17Mn-5Si-5Cr Alloy in concrete structures *Materials (Basel).* **13** 5531
[27] Farhat H, Griffin R, Oguocha I N A and Evitts R 2018 An investigation of the corrosion behaviour of a FeNiCoAlTa shape memory alloy in 3.5 wt-% NaCl solution *Corros. Eng. Sci. Technol.* **53** 611–6
[28] Zhang X, Cui T, Liu Q, Dong Z and Man C 2021 Effect of Nd addition on the microstructure,

mechanical properties, shape memory effect and corrosion behaviour of Cu–Al–Ni high-temperature shape memory alloys *J. Alloys Compd.* **858** 157685

[29] Dawood N M and Abid Ali A R 2021 Influence of Titanium Additions on the Corrosion Behavior of Cu-Al-Ni Shape Memory Alloys *Materials Science Forum* vol 1021 (Trans Tech Publ) pp 55–67

[30] Mishra R K, Sahay P P and Shahi R R 2019 Alloying, magnetic and corrosion behavior of AlCrFeMnNiTi high entropy alloy *J. Mater. Sci.* **54** 4433–43

[31] Zhang Y, Jiang X, Sun H and Shao Z 2020 Effect of annealing heat treatment on microstructure and mechanical properties of nonequiatomic CoCrFeNiMo medium-entropy alloys prepared by hot isostatic pressing *Nanotechnol. Rev.* **9** 580–95

[32] Nawaz T, Ali A, Ahmad S, Piatkowski P and Alnaser A S 2023 Enhancing Anticorrosion Resistance of Aluminum Alloys Using Femtosecond Laser-Based Surface Structuring and Coating *Nanomaterials* **13**

[33] Ali A, Piatkowski P, Nawaz T, Ahmad S, Ibrahim T, Khamis M and Alnaser A S 2023 A Two-Step Femtosecond Laser-Based Deposition of Robust Corrosion-Resistant Molybdenum Oxide Coating *Materials (Basel).* **16**

[34] Mohamed O, Hassan M, Egilmez M, Abuzaid W, Ibrahim T and Khamis M 2022 Corrosion behavior of CoCrNi/mild steel medium entropy alloy thin films *Mater. Today Commun.* **30** 103015

[35] Ahmad S, Ahmad W, Abu Baker A, Egilmez M, Abuzaid W, Orhan M F, Ibrahim T, Khamis M and Alnaser A S 2022 Enhancement of the corrosion resistance of mild steel with femtosecond laser- nanostructuring and CrCoNi medium entropy alloy coating *Appl. Surf. Sci. Adv.* **12** 100321

# Light absorption enhancement in dye-sensitized solar cells using thin nanostructured photoelectrodes

**Maitha S Alqaydi**[1,2], **Saad Oboudi**[3], and **Di Zhang**[1,3,]

[1] Sustainable and Renewable Energy Engineering Department, University of Sharjah
[2] Industrial Engineering and Engineering Management Department, University of Sharjah
[3] Centre for Advanced Materials Research, University of Sharjah
Email: zdi@sharjah.ac.ae

**Abstract.** As thin-film solar cells are the attraction of many research works nowadays, dye-sensitized solar cells (DSSCs) among other types hold the promise of excellent cost/performance ratio because of the relatively low material costs and simple processing conditions. In order to improve its relatively low performance, different approaches are being explored to enhance the photon-to-current conversion efficiency. Notably, one of the most accessible approaches is by modifying the optical characteristics of the photoelectrode of DSSCs to improve light trapping within the cell. Here, we introduce a cost-effective nanopatterning process done on the photo-active $TiO_2$ photoelectrode layer for better light absorption. This simple technique resulted in increased cell short-circuit current density (from 4.664 to 5.963 mA/cm$^2$) which consequently enhanced cell efficiency (from 2.243 to 2.819%). Furthermore, without compromising the absorption significantly, nanopatterned photoelectrodes with a smaller thickness can lead to reduced material costs and possibly allowing the construction of semitransparent cells. This work can contribute to an attainable efficient DSSC manufacturing route for practical applications.

## 1. Introduction

Third-generation solar cells based on semiconducting organic macromolecules, inorganic nanoparticles, or hybrids are especially attractive due to their potentially low costs, solution-processed fabrication, and mechanical flexibility. Nevertheless, these technologies have not yet been proven commercially due to the significant limitations in terms of improving efficiency at large scale (modules) and device stability in particular under harsh operating conditions (mainly related to humidity, thermal and light-induced degradations) [1]. Among the third-generation photovoltaic technologies, dye-sensitized solar cell (DSSC) is a promising candidate. The operation of a DSSC depends largely on the optical/electrical properties of the photoelectrode, which typically consists of a mesoporous $TiO_2$ layer (covalently bonded by light-absorbing organic dye) deposited on fluorine-doped tin oxide (FTO) glass.

Aiming for high-efficiency and less costly cells, many approaches are actively being explored to increase device power conversion efficiency (PCE). Light-trapping techniques, where light absorption is increased by providing additional light scattering from nanoparticles or by increased absorption at the desired location from textured surfaces, have shown promise in improving the efficiency of many thin-film solar cells. Notably, the nano-imprint or the nanopatterning method has been reported as an effective light-trapping strategy to generate desired optical nanostructures via applying an external force and thermal energy to a soft polymer transforming the underlying thin film. Using this approach, organic solar cells with nano-imprinted poly (3,4-ethylene-dioxythiophene) polystyrene sulfonate (PEDOT:PSS) nanogratings have been reported with enhanced performance [2]. The patterned nanostructure was further investigated together with silver nanograting arrays as the active layer in inverted polymer solar cells [3,4]. Furthermore, it has been investigated that the nano-imprint method induces light trapping not only in a silicon solar cell [5] and a dye-sensitized solar cell [6] but also in a perovskite solar cell [7]. In DSSCs, given the much thicker photoelectrode compared to the active layer in e.g. organic cells (~ 10 μm vs. a few hundred nm), light trapping by nanopatterning will be even more appealing as the material consumption for $TiO_2$ can be considerably reduced using a thinner nanostructured photoelectrode. In addition, thin photoelectrodes with a thickness of a few μm would allow the construction of semi-transparent DSSCs.

In this work, we propose implementing a nanograting structure on the DSSC's photoelectrode layer using an accessible and low-cost nanoimprint technique (instead of more costly methods such as photolithography) [8,9]. Increased scattering effect from the nanopatterned mesoporous $TiO_2$ is expected to enhance the light absorption in the photoelectrode and subsequently the cell efficiency. Meanwhile, DSSCs using thinner photoelectrode can potentially realize semi-transparent photovoltaics at a performance level similar to conventional designs. Finally, the approach of nanopatterning thin films using nanograting for enhancing optical absorption is easily applicable to other types of solar cell technology.

## 2. Experimental Procedure

### *2.1 PDMS stamp preparation*

Polydimethylsiloxane (PDMS) nanopatterning stamp was prepared by mixing Sylgard 184 silicone elastomer components of a base and curing agent with a ratio of (10:1). The mixed elastomer was placed in a vacuum chamber for ~ 15 min to remove air bubbles. A digital versatile disc (DVD) sample was cut into smaller pieces and washed a few times with ethanol to remove the protective layer until becoming completely transparent. The liquid PDMS was poured onto the cleaned DVD pieces placed in a petri dish, which made contact with the patterned side of the DVD. The petri dish was then carefully placed again into the vacuum chamber at 80°C for 2 hours for the PDMS to harden. Finally, the PDMS stamp was peeled and separated from the DVD, and cut to the desired size.

### *2.2 DSSC fabrication*

To prepare the nanostructured photoelectrode, FTO glass substrates (2.2 mm glass thickness and 7 ohm/sq, purchased from Solaronix) were coated with a mesoporous $TiO_2$ paste (purchased from Solaronix) by blade coating. The $TiO_2$ layer was then patterned by the prepared PDMS stamp placed on top for a few minutes to get rid of trapped air bubbles. The pressure was carefully applied between small magnets placed on the top of the stamp and the bottom of the petri dish, which was placed in an oven and heated for 20 minutes at 80°C. After cooling the sample and carefully detaching the magnets and the PDMS stamp, the $TiO_2$ layer was successfully nanopatterned, proven by the multicolored reflection as demonstrated in Figure 1.

For DSSC preparation, platinum-coated FTO glass (purchased from Solaronix) functioning as the counter electrode and the $TiO_2$ photoelectrode (patterned or non-patterned) were annealed at ~ 250°C for 30 minutes and then the temperature was ramped up to 500°C for 45 minutes. The cooled photoelectrodes were stored in a container and dipped in an ethanol dye solution (N719 purchased from Solaronix) for 24 hours in dark. The electrodes were then washed with ethanol to remove the excess dye and dried. The photoelectrode and counter electrode were assembled and held together by paperclips while being separated by a square shaped-hole spacer to avoid short-circuit. Iodine electrolyte (purchased from Solaronix) was applied between the two electrodes by capillary force. Copper tapes were attached to the cell contacts for increasing conductivity during testing.

### *2.3 Characterizations*

Scanning electron microscope (SEM) images were taken by a VEGA3 TESCAN SEM. Maya 2000-Pro high-resolution spectrometers (Ocean Optics) were used for the optical spectrometry measurements. The current density-voltage (J-V) measurements on DSSCs were conducted by an ABET Sunlite solar simulator and a Keithley 2400 SourceMeter.

## 3. Results and Discussion

Nanostructured patterns transferred from DVD gratings to the PDMS stamp and then to the $TiO_2$ photoelectrode are shown in Figure 1 **Error! Reference source not found.**(a) and (b). The multicolored r eflection proves the complete transfer of the nanopattern onto both samples. As a reference, the opposite side of the PDMS stamp was used to create a flat (non-patterned) $TiO_2$ layer under identical conditions. Figure 1 (c) shows the cross-sectional SEM of a $TiO_2$ photoelectrode. The thickness of the $TiO_2$ layer was found to be 3.57 μm, which demonstrates the successful fabrication of thin $TiO_2$ photoelectrode. The top-down SEM in Figure 1 (d) exhibits a ~350 nm grating period of the imprinted grating pattern with ~200 nm line width.

Figure 1. (a)(b) Photographs of imprinted DVD patterns of a PDMS stamp and a $TiO_2$ photoelectrodes, respectively. (c) Cross-sectional SEM and (d) top-down SEM images of a patterned $TiO_2$ photoelectrode. The thickness of the $TiO_2$ photoelectrode layer is indicated.

To study the effects of the nanopattern on the optical properties of the electrodes, optical transmission in the visible spectral range was studied using spectrometry measurements. Figure *2* shows the results of the measurements. The transmission of the patterned photoelectrode was found to be overall lower compared to the reference flat sample. The decreased transmission indicates a direct optical improvement resulting from increased light trapping by the nanostructured surface morphology. Related work by Usami demonstrated that the incident solar energy is absorbed effectively in the thinner sensitized porous film due to multiple scattering at the bottom and total reflection at the surface [10]. This optical improvement is expected to reflect on the cell performance and therefore on increasing the cell efficiency.

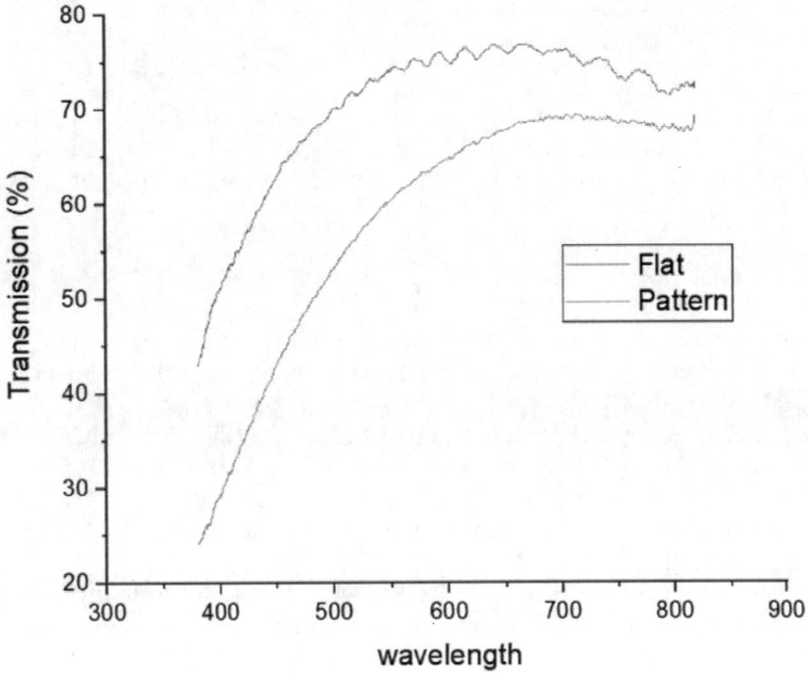


Figure 2. Optical transmission of flat and patterned $TiO_2$ photoelectrodes in the visible spectral range.

As mentioned above, the nanopattern enhances light trapping. We measured the photovoltaic performance of DSSCs using patterned and flat photoelectrodes. The patterned cell showed an enhanced performance. The photovoltaic parameters of the DSSCs are summarized in Table 1, with $J_{SC}$ of 4.664 and 5.963 mA/cm$^2$ and $V_{OC}$ of 0.734 and 0.741 V for the reference and grating imprinted cells, respectively. The increase of $J_{SC}$ in the patterned cell confirmed the role of the nanopatterned photoelectrode in improving light trapping, as is also supported by the spectrometry measurements in Figure 2. This resulted in an increased cell PCE from 2.243% to 2.819% with comparable fill factors (FFs) observed (65.5% vs. 63.8%). Further research and experimentation are required to optimize the DSSC performance with thinner nanopatterned photoelectrodes.

| | **Reference cell (flat)** | **Patterned cell** |
|---|---|---|
| $J_{SC}$ (mA/cm$^2$) | 4.664 | 5.963 |
| $V_{OC}$ (V) | 0.734 | 0.741 |
| FF | 0.655 | 0.638 |
| Efficiency (%) | 2.243 | 2.819 |

Table 1. DSSC performance with flat and nanopatterned photoelectrodes.

## 4. Conclusions

In conclusion, the nano-imprint method used for the fabrication of DSSCs with enhanced light trapping yields better photovoltaic performance. The approach of fabricating the patterned photoelectrode is cost-efficient since only soft polymer stamps were used to replicate the gratings from existing DVD masters, instead of the costly lithography process. The obtained results indicated that efficient light-trapping within the cell was the reason for the cell performance enhancement. This work provides a simple and fast approach and opens many opportunities to fabricate and investigate the effect of light-harvesting with a direct nanopatterning technique, which is easy to up-scale to fabricate large-area DSSCs inexpensively.

## 5. Acknowledgments

The authors would like to acknowledge Muhammad Adil Abbasi and Mohamed Shameer from the Centre for Advanced Materials Research at the University of Sharjah. This research is financially supported by the Competitive Research Project (18020406113) of the University of Sharjah.

## References

[1] William R. Mateker, Michael D. McGehee 2016 Progress in Understanding Degradation Mechanisms and Improving Stability in Organic Photovoltaics Adv. Mater. 29, 1603940. DOI:10.1002/adma.201603940

[2] Lee K S, Lee H K, Wang D H, Park N-G, Lee J Y, Park O O and Park J H 2010 Dye-sensitized solar cells with Pt- and TCO-free counter electrodes *Chem. Commun.* 46 4505

[3] Zhu X, Choy W C H, Xie F, Duan C, Wang C, He W, Huang F and Cao Y 2012 A study of optical properties enhancement in low-bandgap polymer solar cells with embedded PEDOT:PSS gratings Solar Energy Materials and Solar Cells 99 327–32

[4] Li X H, Sha W E I, Choy W C H, Fung D D S and Xie F X 2012 Efficient Inverted Polymer Solar Cells with Directly Patterned Active Layer and Silver Back Grating J. Phys. Chem. C 116 7200–6

[5] Erik Garnett and Peidong Yang 2010 Light trapping in silicon nanowire solar cells, Nano Lett. 10 1082–1087

[6] Kai Zhu, Nathan R. Neale, Alexander Miedaner, and Arthur J. Frank 2007 Enhanced charge-collection efficiencies and light scattering in dye-sensitized solar cells using oriented $TiO_2$ nanotubes arrays, Nano Lett. **7** 69–74.

[7] Dong In Kim , Ji Won Lee, Rak Hyun Jeong, Jung-Hoon Yu, Ju Won Yang, Sang-Hun Nam, Jin-Hyo Boo 2019 Enhancing the optical properties using hemisphere $TiO_2$ photonic crystal as the electron acceptor for perovskite solar cell, Applied Surface Science 487 409-415. https://doi.org/10.1016/j.apsusc.2019.05.092

[8] Jingbi Y, Xuanhua Li, Feng-xian Xie, Wei E. I. Sha , Johnson H. W. Kwong, Gang Li, Wallace C. H. Choy, and Yang Y 2012 Surface Plasmon and Scattering-Enhanced Low-Bandgap Polymer Solar Cell by a Metal Grating Back Electrode Adv. Energy Mater. 2 1203–7

[9] Jeonghun K, Jong K K, Byeonggwan K, Jong H K, Eunkyoung K 2012 Nanopatterning of Mesoporous Inorganic Oxide Films for Efficient Light Harvesting of Dye-Sensitized Solar Cells Angew. Chem. Int. Ed. 51 6864–9. DOI: 10.1002/anie.201202428

[10] Akira U 1997 Theoretical study of the application of multiple scattering of light to a dye-sensitized nanocrystalline photoelectrochemical cell, Chem. Phys. Lett. 277 105–108. DOI:10.1016/s0009-2614(97)00878-6

# New 2D penta-SiPN: A wide and indirect bandgap semiconductor

**I. A. Qattan[1*], Shambhu Bhandari Sharma[1], Santosh KC[2], and Sufian Abedrabbo[1]**

[1]Department of Physics, Khalifa University of Science and Technology, Abu Dhabi, P.O. Box 127788, UAE.
[2]Chemical and Materials Engineering, San Jose State University, San Jose, CA 95112, USA.

E-mail: *issam.qattan@ku.ac.ae

**Abstract.** In recent years, two-dimensional (2D) pentagonal ternary monolayers have attracted much attention and emerged as a new class of materials because of their new feature and extensive applicability. Using first-principles density functional theory (DFT) calculations, we predict a new 2D pentagonal-SiPN or p-SiPN monolayer material. The new monolayer has shown to be structurally, thermodynamically, and dynamically stable. Our findings imply that p-SiPN is a wide and indirect bandgap semiconductor, with a highly tunable bandgap with applied equ-biaxial strain. This makes p-SiPN a promising candidate for futuristic optoelectronics and nanomechanics device applications.

## 1. Introduction

The first two-dimensional (2D) hexagonal honeycomb lattice-based material, Graphene [1], has shown extraordinary physiochemical properties, however, it lacks semiconducting functionality with zero bandgap, restricting its applications in optoelectronics and piezoelectric industries. Meanwhile, the newly discovered penta-Graphene [2, 3], with semiconducting behavior, mechanical strength equivalent to that of Graphene, and very rare auxetic mechanics, has shed light into the importance of the new 2D Cairo-pentagonal lattice-based sheets (elementary, binary compounds) as promising future materials, which demonstrated exceptional physical and chemical properties. These properties have been identified mostly with theoretical and computational approaches [4, 5, 6, 7, 8], with promising possibility for experimental synthesis and experimental feasibility [9, 10, 11, 12, 13]. They exhibit buckling/puckering nature combined with three virtual layers geometry, which enable them to parade phenomenal optoelectronic, mechanical, and thermal conductivity properties. In particular, period I-II group light-weight penta materials have garnered a lot of attention lately because of their tremendous abundance, light-mass, and environmental traits. Moreover, binary pentagonal monolayers [14, 15, 16] among many others have demonstrated multi-functional responses and exceptional mobility, thermal conductivity, transport, and optoelectronic properties, which make them excellent candidates for several applications involving thermoelectric devices and catalysis. In addition, the recently predicted ternary penta compounds [17, 18] have demonstrated extraordinary auxetic mechanics, piezoelectricity, and kinetic stability properties superior to that of their binary counterpart. Motivated in part by the buckled configuration of the N, P, and Si atoms, together with the adaptable structure of the electronic configuration of the P atom, and similar bond lengths and bond angles, we design a new 2D p-SiPN monolayer material, which we find to be structurally, dynamically, and thermodynamically stable.

## 2. Computational Details

In our calculations, the SIESTA code [19] was used to carry-out all DFT calculations. The GGA approximation based on the PBE parametrization was used to account for the electronic exchange and correlation potentials. The electrons-core interactions as implemented in the Kleinman-Bylander form were calculated using the Troullier-Martins norm-conserving pseudopotentials within the semi-local form. Moreover, the electronic distribution of each atom was described using the double-$\zeta$ polarized basis based on the numerical atomic orbitals. To establish the convergence criterion, we used a mesh of 20×20×1 k-points within the Monkhrost pack scheme, and a cutoff energy = 350 Ry. For geometry optimization, atomic forces < 0.01 eV/Å, self-consistent field = $10^{-6}$ eV, and a vacuum-gap = 25 Å along the out-of-plane (z-direction) are used. We ensured chemical stability by calculating both the cohesive ($E_{\mathrm{coh}}$) and formation ($E_f$) energies using:

$$E_{coh} = (E_{p-SiPN} - 2E_{Si} - 2E_P - 2E_N)/6, \quad (1)$$

where $E_{p-SiPN}$, $E_{Si}$, $E_P$, and $E_N$ represent the energy of p-SiPN, and free/isolated Si, P, and N atoms, respectively, and:

$$E_f = (E_{p-SiPN} - 2E_{Si}^{Bulk} - 2E_P^{Bulk} - 2E_N^{Bulk})/6, \quad (2)$$

with $E_{Si}^{Bulk}$, $E_P^{Bulk}$ and $E_N^{Bulk}$ being the energy at the most stable bulk geometry of Si, P, and N atoms, respectively. The finite displacement method implemented in the PHONOPY package [20] within the Viena Ab-initio Simulation Package (VASP) is utilized to calculate the phonon bands needed to ensure the p-SiPN dynamical stability. Moreover, we carried ab-initio molecular dynamics (AIMD) simulations at T = 300 K, in a time step = 1fs up to 2000 fs to investigate the p-SiPN thermodynamical stability. Both stability calculations were performed using a supercell of 7×7×1 in order to achieve convergence.

## 3. Results and Discussion

### 3.1. Structural Properties

Based on the arrangement of three and four carbon atoms in the bottom, upper, and middle layers of penta-Graphene, the p-SiPN shape is formed with a central layer of 4-coordinated Si ($Si_{4c}$) and bottom/upper layers of 3-coordinated P/N ($P/N_{3c}$). Moreover, we experimented with various versions of the 2D pentagonal structure and found that the $Si_{4c}$ structure was the most stable, as shown in Figure (1a). The final p-SiPN unit cell is composed of two atoms of Si, P, and N each, organized in three virtual layers within one monolayer. Next, we fine-tuned the p-SiPN configuration to achieve the minimum energy level, inter-atomic forces, and stresses. The resulting structure exhibits pentagonal symmetry in two dimensions, possessing $P-2_1$ crystal symmetry with space group No. 4. The supercell's aerial perspective reveals four well-defined and unique pentagonal Cairo structures, featuring optimized lattice parameters of $a$ = 4.41 Å and $b$ = 4.43 Å.

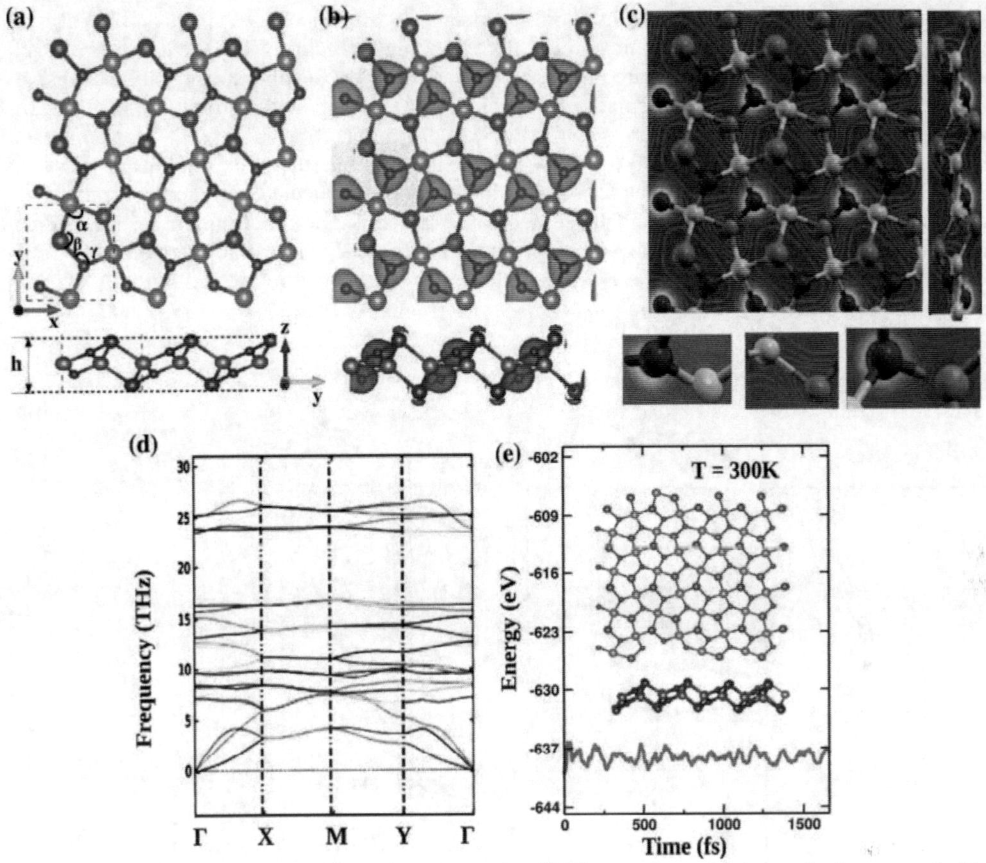


**Figure 1.** (Color online) *(a)* The final optimized p-SiPN monolayer is shown in the top and side views. The Si, P, and N atoms are shown in green, purple, and blue balls, respectively, (b) the valence charge-density iso-surface (light-magenta) with 0.11 e/Å$^3$, (c) the 2D charge density contour plots of the inter-atomic charge distribution are shown in the top/right and side/left panel, (d) phonon band, and (e) energy fluctuation during the NVT ensemble at T = 300 K are based on AIMD simulation. The diagram displays the optimized geometry of p-SiPN from both top-view and side-view perspectives.

The bond lengths measured for Si–P, P–N, and Si–N are 2.32 Å, 1.81 Å, and 1.78 Å, in that order. The thickness that was measured is $h$ = 2.85 °A, which is greater than the thickness found in previous studies on 2D penta-monolayers ($1.34 \leq h \leq 2.50$) Å, because of the longer bond-lengths observed in p-SiPN. The bond angles for α, β, and γ in the P-Si-P, Si-P-N, and P-N-Si structures are 109.30$^o$, 97.21$^o$, and 119.67$^o$, respectively. To further comprehend the atomic charge distribution and bonding mechanism, we have conducted Mulliken charge density analysis. The valence charge density iso-surface plot in Fig. (1b) indicates that the charge is primarily concentrated around the N atom with some distribution along the Si-N and N-P bonds. This is clearly due to the increased N atom electronegativity. Figure (1c) displays a contour plot of 2D charge density, with green representing the highest intensity and red representing the lowest intensity. The presence of a greater electron charge in Si-N and P-N bonds, the green color, and the intersection of concentric lines with the merging of electronic wave functions validate covalency. The distinctive smaller wave functions overlapping along with the deformation of the conjoining contour lines, shown as dumbbell-shaped with smaller charge spread, faded-green, between P-N clearly suggest the ionic and covalent bonding formation, with the bonding of P-N and Si-N being the strongest compared to the Si-P bond. Further, the obtained negative cohesive energy $E_{coh} = -5.28$ eV/atom, calculated based on Eq. (1), confirms the structural stability. In addition, we found formation energy $E_f = +1.56$ eV/atom, based on Eq. (2), which suggests the possible synthesis of the 2D p-SiPN monolayer through endothermic experimental process. The phonon band spectrum along the first Brillouin zone is also calculated. It contains only real frequencies with fifteen optical and three acoustic branches phonon frequencies, Fig. (1*d*), which further confirms dynamical stability. The shown AIMD simulation performed at $T$ = 300 K, Fig. (1*e*), suggests that the p-SiPN monolayer is highly thermodynamically stable, and does not exhibit any distortion.

*3.2. Electronic Properties*

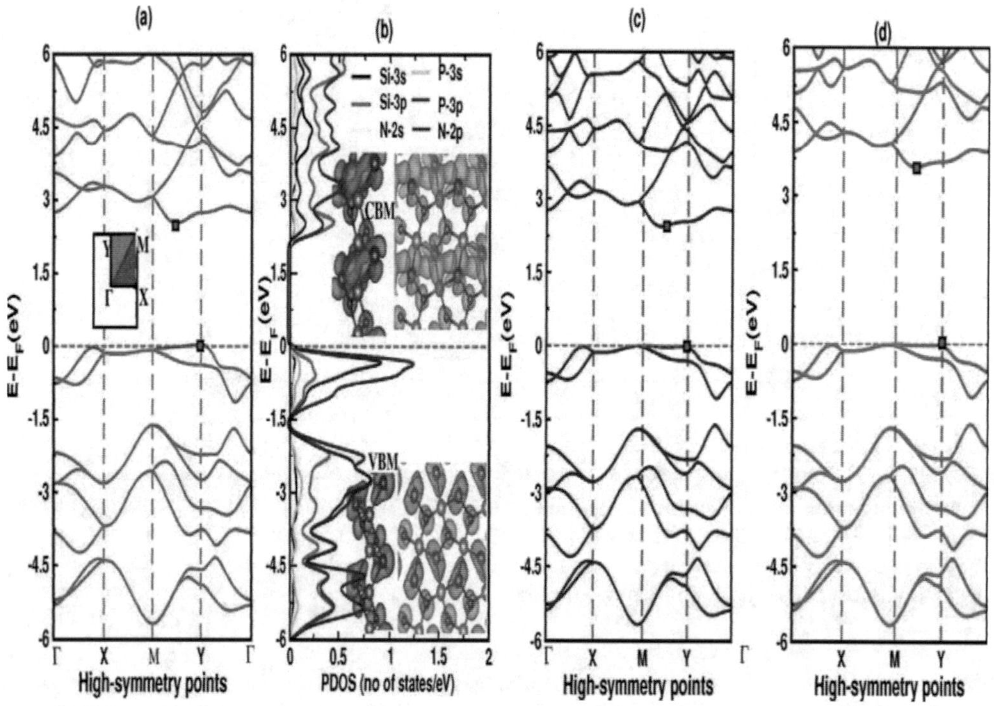


**Figure 2.** (Color online) (a) The band structure (electronic), (b) the PDOS with the VBM/CBM calculated using GGA approximation displayed in the top/bottom panels, (c) electronic band structure acquired from GGA approximation, and (d) comparison of band structure using HSE06 functional. The green box indicates the endpoint locations of the VBM and CBM bands.

To understand the electronic structure of p-SiPN, the spin-polarized/unpolarized electronic bands, and PDOS are calculated, Figs. (2a-2b). Both the spin-up and spin-down bands show clear symmetry suggesting a zero total magnetic moment with a non-magnetic ground state. There is a high-symmetry observed in the first Brillouin zone (BZ) along the $\Gamma\text{–}X\text{–}M\text{–}Y\text{–}\Gamma$ path as calculated using the NAO-basis implemented in SIESTA, Fig. (2*a*), and the plane-wave approach used by VASP, Fig. (2*c*). Both the GGA and HSE06 approximations yield similar electronic band structure, and confirm the indirect bandgap semiconducting behavior of p-SiPN. While the valence-band-maximum (VBM) occurs at the Y points, the conduction-band-minimum (CBM) occurs along the M–Y path with a bandgap energy of $E_g = 2.43$ eV, and a flat band near the Fermi level (FL) of the valance band. It is known that the GGA approximation seems to underestimate the bandgap value. For a better calculation of $E_g$, we utilized the hybrid functional HSE06 form built in VASP and obtained $E_g = 3.43$ eV, Fig. (2*d*). Additionally, the PDOS in Fig. (2b) indicates that the P-*3p*, N-*2p*, and Si-*3p* electronic orbitals with P*3s* orbitals ($sp^3$-hybridized) have a decreasing contribution to the VBM. Meanwhile, the main contributors to the CBM appear to be the Si-*3p*, P-*3p*, and N-*2p* in combination with N-*2s*. This is further supported by the partial charge densities of the VBM/bottom right and CBM/top right, showing similar atomic charge distribution in both the VBM and CBM regions.

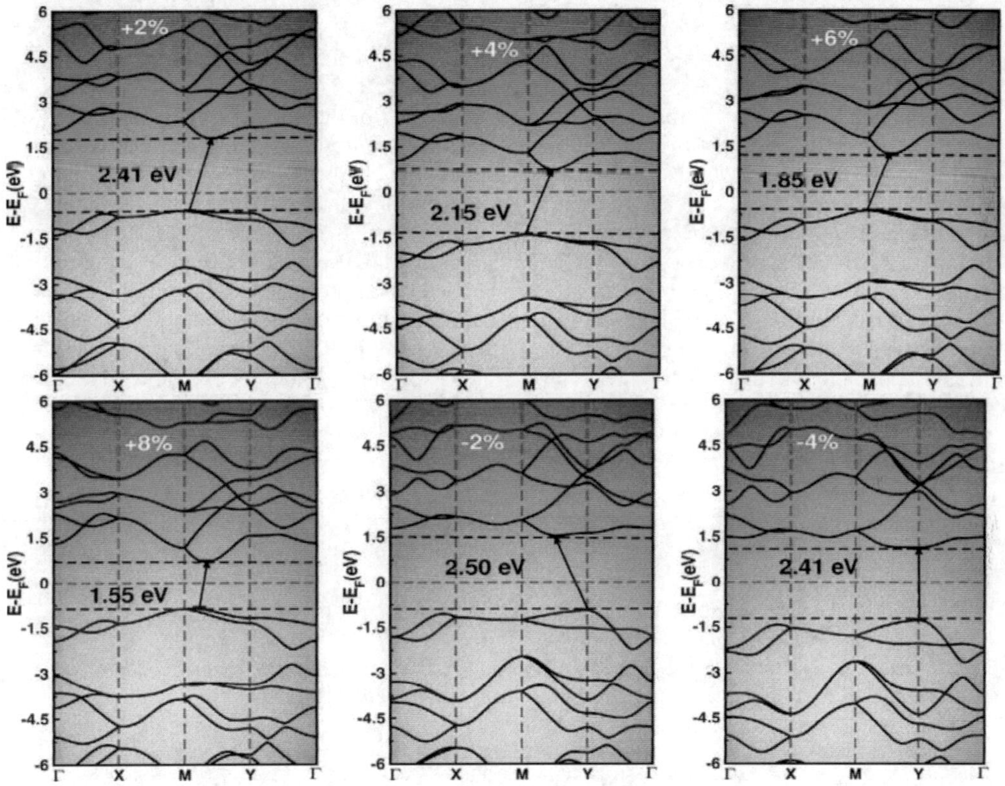


**Figure 3.** (Color online) The change in the electronic band due to applied equi-biaxial tensile and compressive strain $\varepsilon_{xy}$. The shown dotted horizontal lines display the valence band maximum and conduction band minimum, and the solid arrow represents the nearest inter-band transition.

We further investigate the modulation of the electronic bandgap and the tuning mechanism of p-SiPN, Fig. 3($a$ – $b$), when both an equ-biaxial tensile and compressive strain ($-4\% \leq \varepsilon_{xy} \leq +8\%$) are applied. In order to reach the state of minimum energy, we permitted the atomic positions to relax during the process. When subjected to a compressive strain of $\varepsilon_{xy} = -2\%$, the bandgap expanded to 2.50 eV while maintaining its indirect nature. When the strain was at $\varepsilon_{xy} = -4\%$, the bandgap reduced to 2.41 eV and exhibited a direct bandgap behavior at the Y point,

Journal of Physics: Conference Series **2751** (2024) 012013 doi:10.1088/1742-6596/2751/1/012013

likely attributed to the meta-stability of p-SiPN under applied strain. However, applying a tensile strain of $\varepsilon_{xy}$ = + 2% results in obtaining an indirect bandgap of 2.41 eV, leading to the VBM shifting from the Y to the M point. By further increasing the tensile strain to a range of 4% $\leq \varepsilon_{xy} \leq$ 8%, the bandgap decreases even more to 2.15 eV, 1.85 eV, and 1.55 eV for strain levels of 4%, 6%, and 8% respectively. Compared to being strain-free, the electronic bandgap value decreases by approximately 36% when stretched by 8%. This indicates that the electronic performance of p-SiPN can be easily adjusted by changing the applied strain, which is essential for various optoelectronic, nanomechanical, energy storage, and light harvesting devices [22, 23, 24].

## 4. Conclusions

In this work, DFT calculations were carried-out to predict a new 2D p-SiPN material. The new p-SiPN monolayer is verified to be structurally, dynamically, and thermodynamically stable. The p-SiPN obtained negative cohesive energy $E_{coh} = -5.28$ eV/atom and formation energy $E_f$ = 1.56 eV/atom confirm the structural stability of the 2D p-SiPN monolayer suggesting its possible experimental synthesis using endothermic process. In addition, the performed AIMD simulation carried-out at $T$ = 300 K confirms that the 2D p-SiPN monolayer is highly thermodynamically stable without any distortion. The plots of the valence charge density iso-surface for p-SiPN reveal that the charge mostly clusters around the N atom, while also being widely distributed along the Si-N and N-P bonds, due to the N atom's greater electronegativity. Our calculations suggest that both the spin-up and spin-down bands show clear symmetry indicating a zero total magnetic moment with a non-magnetic ground state of the 2D p-SiPN. Moreover, p-SiPN seems to exhibit a wide and indirect bandgap semiconducting behavior with a bandgap value of $E_g = 2.43$ eV (3.43 eV) as calculated based on the GGA (hybrid functional HSE06) approximation. It also shows high tunable bandgap against both applied biaxial and compressive strains in the range of $-4\% \leq \varepsilon_{xy} \leq +8\%$, making p-SiPN an attractive material for various applications such as energy-storage, optoelectronics, nanomechanics, and light-harvesting devices [22,23,24].

## References


[1] Novoselov K S, Geim A K, Morozov S V, Jiang D, Zhang Y, Dubonos S V, Grigorieva I V, and Firsov A A 2004 *Science* **306** (5696) 666-669.

[2] Zhang S, Zhou J, Wang Q, Chen X, Kawazoe Y, and Jena P 2015 *Proc. Natl. Acad. Sci. USA* **112** 2372-2377.

[3] Nazir M A, Hassan A, Shen Y, and Wang Q 2022 *Nano Today* **44** 101501.

[4] Shen Y, and Wang Q 2022 *Phys. Rep.* **964** 1-42.

[5] Sharma S B, Qattan I A, Kc S, and Alsaad A M 2022 *ACS Omega* **7** (41) 36235-36243.

[6] Sharma S B and Qattan I 2022 *Applied Surface Science* **599** 153997.

[7] Sharma S B, Qattan I A, Jaishi M, and Paudyal D 2022 *ACS Applied Electronic Materials* **4** 5 2561-2569.

[8] Sharma S B, Qattan I A, KC S, and Abedrabbo S 2022 *Nanomaterials* **12** (22) 4068.

[9] Oyedele A D, Yang S, Liang L, Puretzky A A, Wang K, Zhang J, Yu P, Pudasaini P R, Ghosh A W, Liu Z *et al.* 2017 *J. Am. Chem. Soc.* **139** 14090-14097.

[10] Li E, Wang D, Fan P, Zhang R, Zhang Y Y, Li G, Mao J, Wang Y, Lin X, Du S *et al.* 2018 *Nano Res.* **11** 5858-5865.

[11] Alsaad A M, Al-Bataineh Qais M, Qattan I A, Ahmad Ahmad A, Ababneh A, Albataineh Zaid, Aljarrah Ihsan A, and Telfah Ahmad 2020 *Frontiers in Physics* **8** 115.

[12] Ahmad A A, Alsaad A M, Albiss B A, Al-Akhras M-Ali, El-Nasser H M, and Qattan I A 2015 *Physica B: Condensed Matter* **470-471** 21-32.

[13] Ahmad A A, Alsaad A M, Albiss B A, Al-Akhras M-Ali, El-Nasser H M, and Qattan I A 2016 *Thin Solid Films* **606** 133-142.

[14] Zhang S, Zhou J, Wang Q, and Jena P 2016 *J. Phys. Chem.* C **120** 3993-3998.

[15] Wang H, Chen Z, and Liu Z 2022 *Appl. Surf. Sci.* **585** 152536.
[16] Wang H, Qin G, Qin Z, Li G, Wang Q, and Hu M 2018 *J. Phys. Chem. Lett.* **9** 2474-2483.
[17] Zhao K, Guo Y, Shen Y, Wang Q, Kawazoe Y, and Jena P 2020 *J. Phys. Chem. Lett.* **11** 3501-3506.
[18] Sun W, Shen Y, Guo Y, Chen Y, and Wang Q 2021 *Appl. Surf. Sci.* **554** 149499.
[19] Soler J M, Artacho E, Gale J D, Garc´ıa A, Junquera J, Ordej´on P, and S´anchez-Portal D 2002 *J. Phys. Condens. Matter* **14** 2745.
[20] Togo A, Oba F, and Tanaka 2008 *Phys. Rev. B* **78** 134106.
[21] Kresse G, and Joubert D 1999 *Phys. Rev. B Condens. Matter. Mater Phys.* **59** 1758.
[22] Lokhande A C, Babar P T, Karade V C, Jang J S, Lokhande V C, Lee D J, Kim I C, Patole S P, Qattan I A, Lokhande C D, and Kim J H 2019 *Materials Today Chemistry* **14** 100181.
[23] Lokhande A C, Teotia S, Shelke A R, Hussain T, Qattan I A, Lokhande V C, Patole Shashikant, Kim J H, and Lokhande C D 2020 *Chemical Engineering Journal* **399** 125711.
[24] Lokhande A C, Qattan I A, Lokhande C D, and Patole S P 2020 *J. Mater. Chem. A* **8** 918.

# Electronic, Mechanical, and thermoelectric properties of Ge/Zn co-doped SnSe: first-principles calculations

**N Al Bouzieh[1], M Benkraouda[1] and N Amrane[1,2]**

[1]Physics Department, College of Science, United Arab Emirates University (UAEU), 15551, Al Ain, UAE
[2]Corresponding author's Email: namrane@uaeu.ac.ae

**Abstract.** Materials based on tin selenide have attracted significant attention due to their unique properties, particularly their high ZT value. This study investigates the impact of co-doping Germanium and Zinc on the electronic, mechanical, and thermoelectric properties of SnSe crystal using first-principles calculations. The doped structure demonstrated a p-type semiconducting behavior with a triclinic stable structure, which was predicted by the calculated elastic constants. Thermoelectric properties were studied for both doped and undoped systems across a wide range of carrier concentrations and temperatures. Results showed that co-doping SnSe with Ge/Zn reduced electronic thermal conductivity at room temperature while simultaneously doubling the Seebeck coefficient. This promising combination of features suggests high thermoelectric performance for the material.

## 1. Introduction
As traditional fossil fuel resources dwindle, thermoelectric materials are gaining popularity as a viable alternative energy source. Tin chalcogenides-based thermoelectric materials are especially promising due to their outstanding electrical, optical, and catalytic properties. In this context, tin selenide (SnSe) stands out as it has the smallest band gap of all the tin chalcogenides, making it a desirable candidate for thermoelectric applications [1]. Recent years have seen considerable efforts to enhance its thermoelectric properties to rival other existing thermoelectric materials. SnSe's versatility is another major advantage, as it can be utilized in various applications such as waste heat recovery and refrigeration systems. Additionally, SnSe exhibits good mechanical and chemical stability, which ensures its durability over long-term use. However, there are still challenges to be addressed before SnSe can be widely commercialized as a thermoelectric material. For instance, its stability at high temperatures needs to be improved, and its thermal conductivity must be reduced.

In the field of thermoelectric materials, the figure of merit (ZT) is widely used to assess their performance. The low thermal conductivity of SnSe makes it a promising candidate in this regard [2], with a figure of merit of 2.6 at 923K along the b-axis [3]. However, tin selenide can be further enhanced by doping with appropriate elements, such as Ge and Zn. Prior research indicates that introducing Ge as a dopant notably enhances the Seebeck coefficient [4], Zhang and colleagues found that when SnSe was doped with Zn, it yielded a ZT value of 0.96 at 873 K, marking a 41% increase compared to undoped SnSe [5]. Additionally, Sheini noted that Zn-doped SnSe films outperformed those without doping in terms of efficiency [6]. In this work, the effects of co-doping with Ge and Zn on the electrical, mechanical, and thermoelectric properties of SnSe were examined using first-principles calculations. The results showed that Ge/Zn co-doping led to higher Seebeck coefficients and electrical conductivity, as well as extremely low electronic thermal conductivitie. The study also provided a comprehensive characterization of the doped SnSe structures, including crystal structure refinement, DFT band-structure calculations, and transport properties analysis.

## 2. Computational Details
Using the Vienna Ab-initio Simulation Package (VASP) for first-principle computations, we explored how the properties of SnSe are influenced by the joint incorporation of germanium and zinc. To compute

the exchange-correlation energy, we applied the Perdew-Burke-Ernzerhof (PBE) generalized gradient approximation (GGA) [7] in conjunction with the projector augmented plane wave method (PAW) [8] and, for greater accuracy in calculating we employed the SCAN functional to examine the electronic band gap [9]. Our research focused on a supercell with the dimensions of 2 × 3 × 3 where we randomly changed three Sn atoms with three Ge atoms and one Sn atom with one Zn atom. Structural parameters were calculated for both compounds using a plane wave energy cut-off of 450 eV and a Monkhorst-Park k-point mesh [10]. We employed a 4 × 12 × 12 mesh grid for pure SnSe, and a 2 × 6 × 6 grid for the doped system to conduct crystal structure optimization and band structure calculations. We performed computations for the Electronic Density of States (DOS) with a more finely spaced K-mesh grid, using 8 × 24 × 24 for the pristine SnSe and 4 × 12 × 12 for the doped system. We relaxed the lattice parameters and internal ion coordinates using a force tolerance of 0.02 eV/Å and kept the electronic convergence threshold for energy at $1 \times 10^{-6}\ eV$. To compute elastic properties we used the energy-strain method [11]. Finally, we evaluated the thermoelectric properties using Boltzmann Transport Equation (BTE) and BoltzTrap2 code [12], considering a constant relaxation time.

## 3. Structural properties

The crystal structures are presented in Figure 1 for the pure and doped SnSe samples. The lattice parameters for the undoped SnSe system were determined, and their values were consistent with previous theoretical and experimental results **[13,14]**. The measured values for a, b, and c were 11.78 **A°**, 4.22 **A°** , and 4.52 **A°**, respectively for the pure structure. However, the parameters of the doped system showed slight differences, with a, b, and c values 11.75 **A°**, 4.19 **A°** , and 4.55 **A°**, respectively, as shown in Table 1. Due to the smaller atomic sizes of Ge and Zn compared to Sn, the overall unit-cell volume of the doped system was decreased.

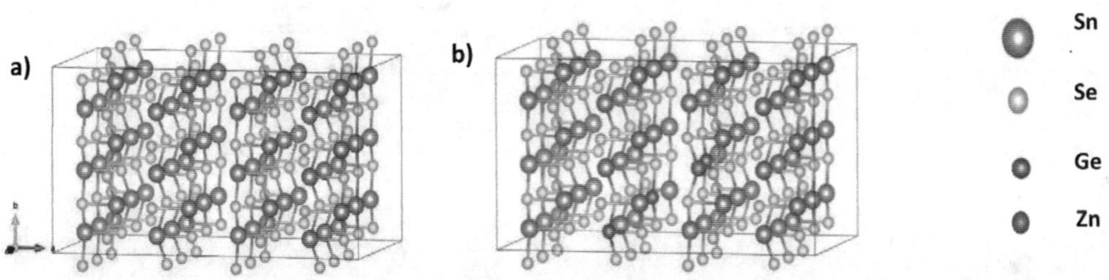


**Figure 1.** The structures of crystalline 2 × 3 × 3 SnSe supercell systems, which includes (a) the undoped α-SnSe, and (b) the structure doped with both Ge and Zn.

**Table 1.** DFT Calculations of Lattice Parameters for Undoped and Doped SnSe.

| | $a(A^{\circ})$ | $b(A^{\circ})$ | $c(A^{\circ})$ |
|---|---|---|---|
| $\alpha - SnSe$ | 11.78 | 4.22 | 4.52 |
| **Other DFT calculations[a]** | 11.72 | 4.20 | 4.55 |
| **Experiment[b]** | 11.50 | 4.15 | 4.45 |
| $Sn_{0.95}Ge_{0.04}Zn_{0.01}Se$ | 11.75 | 4.19 | 4.55 |

[a] Ref [13].
[b] Ref [14].

## 4. Electronic properties

Previous studies extensively reported on DFT calculations of the electronic structure of orthorhombic SnSe crystal [4,15–17]. However, it is known that DFT band-structure calculations tend to underestimate energy gaps in semiconductors and insulators [18]. Therefore, we computed the electronic band gap for both pure and doped SnSe systems using GGA-PBE and SCAN functionals. Table 2 presents the determined band gaps, while Figure 2 illustrates the electronic band structure of the doped compound. The PBE-GGA potential yielded a DFT gap for undoped SnSe of 0.601 eV, which aligns with earlier calculations as referenced [19–21]. Nonetheless, this is smaller than the values obtained from experimental values, which ranged between 0.9 and 1.1 eV [22], as predicted due to the PBE approach's tendency to underestimate semiconductors' band gap. We determined energy band gap values of 0.780 eV utilizing the SCAN functional, which aligns more closely with the experimental results presented in Table 2. SnSe doped structure was found to be a p-type semiconductor, with a similar indirect bandgap of 0.821 eV, using the SCAN functional. Sn p-states were the largest contributors among the conduction bands, while Se p-states dominated in the valence bands, as shown in Figure 2. The CBM of bulk SnSe was found in the Y-Γ region, and the VBM was located in the Γ-Z region, while the doped system exhibited VBM at the Γ point and CBM in the Y-Γ direction. Zn (3d) state exhibited strong hybridization with Sn (5s) state at around -7 eV, while Zn (4s) state contributed to the valence band energy at around -4 eV. All doped and undoped systems exhibited semiconductor behavior.

**Table 2.** Computed Bandgap Values for Pristine and Doped Systems.

| **Compound** | functional | Band gap (eV) |
|---|---|---|
| $\alpha - SnSe$ | PBE | 0.601 |
| | SCAN | 0.780 |
| | EXP[a] | 0.860 |
| $Sn_{0.95}Ge_{0.04}Zn_{0.01}Se$ | PBE | 0.613 |
| | SCAN | 0.821 |

[a] Ref [22]

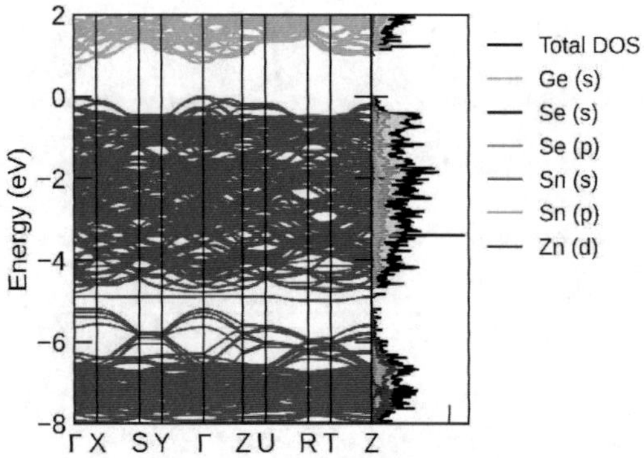


**Figure 2.** Electronic band structure and the corresponding partial density of states (PDOS) obtained with SCAN functionals for $Ge/Zn$ doped structure.

## 5. Mechanical properties

Mechanical properties play a vital role in the longevity of thermoelectric materials. To ensure mechanical stability, we begin by investigating the elastic constants of both doped and undoped crystals. Our DFT calculations show that the doped SnSe structure is a triclinic crystal system with 21 independent elastic constants that describe its infinitesimal elastic behavior. The calculated values of the elastic parameters are listed in Table 3, where $C_{44}$, $C_{55}$ and $C_{66}$ correspond to the deformations of the lattice angles α, β, and γ [23]. From the provided data, we can observe that the angle β holds the most stability since the value of $C_{55}$ surpasses those of $C_{44}$and $C_{66}$. According to Hooke's law [24], the stress vector $\sigma_i$ is equal to the elastic stiffness matrix $C_{ij}$ multiplied by the strain vector $\varepsilon_i$. $C_{11}$and $C_{22}$ has higher values than $C_{33}$, demonstrating that the anti-deformation capabilities of the c-axis is inferior to that of the a-axis and b-axis. The value of $C_{55}$ exceeds those of $C_{44}$ and $C_{66}$, suggesting that a shear force applied on the ac plane would cause more strain than when it's solely exerted on the ab plane.

**Table 3:** Elastic Stiffness Constants of Pristine and Doped Systems Calculated via DFT.

| | $C_{11}$ | $C_{22}$ | $C_{33}$ | $C_{44}$ | $C_{55}$ | $C_{66}$ |
|---|---|---|---|---|---|---|
| $\alpha - SnSe$ | 42.7 | 65.0 | 32.5 | 14.0 | 29.6 | 10.8 |
| **Other DFT calculations**[a] | 41.1 | 65.7 | 29.6 | 12.0 | 29.6 | 14.7 |
| **Experiment**[b] | 41.8 | 59.7 | 32.5 | 13.2 | 24.5 | 20.5 |
| $Sn_{0.95}Ge_{0.04}Zn_{0.01}Se$ | 41.9 | 65.2 | 28.4 | 13.1 | 28.1 | 9.8 |

[a] Ref [25]

[b] Ref [26]

During heat transfer operations, materials often encounter challenges like thermal stress and expansion. These issues can be mitigated by reducing certain thermodynamic properties, including Young's modulus, the linear coefficient of thermal expansion, and Poisson's ratio, while maintaining a steady temperature gradient [27]. We assessed the thermodynamic characteristics for both the doped and pristine SnSe systems, and the results are displayed in Table 4. Bulk modulus and shear modulus are known to indicate the hardness of the material, where the bulk modulus represents the resistance to fracture and the shear modulus represents the resistance to plastic deformation [28]. From Table 4, and by applying Pugh relation [29] B/G, which is analytically linked to the ductile/brittle properties of the material, both SnSe structures lie in the brittle region. Young's modulus, which is the ratio of principal stress to corresponding strain in the elastic range in the same direction, and Poisson's ratio, a parameter that depicts how a material reacts to a load applied in one direction [30], are essential for technology and engineering applications [31]. The doped structure has a lower Young's modulus, indicating that it is less stiff than the undoped SnSe.

**Table 4.** Elastic moduli: Bulk Modulus (B), Shear Modulus (G), Young's Modulus (E), Poisson's Ratio ($v$), and Ductility Index (B/G) along with the values of longitudinal wave velocity ($v_l$), average sound velocity ($v_m$) of the pristine and doped systems.

| | B (GPa) | G (GPa) | E (GPa) | $v$ | B/G | $v_l$ $(m/s)$ | $v_m$ $(m/s)$ |
|---|---|---|---|---|---|---|---|
| $\alpha - SnSe$ | 23.9 | 15.7 | 38.7 | 0.23 | 1.52 | 2771 | 1817 |
| **Other DFT calculations**[a] | 24.1 | 15.2 | 37.6 | 0.24 | 1.59 | 2711 | 1759 |
| **Experiment**[b] | 23.2 | 16.0 | 39.0 | 0.22 | 1.45 | 2717 | 1803 |
| $Sn_{0.95}Ge_{0.04}Zn_{0.01}Se$ | 24.0 | 14.1 | 35.3 | 0.25 | 1.70 | 2719 | 1733 |

[a] Ref [25]

[b] Ref [26]

## 6. Thermoelectric properties

The Seebeck coefficient (S), electrical conductivity ($\sigma$), and electrical thermal conductivity ($\kappa_e$) were calculated for various carrier densities and temperatures in both the pure and doped systems in order to assess the thermoelectric properties of SnSe with Ge/Zn doping. The results are presented in Figures 3 and 4, where the parameters are expressed in (T, n) plane. The left column of the figures shows pure SnSe, while the right column shows doped SnSe. Figure 3 (a & b) shows that the Seebeck coefficient has much higher values for the doped structure, particularly in the region of relatively low and mid-range $n$ at low temperatures. This increase may be due to the addition of a small percentage of Ge and Zn atoms, which may cause lower mobility of the charge carriers [4].

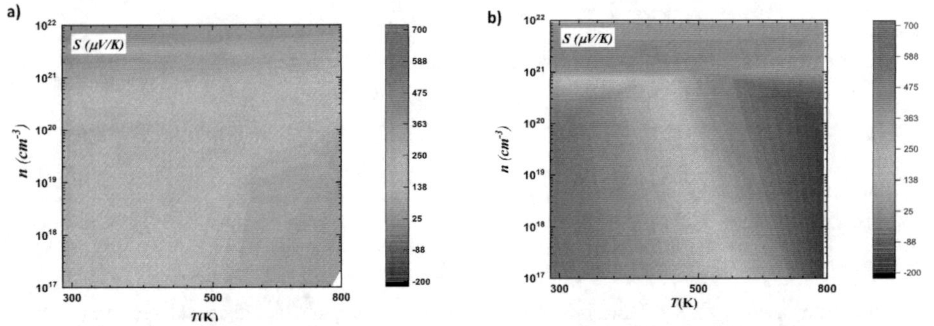


**Figure 3.** Seebeck coefficient values on the (T, n) plane for (a) pure SnSe and (b) Ge/Zn doped SnSe.

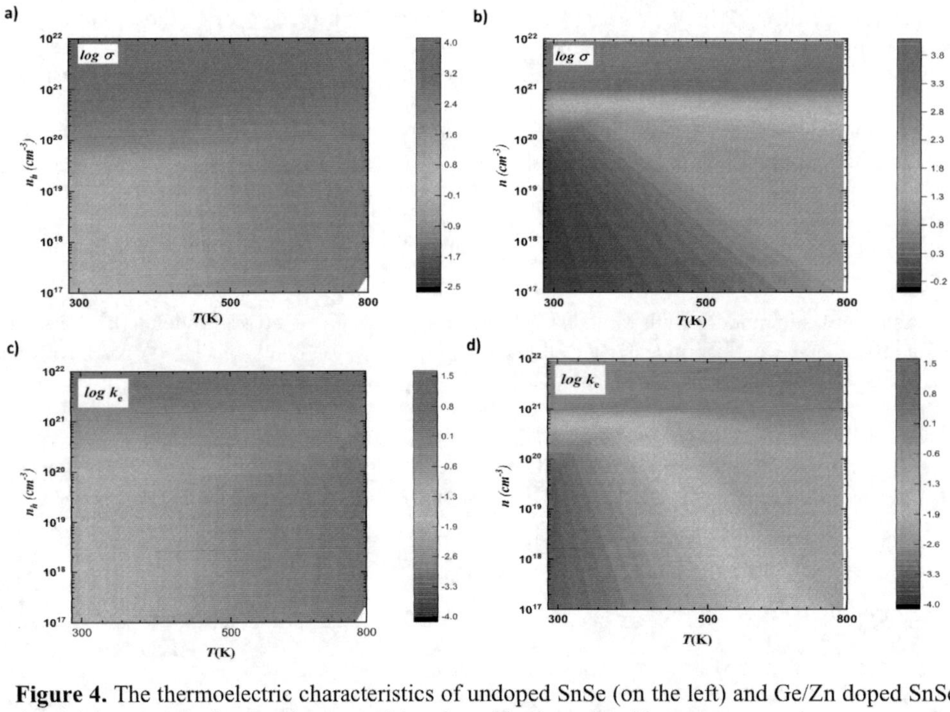


**Figure 4.** The thermoelectric characteristics of undoped SnSe (on the left) and Ge/Zn doped SnSe (on the right) were calculated and presented in a logarithmic scale with respect to carrier concentration ($n$) and temperature (T) from 300 to 800 Kelvin. The top row shows the electrical conductivity $\sigma$ ($S\ cm^{-1}$), while the bottom row presents the electronic thermal conductivity $k_e$ ($WK^{-1}m^{-1}$).

The electrical conductivity σ and electronic thermal conductivity $\kappa_e$ were examined and shown in Figure 4 to better understand the thermal behavior of the doped system. Because of the linear temperature dependence predicted by the Wiedemann-Franz law $\kappa_e = L\sigma T$, where $L$ is the Lorentz constant, $\sigma$ and $\kappa_e$ values increase with increasing T in the low *n* region, and have maximum values in the region of relatively high *n*.

## 7. Conclusion

In this study, we utilized advanced density-functional theory to investigate the electronic, mechanical, and thermal characteristics of SnSe that was co-doped with Germanium and Zinc across various temperatures and carrier concentrations. Our calculations revealed that the doped structure exhibited a triclinic configuration with an indirect bandgap of 0.824 eV, which is higher than the 0.780 eV bandgap observed in the pristine α-SnSe using the SCAN functional. The mechanical stability of the doped structure was verified through the evaluation of elastic constants. Furthermore, we investigated the thermoelectric properties of the Ge/Zn-doped system across a temperature range of 300-800 K and various carrier concentrations, and our results indicated that the doping substantially increased the values of the Seebeck coefficient.

## References

[1] Sun Q, Yang M, Zeng G, Li J, Hu Z, Li D, Wang S, Si P, Tian Y and Ci L 2022 Insights into the Potassium Ion Storage Behavior and Phase Evolution of a Tailored Yolk–Shell SnSe@C Anode *Small* **18** 2203459

[2] Sheng Z, Guo C, Hu S, Xiao Z and Guo L 2022 Routes to High-Carrier-Density Doping in Thermoelectric SnSe *J. Phys. Chem. C* **126** 1230–41

[3] Zhao L-D, Lo S-H, Zhang Y, Sun H, Tan G, Uher C, Wolverton C, Dravid V P and Kanatzidis M G 2014 Ultralow thermal conductivity and high thermoelectric figure of merit in SnSe crystals *Nature* **508** 373–7

[4] Gharsallah M, Serrano-Sánchez F, Nemes N M, Mompeán F J, Martínez J L, Fernández-Díaz M T, Elhalouani F and Alonso J A 2016 Giant Seebeck effect in Ge-doped SnSe *Sci Rep* **6** 26774

[5] Li J C, Li D, Qin X Y and Zhang J 2017 Enhanced thermoelectric performance of p-type SnSe doped with Zn *Scripta Materialia* **126** 6–10

[6] Jamali-Sheini F, Cheraghizade M and Yousefi R 2018 Electrochemically synthesis and optoelectronic properties of Pb- and Zn-doped nanostructured SnSe films *Applied Surface Science* **443** 345–53

[7] Perdew J P, Burke K and Ernzerhof M 1996 Generalized Gradient Approximation Made Simple *Phys. Rev. Lett.* **77** 3865–8

[8] Blöchl P E 1994 Projector augmented-wave method *Phys. Rev. B* **50** 17953–79

[9] Sun J, Ruzsinszky A and Perdew J P 2015 Strongly Constrained and Appropriately Normed Semilocal Density Functional *Phys. Rev. Lett.* **115** 036402

[10] Pack J D and Monkhorst H J 1977 "Special points for Brillouin-zone integrations"---a reply *Phys. Rev. B* **16** 1748–9

[11] Nielsen O H and Martin R M 1985 Stresses in semiconductors: Ab initio calculations on Si, Ge, and GaAs *Phys. Rev. B* **32** 3792–805

[12] Madsen G K H, Carrete J and Verstraete M J 2018 BoltzTraP2, a program for interpolating band structures and calculating semi-classical transport coefficients *Computer Physics Communications* **231** 140–5

[13] Dewandre A, Hellman O, Bhattacharya S, Romero A H, Madsen G K H and Verstraete M J 2016 Two-Step Phase Transition in SnSe and the Origins of its High Power Factor from First Principles *Phys. Rev. Lett.* **117** 276601

[14] Lee Y K, Ahn K, Cha J, Zhou C, Kim H S, Choi G, Chae S I, Park J-H, Cho S-P, Park S H, Sung Y-E, Lee W B, Hyeon T and Chung I 2017 Enhancing p-Type Thermoelectric Performances of Polycrystalline SnSe via Tuning Phase Transition Temperature *J. Am. Chem. Soc.* **139** 10887–96

[15] Zhang Q, Chere E K, Sun J, Cao F, Dahal K, Chen S, Chen G and Ren Z 2015 Studies on Thermoelectric Properties of n-type Polycrystalline SnSe1-xSx by Iodine Doping *Advanced Energy Materials* **5** 1500360

[16] Peng K, Lu X, Zhan H, Hui S, Tang X, Wang G, Dai J, Uher C, Wang G and Zhou X 2016 Broad temperature plateau for high ZTs in heavily doped p-type SnSe single crystals *Energy Environ. Sci.* **9** 454–60

[17] Al Bouzieh N, Sattar M A, Benkraouda M and Amrane N 2023 Optoelectronic, mechanical, and thermoelectric properties of Na/I co-doped SnSe via ab initio calculations *Journal of Solid State Chemistry* **320** 123858

[18] Vu T V, Lavrentyev A A, Gabrelian B V, Parasyuk O V, Ocheretova V A and Khyzhun O Y 2018 Electronic structure and optical properties of Ag2HgSnSe4: First-principles DFT calculations and X-ray spectroscopy studies *Journal of Alloys and Compounds* **732** 372–84

[19] Carrete J, Mingo N and Curtarolo S 2014 Low thermal conductivity and triaxial phononic anisotropy of SnSe *Appl. Phys. Lett.* **105** 101907

[20] González-Romero R L, Antonelli A and Meléndez J J 2017 Insights into the thermoelectric properties of SnSe from ab initio calculations *Phys. Chem. Chem. Phys.* **19** 12804–15

[21] Kim H S, Choi G, Ha M Y, Kim D H, Park S H, Chung I and Lee W B 2019 Thermoelectric transport properties of Pb doped SnSe alloys (PbxSn1-xSe): DFT-BTE simulations *Journal of Solid State Chemistry* **270** 413–8

[22] Meléndez J J, González-Romero R L and Antonelli A 2018 Quasiparticle bands and optical properties of SnSe from an ab initio approach *Computational Materials Science* **152** 107–12

[23] Qin H, Yan B-L, Zhong M, Jiang C-L, Liu F-S, Tang B and Liu Q-J 2019 First-principles study of structural, elastic, and electronic properties of triclinic TATB under different pressures *Physica B: Condensed Matter* **552** 151–8

[24] Yu C-F, Cheng H-C and Chen W-H 2016 An Ab-initio study of structural, elastic, electronic and thermodynamic properties of triclinic Cu7In3 *Materials Chemistry and Physics* **174** 70–80

[25] Li G, Aydemir U, Wood M, Goddard W A, Zhai P, Zhang Q and Snyder G J 2017 Ideal Strength and Deformation Mechanism in High-Efficiency Thermoelectric SnSe *Chem. Mater.* **29** 2382–9

[26] Wu Z, Zhao E, Xiang H, Hao X, Liu X and Meng J 2007 Crystal structures and elastic properties of superhard $\mathrm{Ir}{\mathrm{N}}_{2}$ and $\mathrm{Ir}{\mathrm{N}}_{3}$ from first principles *Phys. Rev. B* **76** 054115

[27] Weller D P and Morelli D T 2022 Tetrahedrite Thermoelectrics: From Fundamental Science to Facile Synthesis *Frontiers in Electronic Materials* **2**

[28] Teter D M 1998 Computational Alchemy: The Search for New Superhard Materials *MRS Bulletin* **23** 22–7

[29] Pugh S F 1954 XCII. Relations between the elastic moduli and the plastic properties of polycrystalline pure metals *The London, Edinburgh, and Dublin Philosophical Magazine and Journal of Science* **45** 823–43

[30] Yu L, Yan Q and Ruzsinszky A 2017 Negative Poisson's ratio in 1T-type crystalline two-dimensional transition metal dichalcogenides *Nat Commun* **8** 15224

[31] Lakel S, Ibrir M and Guibadj A 2016 Structural and elastic properties of BiOCu1-xS with Cu vacancies *Materials Today: Proceedings* **3** 2877–82

Journal of Physics: Conference Series **2751** (2024) 012015 doi:10.1088/1742-6596/2751/1/012015

# Comparing the plasmon dispersion in graphene and $MoS_2$ nanoribbons array under Electromagnetic excitation

**Omnia Samy[1], Amine El Moutaouakil[1,*]**

[1]Department of Electrical and Communication Engineering, College of Engineering United Arab Emirates University, Al Ain UAE

a.elmoutaouakil@uaeu.ac.ae

**Abstract**. Terahertz properties of different materials have been recently studied due to their wide applications in optoelectronics, industry, product inspection, and spectroscopy. Terahertz frequency applications are promising for the medical field as they are considered safe frequencies. Previous terahertz plasma response focused on 2D materials like graphene and transition metal dichalcogenides (TMDs) due to their favourable electronic properties, high electric conductivity, and their band gap characteristics, so they can be used in electronic devices. Some of these materials showed good biocompatibility so they can be used in biomedical applications. Since graphene has zero band gap, researchers are continuously exploring methods to increase its band gap to be used in electronics. Graphene heterostructures or metamaterials are ways to enhance graphene characteristics for specific applications. This work investigates the possibility of using $MoS_2$ with graphene in THz applications. The plasmon dispersion for graphene and $MoS_2$ nanoribbon array structure is compared. Both graphene and $MoS_2$ behave differently in response to terahertz radiation due to their different band gaps. The results showed that $MoS_2$ exhibits a plasmonic response in the THz region at high carrier concentrations. This opens up opportunities for $MoS_2$ to be employed in THz sensors, both independently and in conjunction with graphene within heterostructures or metamaterials for power sources and detectors. These advancements hold significant potential for the future THz imaging and communication technologies.

## 1. Introduction

2D materials like graphene and transition metal dichalcogenides (TMDs) showed desirable electrical, optical, and chemical properties that permitted their use in a wide range of applications [1–5]. The fact that their band gap changes for different structures (from bulk to several layers) grabbed the attention to study their behavior to use them in electronic and optical applications. $MoS_2$ band gap changes from an indirect band gap of 1.29 eV in bulk structure to a direct one of 1.8-1.9 eV in monolayer $MoS_2$ [6–11]. A bulk structure of graphene (graphite) has zero band gap, while a bilayer or a monolayer structure has a band gap of 0.5 eV. At the same time, studies are directed towards the terahertz frequency response of these materials to be used in optoelectronic applications especially because the terahertz frequency is a low energy frequency which is considered safe in medical applications when compared to other frequency ranges like X-rays or Gamma rays [12–16]. Since semiconductors showed plasmonic frequency response at low dimensions [17–23], the study of the terahertz frequency response of these materials became a must. Most of $MoS_2$ research work was done in the visible or infra-red region due to its relatively large band gap [14,24] but very few discusses its behavior in the THz range which we seek to investigate in this work. The work in [25] showed very small absorption peaks for $MoS_2$ in the THz region and the work in [26] showed the sensing capabilities of a D-shaped complex structure based on graphene and

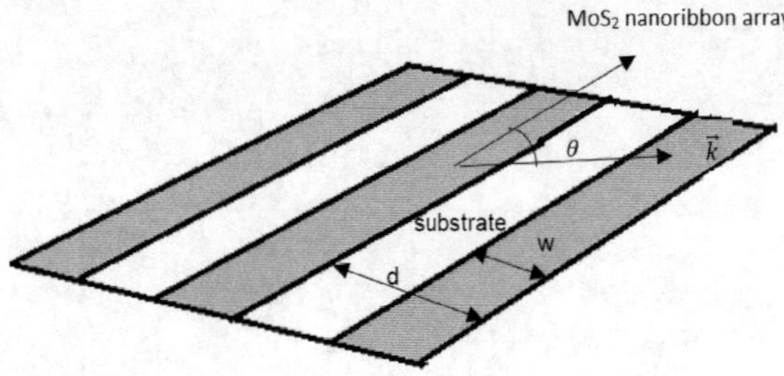


**Figure 1.** Schematic of $MoS_2$ nanoribbon array

$MoS_2$. In this work, we investigate the sensitivity of a simple nanoribbon structure of $MoS_2$ in response to THz waves. Obtaining simple structures with good characteristics facilitates its implementation and decreases the costs of manufacturing. The response of n-doped $MoS_2$ nanoribbon array structure when excited with external electric field, applied in the $MoS_2$ nanoribbon plane, that has an angle $\theta$ with the $MoS_2$ nanoribbon direction, is compared to that of graphene. The work is based on the equations published in [27] for graphene.

## 2. Plasmons' dispersion relation for $MoS_2$ nanoribbons

The nanoribbon structure of $MoS_2$ is shown in Figure 1 with a nanoribbon width w. Both $MoS_2$ and graphene optical conductivity in the THz range can be represented by the Drude conductivity model in (1)

$$\boldsymbol{\sigma = i\,\frac{e^2 N_{2D}}{m^*(\omega+iv)}} \tag{1}$$

Where $\boldsymbol{N_{2D}}$ is the 2D electron concentration, $\boldsymbol{v}$ is the electron relaxation rate, and $\boldsymbol{m^*}$ is the electron effective mass. For graphene we took $\boldsymbol{N_{2D} = 10^{12} cm^{-2}}$ and $\boldsymbol{v = 1.32 \times 10^{13} s^{-1}}$ and $\boldsymbol{m^* = 0.002 m_o}$ where $\boldsymbol{m_o}$ is the electronic mass *[27]* and for $MoS_2$, $\boldsymbol{N_{2D} = 2.5 \times 10^{13} cm^{-2}}$ as n-doped, $\boldsymbol{v = 5.88 \times 10^{12} s^{-1}}$ and $\boldsymbol{m^* = 0.53 m_o}$ [17], e is the electronic charge The external wave has a propagation vector $\boldsymbol{\vec{k}}$ and an angle $\boldsymbol{\theta}$ between the nanoribbon array direction and the wave vector $\boldsymbol{\vec{k}}$ (Figure 1). Using the same procedure as in *[27]* to solve for the dispersion relation for plasmons in $MoS_2$ nano array we can reach the same equation for the dispersion frequency for $MoS_2$ nanoribbon arrays (2).

$$\boldsymbol{\omega = \sqrt{2\pi \frac{\vec{k}}{\epsilon} \frac{e^2 N_{2D}}{m^*} \cos^2\theta - \frac{v^2}{4}} - \frac{iv}{2}} \tag{2}$$

Where $\epsilon$ is the substrate permittivity, usually Silicon ($\epsilon_{Si}$ = 11.7). The real part of $\boldsymbol{\omega}$ represents the dispersion frequency and the imaginary part is the decay rate. Figure 2 shows a comparison between graphene and $MoS_2$ nanoribbon arrays, where the dispersion frequency and the decay rate of $MoS_2$ are less than graphene. From Figure 2a, it is clear that, the dispersion frequency is higher than the excitation frequency, but it is much less for $MoS_2$, also as $\boldsymbol{\theta}$ increases the dispersion frequency decreases for both graphene and $MoS_2$ and a cut off frequency appears which is higher in the case of $MoS_2$. We have to excite the nanoribbon with a higher frequency than the cut off frequency to have a dispersion frequency in the THz range. In Figure 2b, the decay for $MoS_2$ is less than that of graphene.

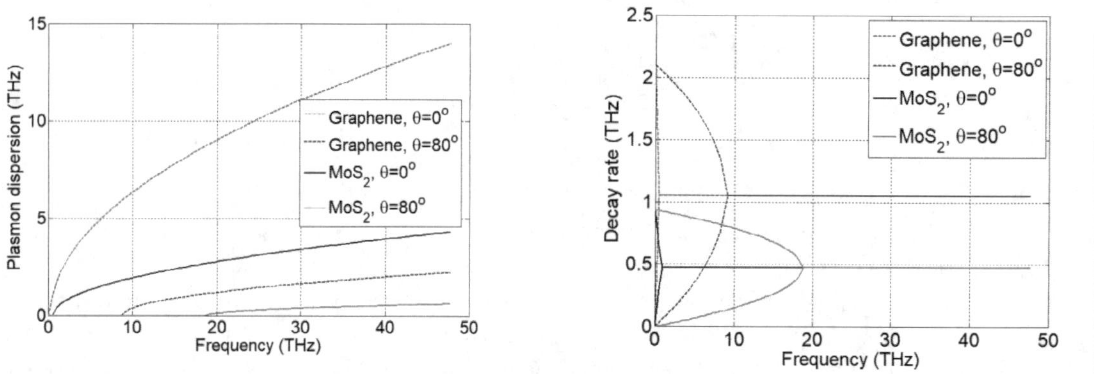


(a) (b)

**Figure 2**. Comparison between graphene (light lines) and $MoS_2$ (bold lines); **(a)** Plasmon dispersion; **(b)** decay rate. For $MoS_2$ $v = 5.88235\times \mathbf{10^{12}}\ s^{-1}, m^* = 0.53m_0, N_{2D} = 2.5\times 10^{13}\ cm^{-2}$, for graphene $v = 1.32\times 10^{13}\ s^{-1}, m^* = 0.002m_0, N_{2D} = 10^{12}\ cm^{-2}$ *[25,27]*

## 3. Conclusion

In conclusion, in this work, we investigated the behavior of n-doped $MoS_2$ nanoribbons when excited with terahertz EM waves in the same plane of the nanoribbon array. Both graphene and $MoS_2$ have a dispersion frequency in the THz range, that increase with increasing the excitation frequency. Although $MoS_2$ has much lower dispersion frequencies than graphene, and it needs to be excited with higher frequencies than graphene, it still has dispersion frequency in the THz range. The dispersion frequency of both of them decreases with increasing the angle $\theta$ which is the angle between the direction of the nanoribbon and the wave vector. $MoS_2$ has higher cut off frequencies that appears clearly at larger angles of excitation. The behaviour of $MoS_2$ paves the way for its integration with graphene in THz sensors and power sources that are crucial for the next THz era. This combination holds an immense promise in applications such as imaging and high-speed communication systems.

## Acknowledgments

This work was supported partly by funds from the UAE University Startup project No. 31N312 and the UPAR project No. 31N393.

## References

[1] Yadav V, Roy S, Singh P, Khan Z and Jaiswal A 2019 2D MoS $_2$ -Based Nanomaterials for Therapeutic, Bioimaging, and Biosensing Applications *Small* **15** 1803706

[2] Nawz T, Safdar A, Hussain M, Sung Lee D and Siyar M 2020 Graphene to Advanced MoS2: A Review of Structure, Synthesis, and Optoelectronic Device Application *Crystals* **10** 902

[3] Samy O and El Moutaouakil A 2024 A graphene-based THz selective absorber with absorptivity 95 % and wide-range electrical tunability *Results in Physics* **58** 107490

[4] Samy O, Belmoubarik M, Otsuji T and El Moutaouakil A 2023 A Voltage-Tuned Terahertz Absorber Based on MoS2/Graphene Nanoribbon Structure *Nanomaterials* **13** 1716

[5] Moutaouakil A E, Kang H-C, Handa H, Fukidome H, Suemitsu T, Sano E, Suemitsu M and Otsuji T 2011 Room Temperature Logic Inverter on Epitaxial Graphene-on-Silicon Device *Jpn. J. Appl. Phys.* **50** 070113

[6] Hong J, Li K, Jin C, Zhang X, Zhang Z and Yuan J Dimensional cross-over of the bandgap transition in quasi-two-dimensional MoS2 13

[7] Yan X, Zhu L, Zhou Y, E Y, Wang L and Xu X 2015 Dielectric property of MoS_2 crystal in terahertz and visible regions *Appl. Opt.* **54** 6732

[8] Rahman I A and Purqon A 2017 First Principles Study of Molybdenum Disulfide Electronic Structure *J. Phys.: Conf. Ser.* **877** 012026

[9] Samy O and El Moutaouakil A 2021 A Review on MoS2 Energy Applications: Recent Developments and Challenges *Energies* **14** 4586

[10] Samy O, Zeng S, Birowosuto M D and El Moutaouakil A 2021 A Review on MoS2 Properties, Synthesis, Sensing Applications and Challenges *Crystals* **11** 355

[11] Hijazi A and Moutaouakil A E 2019 Graphene and MoS2 Structures for THz Applications *2019 44th International Conference on Infrared, Millimeter, and Terahertz Waves (IRMMW-THz)* 2019 44th International Conference on Infrared, Millimeter, and Terahertz Waves (IRMMW-THz) (Paris, France: IEEE) pp 1–2

[12] Li Y, Li Z, Chi C, Shan H, Zheng L and Fang Z 2017 Plasmonics of 2D Nanomaterials: Properties and Applications *Adv. Sci.* **4** 1600430

[13] Mikhailov S A 2018 Graphene for Emission of Terahertz Radiation 3

[14] Moynihan E, Rost S, O'Connell E, Ramasse Q, Friedrich C and Bangert U 2020 Plasmons in $MoS_2$ studied via experimental and theoretical correlation of energy loss spectra *Journal of Microscopy* **279** 256–64

[15] Andersen K, Jacobsen K W and Thygesen K S 2014 Plasmons on the edge of MoS 2 nanostructures *Phys. Rev. B* **90** 161410

[16] Luo X, Qiu T, Lu W and Ni Z Plasmons in graphene: Recent progress and applications 84

[17] Dyakonov M and Shur M 1996 Detection, mixing, and frequency multiplication of terahertz radiation by two-dimensional electronic fluid *IEEE Trans. Electron Devices* **43** 380–7

[18] Deng X-Y, Deng X-H, Su F-H, Liu N-H and Liu J-T 2015 Broadband ultra-high transmission of terahertz radiation through monolayer $MoS_2$ *Journal of Applied Physics* **118** 224304

[19] Moutaouakil A E 2018 Two-Dimensional Electronic Materials for Terahertz Applications : Linking the Physical Properties with Engineering Expertise *2018 6th International Renewable*

*and Sustainable Energy Conference (IRSEC)* 2018 6th International Renewable and Sustainable Energy Conference (IRSEC) pp 1–4

[20] Moutaouakil A E, Suemitsu T, Otsuji T, Coquillat D and Knap W 2010 Room temperature terahertz detection in high-electron-mobility transistor structure using InAlAs/InGaAs/InP material systems *35th International Conference on Infrared, Millimeter, and Terahertz Waves* 2010 35th International Conference on Infrared, Millimeter, and Terahertz Waves (IRMMW-THz 2010) (Rome, Italy: IEEE) pp 1–2

[21] Moutaouakil A E, Komori T, Horiike K, Suemitsu T and Otsuji T 2010 Room Temperature Intense Terahertz Emission from a Dual Grating Gate Plasmon-Resonant Emitter Using InAlAs/InGaAs/InP Material Systems *IEICE Transactions on Electronics* **E93.C** 1286–9

[22] Moutaouakil A E, Suemitsu T, Otsuji T, Coquillat D and Knap W 2012 Nonresonant Detection of Terahertz Radiation in High-Electron-Mobility Transistor Structure Using InAlAs/InGaAs/InP Material Systems at Room Temperature *j nanosci nanotechnol* **12** 6737–40

[23] El Moutaouakil A, Suemitsu T, Otsuji T, Videlier H, Boubanga-Tombet S-A, Coquillat D and Knap W 2011 Device loading effect on nonresonant detection of terahertz radiation in dual grating gate plasmon-resonant structure using InGaP/InGaAs/GaAs material systems *Phys. Status Solidi (c)* **8** 346–8

[24] Tao Z H, Dong H M, Duan Y F, Huang F and Liu J L 2019 Four Terahertz Plasmon Modes of Hole Gas in Monolayer MoS2 *Plasmonics* **14** 839–43

[25] Reshetnyak V Yu, Zadorozhnii V I, Pinkevych I P, Bunning T J and Evans D R 2018 Surface plasmon absorption in $MoS_2$ and graphene-$MoS_2$ micro-gratings and the impact of a liquid crystal substrate *AIP Advances* **8** 045024

[26] Liu S, Ma R, Li Y, Zhao L, Xia Y, Dong X and Pang Y 2021 D-shaped surface plasmon resonance biosensor based on MoS 2 in terahertz band *Optical Fiber Technology* **66** 102631

[27] Popov V V, Bagaeva T Yu, Otsuji T and Ryzhii V 2010 Oblique terahertz plasmons in graphene nanoribbon arrays *Phys. Rev. B* **81** 073404

# Upper Subcell Impacts on Perovskite/u-CIGS Tandem Solar Cell Performance

**Nour Boukortt[1,*], Yaser M. Abdulraheem[2], Salvatore Patanè[3], Trupti Ranjan Lenka[4], Rabin Paul[4], Amal M. AlAmri[5], Malak Wafi[1], Mariam Ali Mohammad[1], Anfal Almuraished[1], Noor Alenezi[1], Latifa Alameeri[1], Zainab Alenezi[6], Wissem Nasri[7], and Baghdad Hadri[7]**

[1]Electronics and Communication Engineering Department, Kuwait College of Science and Technology, Road N. 7, Doha 13113, Kuwait
[2]Electrical Engineering Department, College of Engineering and Petroleum, Kuwait University, 13060 Kuwait, Kuwait
[3]Department of Mathematics and Computer Science, Physics and Earth Science (MIFT), University of Messina, Viale F. Stagno d'Alcontres, 31, 98166 Messina, Italy
[4]Electronics and Communication Engineering Department, National Institute of Technology Silchar, Assam, 788010, India
[5]Physics Department, College of Science and Art, King Abdulaziz University, POB 344, Rabigh, 21911, Saudi Arabia
[6]Ministry of Electricity and Water and Renewable Energy, Doha East Power Station for Booster and Distillation, Kuwait
[7]Electrical Engineering Department, University of Abd-Elhamid Ibn Badis, Bp227, 27000 Mostaganem, Algeria

*Corresponding author E-mail: n.boukortt@kcst.edu.kw

**Abstract**. In this paper, the two-terminal perovskite/u-CIGS tandem cells have been numerically performed and studied by using TCAD tools. The investigated planar perovskite model has been performed accordingly in order to match the fabricated structure and to be used as a top cell for tandem cells. The ultrathin CIGS solar cell has been also calibrated according to the fabricated model in order to be used as a good substrate (bottom cell) for the top cell (perovskite). These two cells have been simulated individually and they have revealed an efficiency of up to 20 % and 12.51 % for perovskite and u-CIGS models, respectively. The 2-terminal tandem model is then developed by comprising the top cell with the bottom cell by transparent contact that connects the two sub-cells in series electrically. The influence of changing the defect density in the perovskite film (absorber) and at ETL/absorber and absorber/HTL interfaces on tandem cell characteristics are analyzed. Moreover, we show that the power conversion efficiency of this investigated tandem model can reach over 30% by reducing defect densities and adjusting some sub-cell parameters that were already optimized in our previously published work. These approaches allow us to gain a better understanding of the physical dimensions and defect densities' effect on the performance of the tandem devices.

**Keywords:** Perovskite, u-CIGS, tandem cell, efficiency, modeling.

## 1. Introduction

In solar cell technology, different semiconductor materials with various bandgaps and band structures have been investigated to find a suitable candidate that can use sunlight efficiently taking into consideration achieving high power conversion efficiency (PCE), low-cost, and long lifetime [1-4]. According to Shockley-Queisser's single-junction limits, standalone solar cells are still suffering from major power losses such as band edge thermalization, radiative recombination, and transmission (inability to absorb below-bandgap photons) [5, 6]. In solar cells, when the photon energy is equal to or greater than the material bandgap, the electrons from the valence band (VB) rise to the conduction band (CB). Thereafter, the generated electron (hole) in the CB (VB) loses a large amount of its energy in the form of heat (phonon emission) to relax in the CB (VB) edges. Which is well-known as thermalization loss [3]. While the photon with an energy lower than the material bandgap, there is no chance for the photon to excite an electron within the absorber. In this case, the photon will not contribute to the carrier generation process. This loss mechanism is well-known as transmission loss [3]. Figure 1 illustrates the main energy loss mechanisms that can be identified when the device is under illumination. Furthermore, several approaches were employed to fabricate multiple junctions' solar cells in order to surpass the Shockley-Queisser limit [5, 6]. i.e., tandem solar cells [7-13]. Fortunately, tandem solar cell technology came as a good approach to reduce these losses and make a balance between the energy loss from thermalization and transmission in order to achieve high PCE [7-13].

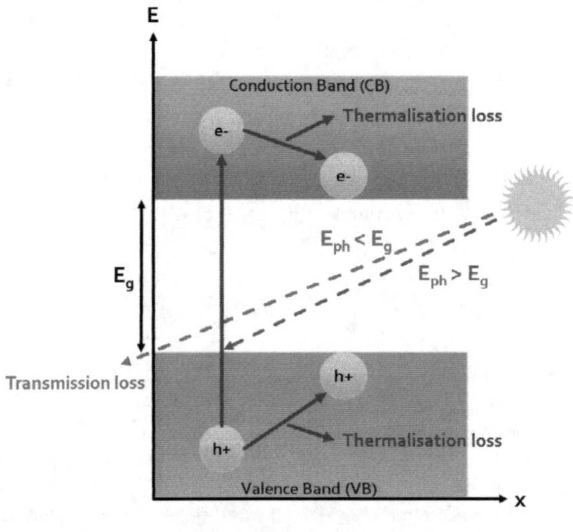


Figure 1. Main energy loss mechanisms can be identified when the device is under illumination.

Tandem cells utilize two or more absorbers each with different properties in order to divide the solar spectrum among them [7-13]. As a rule, one absorber with a large bandgap is going to be responsible to convert UV and blue light into electricity (absorbing lower wavelength photons), while the other with a small bandgap is assigned to convert the red and IR light into electricity (absorbing higher wavelength photons). As stated in the literature, a tandem solar cell can reach an efficiency higher than 30% PCE under the outdoor solar reference spectrum (e.g., AM1.5G) [14-16]. Different tandem configurations have been investigated in order to achieve high efficiencies such as all-perovskite [17-19], perovskite-silicon [20-24], and perovskite-CIGS [24-26] tandem structures, and many more. These combinations have seen massive improvements in recent years [14]. From the above configurations, perovskite materials are mostly used as top cells which exhibit high efficiency due to their unique advantages, large absorption coefficient, and tunable bandgap (1.4 to 1.8 eV), and potentially less production processes due to simpler and less expensive machinery and equipment [19-24]. Several challenges and approaches of scientists have shown many positive attributes in silicon-perovskite tandem solar cells with an efficiency of 32.5% [15], and the battle is still on for further improvement.

Perovskite solar cells (PSCs) have become a promising thin-film photovoltaic (PV) technology due to their unique advantages [27-29]. Perovskite materials generally have cubic structures with the formula $ABX_3$, where A stands for a monovalent cation (MA, FA, $K^+$, $Cs^+$, and $Rb^+$), B stands for a divalent metallic cation ($Sn_2^+$, $Pb_2^+$, and $Ge_2^+$), and X stands for a halogen ($Cl^-$, $I^-$, and $Br^-$). $MAPbI_3$ ($E_g \approx 1.5$ eV) is the first and most commonly used perovskite material for device fabrication at the lab scale using the spin-coating method due to its easy handling and low costs [28]. Poor thermal stability has been shown for $MAPbI_3$ compared to $FAPbI_3$ ($E_g \approx 1.4$ eV) which is metastable by exhibiting better stability at a high temperature (~150 °C). More specifically, a lack of structural stability has been shown for $FAPbI_3$ at ambient temperature [28]. This issue can be solved by modifying the composition of the perovskite using mixed organic cations (MA/FA) which are attracting interest due to their improved stability and capability to tune their properties by changing the composition [28, 29]. Triple cation-based perovskite solar cells use the mixture of Cs/MA/FA and double halide Br/I lead perovskites ($Cs(MAFA)Pb(IBr)_3$, $E_g \approx 1.6$ eV) which offer good moisture tolerance, thermal stability, and high efficiency under inert and ambient conditions [28-30].

In this work, the perovskite/u-CIGS tandem model was performed and optimized using TCAD tools. The influence of the bulk defect density in the absorber layers and interface defect density at heterojunction interfaces has been analyzed on model efficiency for different absorber materials such as $MAPbI_3$, $FAPbI_3$, and $Cs(MAFA)Pb(IBr)_3$. It has been found that they are crucial parameters that affect the performance of the solar cell. Strikingly, efficiencies over 30 % were obtained under moderate simulation conditions for perovskite/u-CIGS tandem cells. Finally, we provide the key challenge of tandem devices in the photovoltaic field.

## 2. Device structure and calibration

The tandem device structure is shown in Figure 2, with perovskite ($CS_{0.05}(MA_{0.10}FA_{0.90})_{0.95}Pb(I_{0.90}Br_{0.10})_3$, 1.6 eV) (top cell) and u-CIGS (bottom cell) with a fixed cell pitch and opening width of 2 μm and 200 nm, respectively. The tandem device has been performed using TCAD tools. The physical parameters of the structure were carefully selected to make the models as realistic as possible, and they have been tabulated in detail in our previous work [14, 31].

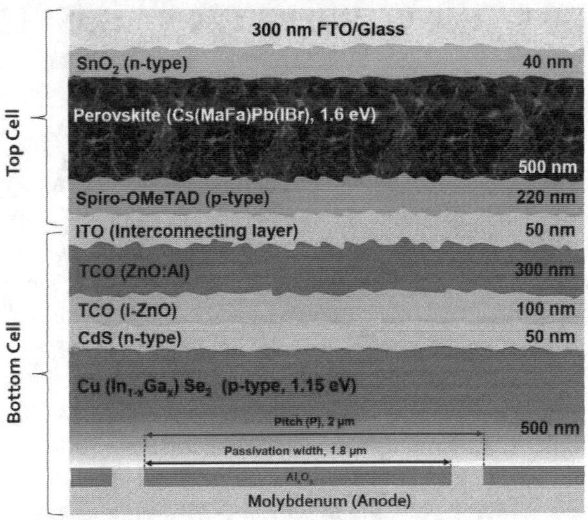


Figure 2. Schematic of monolithically integrated 2T tandem device.

Initially, the subcells were designed, examined, and calibrated individually as per experimental work [30, 32], i.e., with AM1.5G illumination. For the investigated tandem structure, we have proposed the following cell as a good top subcell which consists of $SnO_2$, Perovskite, and Spiro-OMeTAD as an electron extraction layer (ETL), an absorber layer, and a hole extraction layer (HTL), respectively [30]. In contrast, a single u-CIGS bottom sub-cell with a low bandgap absorber (1.15 eV) is utilized to absorb the filtered low-energy photons, full-area front buffer layer, and Molybdenum rear contact with fixed widths grid and full-area on passivation layer using $Al_2O_3$ to reduce the rear recombination [31-35]. Figure 3 illustrates J–V and EQE characteristics of the calibrated u-CIGS, perovskite, and perovskite/u-CIGS tandem models. The physical properties of each layer were taken from reported research articles in order to match the fabricated models [30-35]. Table 1 represents the electrical characteristic of the cell models concerned [30, 32].

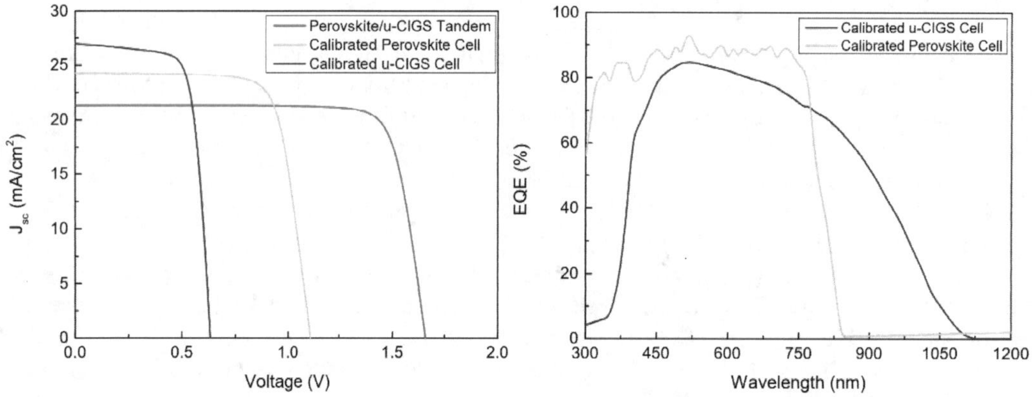


Figure 3. J–V curves and EQE at room temperature for the investigated solar cell models.

Table 1. Device characteristics at room temperature under the AM1.5G spectrum.

| Device | $J_{sc}$ (mA/cm²) | $V_{oc}$ (V) | FF (%) | η (%) |
|---|---|---|---|---|
| Fabricated perovskite cell [30] | 24.1 ± 0.1 | 1.11 ± 0.01 | 75 ± 1 | 20 ± 0.1 |
| Calibrated perovskite cell | 24.17 | 1.111 | 74.85 | 20.10 |
| Fabricated u-CIGS cell [32] | 26.79 | 0.661 | 71.54 | 12.68 |
| Calibrated u-CIGS cell | 26.97 | 0.632 | 73.36 | 12.51 |
| Perovskite/u-CIGS tandem cell (Initial) | 21.30 | 1.659 | 81.33 | 28.76 |

## 3. Results and discussions

Initially, we study three samples with different perovskite absorber materials ($MAPbI_3$, $FAPbI_3$, and $Cs(MAFA)Pb(IBr)_3$) and analyze their impact on tandem cell efficiency while varying the absorber defect density [28]. All-perovskite, perovskite-CIGS, and perovskite-silicon tandem configurations are considered a strategy that has been widely determined as an effective way to achieve high efficiency with a pathway to Low-Cost [1, 25]. After the top and bottom cells were simulated individually, a series connection is formed through a transparent conductive layer (ITO) forming a two-terminal tandem cell. A tunnel recombination junction (TRJ) could be used to ensure the current matching flow across the tandem device [25]. The current output ($J_{sc}$) of the total tandem device is limited by the sub-cell with the smallest current density [25, 26]. Therefore, the total open-circuit voltage ($V_{oc}$), is the sum of the voltages across the subcells. Apart from the absorber defect density effect, the influence of the defect density at heterojunction interfaces (ETL/absorber and absorber/HTL interfaces) of the top cell has also been analyzed. The studied models have a 2D device structure as illustrated in Figure 2. The numerical modeling was evaluated under standard illumination AM1.5G, 100 mW/cm$^2$, 300 K.

### 3.1. *Effect of the absorber defect density*

In solar cell devices, the generation of the charge carriers depends on the film morphology and crystalline quality [28]. Due to the ionic nature of perovskite materials, they can display different states of a defect [28, 36, 37]. Furthermore, the deep defect states in the bandgap of an absorber can crucially limit the cell performance by causing carrier trapping, non-radiative recombination, and hysteresis effect in the J–V curves of PSCs [36-43]. All these factors cause a significant impact on reducing the PCE of PSCs [28, 38]. The latest certified PCE of a standalone perovskite solar cell is 25.5% [42], even though the theoretical efficiency determined by considering the Shockley–Queisser limit is 30.5% [4-6]. This huge efficiency gap is caused by the presence of different types of carrier recombination mechanisms. In our work, the defect density in the perovskite film was adjusted by fitting the device J–V curves of the fabricated model. For the device in Figure 2, the perovskite material defect density is taken as 1 x $10^{15}$ $cm^{-3}$. The SRH recombination can be given as [44]:

$$R_{SRH} = \frac{pn - n_i^2}{\tau\left[n+p+\frac{2\,n_i\,cosh(E_r - E_t)}{KT_L}\right]} \tag{1}$$

$$\tau = \frac{1}{\sigma \times N_t \times V_{th}} \tag{2}$$

where $E_t$ is the trapped energy. $n_i$ is the intrinsic concentration. p and n are the numbers of free carriers (holes and electrons). $T_L$ is the lattice temperature. σ is the capture cross-section, τ is the lifetime, $v_{th}$ is the thermal velocity of carriers, and $N_t$ is the total defect density in the perovskite film.

The defect density in the absorber layers of PSCs is varied from 1 × $10^{12}$ $cm^{-3}$ to 1 × $10^{18}$ $cm^{-3}$ and the optimum $N_t$ value is found within the program to be 1 × $10^{15}$ $cm^{-3}$ for the investigated models (See Figure 4).

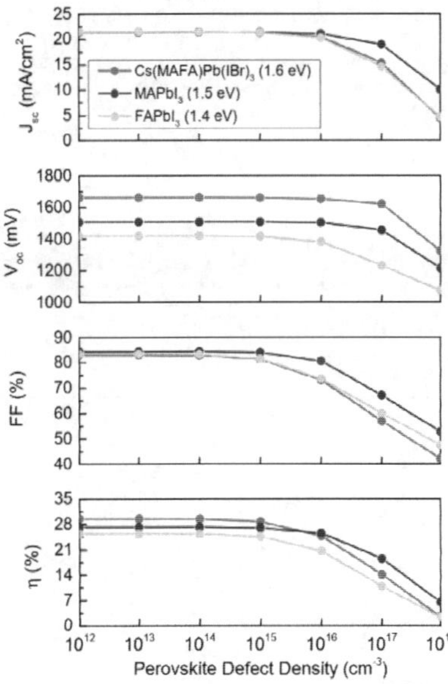


Figure 4. Defect density effects on tandem cell performances.

The defect density affects $J_{sc}$ as it is mostly constant at a fixed value of about 21 mA/cm² for tandem cells, and then decreases beyond the defect density value of $10^{16}$ $cm^{-3}$ for different absorber materials ($MAPbI_3$, $FAPbI_3$, and $Cs(MAFA)Pb(IBr)_3$). Referring to the figure, it can be observed that the $J_{sc}$ behaviour is identical for material $FAPbI_3$ and $Cs(MAFA)Pb(IBr)_3$). The defect density affects $V_{oc}$ as it saturates at a value of about 1111 mV while increasing the defect density and then a slight decrease beyond the $N_t$ value of $10^{17}$ $cm^{-3}$ for three investigated samples. Beyond a certain $N_t$ value ($10^{15}$ $cm^{-3}$) for a sample with $FAPbI_3$, it is observed that the $V_{oc}$ is strongly affected due to strong rate recombination which can affect the cell efficiency. $Cs(MAFA)Pb(IBr)_3$ has been extensively investigated by different researchers as an alternative material for the traditional $MAPbI_3$ seeking high carrier mobility and high device stability. FF and PCE are following the same trend, they start being affected from $10^{15}$ $cm^{-3}$. Increasing the series resistance between the cell's electrodes will cause a decrease in FF, which is observed in Figure 4, thus degradation in cell efficiency. An increase in $N_t$ causes cell degradation due to a negative impact of the recombination mechanism on generated carriers. The carrier diffusion length may increase by lowering the absorber's defect density, thereby enhancing solar cell device performance [36-41].

### 3.2. *Effect of the ETL/Absorber and Absorber/HTL interface defect density*

The energy barrier or poor charge transport capability (high defect and low carrier mobility) of the ETL or HTL causes charge accumulation at the interfaces ETL/absorber or absorber/HTL which limits the charge extraction and increases recombination [38-40]. These issues can reduce the anode/cathode contact charge collection efficiency [38-40]. Therefore, it may be possible to increase J–V hysteresis and light soaking effect in the PSCs [39]. In this part, we will demonstrate in detail how the interfaces affect the efficiency and stability of the investigated models, and at the same time, we will demonstrate how to create progress in interfacial design strategies. The investigated models are simulated to determine the impact of interface defect density ($D_{it}$) at ETL/absorber and absorber/HTL interfaces on the efficiency of tandem models. The $D_{it}$ was varied from $1 \times 10^8$ $cm^{-2}$ to $1 \times 10^{14}$ $cm^{-2}$ keeping other parameters constant. The optimum $D_{it}$ value is found to be $1 \times 10^{10}$ $cm^{-2}$ for the investigated samples (See Figure 5). The $D_{it}$ at the ETL/absorber interface shows a significant impact on $J_{sc}$ because the incident light passes through it to reach the absorber layer and a portion of the generated charges can be trapped and recombined at this interface [39]. The simulated PSC already matched the fabricated model with the optimal $D_{it}$ of $1 \times 10^{10}$ $cm^{-2}$. Further reduction in interface defect density is not expected to improve cell performance as illustrated in Figure 5.

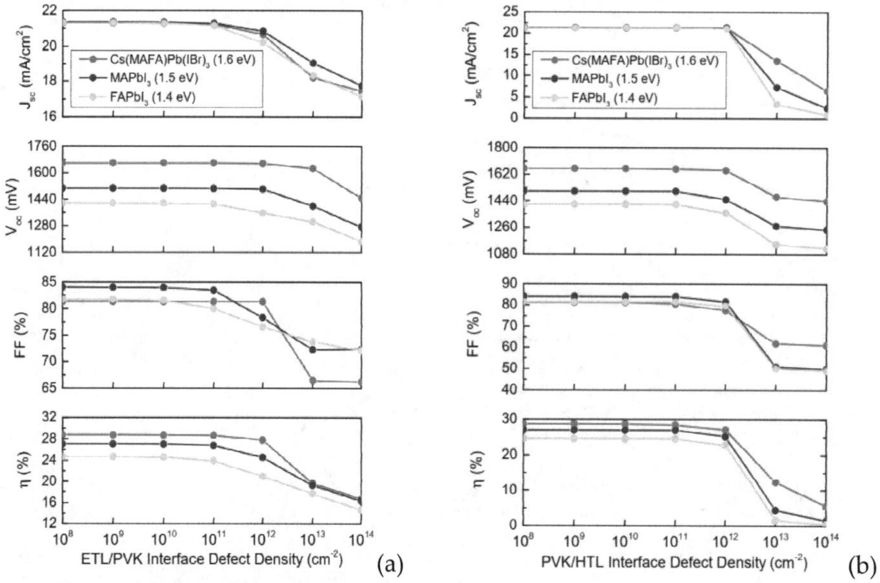


Figure 5. Effects of the interface defect densities on tandem cells performance (a) ETL/perovskite interface (b) perovskite/HTL interface.

The $D_{it}$ at the absorber/HTL interface shows also a significant impact on tandem cell [39]. Interestingly, efficiency becomes almost constant with the entire interface density range until $1 \times 10^{12}$ $cm^{-2}$ as illustrated in Figure 5. Using simulation results, we conclude that the defect density at absorber/HTL and ETL/absorber interfaces has a strong effect on both $J_{sc}$ and $V_{oc}$ [38-40]. Therefore, the interfaces and the interfacial materials need to be carefully engineered and developed to avoid all the problematic aspects [39, 43].

The two-terminal perovskite/u-CIGS tandem solar cell can reach high efficiency by improving the film morphology and crystalline quality, thus enhancing both charge carrier generation and transportation. Several approaches have been focusing on enlarging grain size and enhancing the crystalline quality of the perovskite films in order to reduce the recombination loss and decrease defect densities. The ultrathin CIGS solar cell configuration has been optimized successfully in our previous published work in terms of film thickness, film doping, a rear passivation layer, back contact material, fixed charge density, cell pitch, opening width, and others [33-35]. The perovskite solar cells were also successfully investigated and optimized previously by our research group [26]. Moreover, we have used some of these optimized quantities in the present tandem structure to achieve high efficiency of up to 30%. The optimized tandem structure and J–V curves are shown in Figure 6. Table 2 summarizes the photovoltaic performance of the investigated tandems.

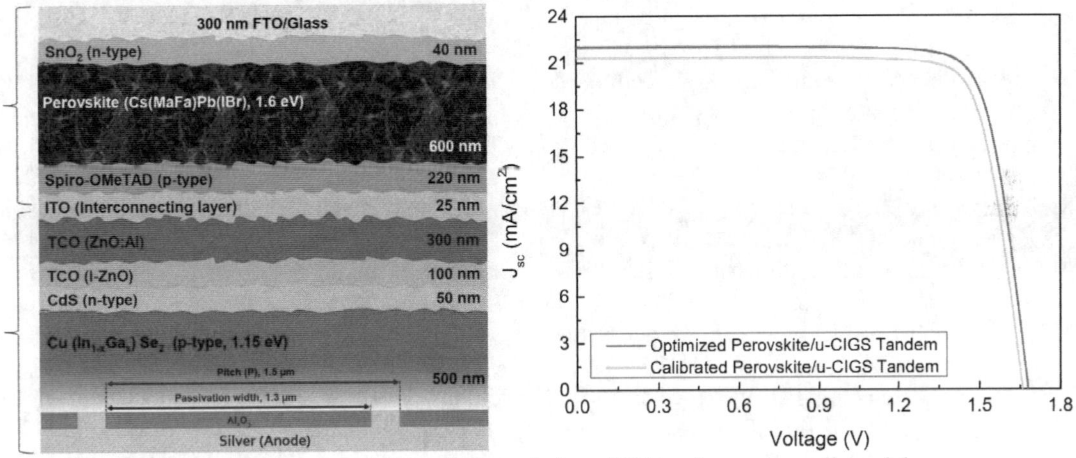


Figure 6. Optimized structure and J–V characteristics of 2T tandem solar cell models.

Table 2. Output characteristics of the optimized solar cell models.

| Cell Characteristics | $J_{sc}$ (mA/cm$^2$) | $V_{oc}$ (V) | FF (%) | η (%) |
|---|---|---|---|---|
| Perovskite/u-CIGS tandem cell (Initial) | 21.30 | 1.659 | 81.33 | 28.76 |
| Perovskite/u-CIGS tandem cell (Optimized) | 22 | 1.680 | 81.24 | 30.03 |

## 4. Conclusions

In this paper, the numerical simulation of perovskite/u-CIGS tandem models was performed using TCAD tools. The impact of the defect density in the perovskite films and at the ETL/perovskite and perovskite/HTL interfaces on tandem solar cells performance was deeply investigated under room temperature, AM1.5G spectrum. It has been analyzed that the defect density in a perovskite absorber has a significant impact on charge carrier generation in the investigated models when it is above $1 \times 10^{16}$ cm$^{-3}$. The defect density at the ETL/absorber and absorber/HTL interfaces with acceptor and donor defects has shown a stronger impact on device efficiencies when it is above $1 \times 10^{11}$ cm$^{-2}$ and $1 \times 10^{12}$ cm$^{-2}$, respectively. The conversion efficiency of 30% and higher can be reached by reducing defect densities in the bulk and at interfaces including the use of some optimized sub-cell parameters that were investigated in our previously published work. These approaches would be useful for further understanding and trends for the near future of ultrathin solar cell technologies.

**Acknowledgments**
This project was supported by the Kuwait Foundation for the Advancement of Sciences (KFAS) under project codes: CN1815EE01 and PN20-35EE-03.

**References**

[1] Gul M, Kotak Y, and Muneer T. Review on recent trend of solar photovoltaic technology [J]. Energy Exploration & Exploitation, 2016, 34(4), pp.485-526.
[2] Olaleru S A, Kirui J K, Wamwangi D, Roro K T, and Mwakikunga B. Perovskite solar cells: The new epoch in photovoltaics [J]. Solar Energy, 2020, 196, pp.295-309.
[3] Meddeb H, Götz-Köhler M, Neugebohrn N, Banik U, Osterthun N, Sergeev O, Berends D, Lattyak C, Gehrke K, and Vehse M. Tunable photovoltaics: adapting solar cell technologies to versatile applications [J]. Advanced Energy Materials, 2022, *12*(28), p.2200713.
[4] Green M A. Third generation photovoltaics: solar cells for 2020 and beyond [J]. Physica E., 2002, 14, 65–70.
[5] Shockley W, Queisser H J. Detailed balance limit of efficiency of p-n junction solar cells [J]. J. Appl. Phys., 1961, 32, 510–519.
[6] Boriskina S. V. and Chen G. Exceeding the solar cell Shockley–Queisser limit via thermal up-conversion of low-energy photons [J]. Optics Communications, 2014, 314, pp.71-78.
[7] Vos AD. Detailed balance limit of the efficiency of tandem solar cells [J]. J. Phys. D Appl. Phys., 1980, 13, 839–846.
[8] Li W, et al. High-performance solar flow battery powered by a perovskite/silicon tandem solar cell [J]. Nat. Mater., 2020, 19, 1326–1331.
[9] Xie M, et al. Two-dimensional BAs/InTe: A promising tandem solar cell with high power conversion efficiency [J]. ACS Appl. Mater. Interfaces, 2020, 12, 6074–6081.
[10] Al-Ashouri A, et al. Monolithic perovskite/silicon tandem solar cell with >29% efficiency by enhanced hole extraction [J]. *Science*., 2020, 370, 1300.
[11] Madan J, Shivani, Pandey R, Sharma R. Device simulation of 17.3% efficient lead-free all-perovskite tandem solar cell [J]. Sol. Energy., 2020, 197, 212–221.
[12] Pandey R, Singla A, Madan J, Sharma R, Chaujar R. Toward the design of monolithic 23.1% efficient hysteresis and moisture free perovskite/c-Si HJ tandem solar cell: A numerical simulation study [J]. J. Micromech. Microeng, 2019, 29, 064001.
[13] Boukortt N E I, Patanè S, Adouane M. Modeling and Investigation of Rear-Passivated Ultrathin CIGS Solar Cell [J]. Electronics, 2023, 12, 758.
[14] https://www.pv-magazine.com "KAUST claims 33.2% efficiency for perovskite/silicon tandem solar cell" APRIL 13, 2023 EMILIANO BELLINI
[15] https://www.pv-magazine.com "group of researchers from Helmholtz-Zentrum Berlin (HZB) has achieved a new world efficiency record for a silicon-perovskite tandem solar cell, with a certified efficiency of 32.5%" DECEMBER 20, 2022 ROLAND ERNST
[16] https://www.pv-magazine.com "Solliance hits 30.1% efficiency on perovskite/silicon tandem solar cell" SEPTEMBER 26, 2022 EMILIANO BELLINI
[17] Eperon G E, et al. Perovskite-perovskite tandem photovoltaics with optimized band gaps [J]. Science, 2016, 354, 861.
[18] Forgács D, et al. Efficient monolithic perovskite/perovskite tandem solar cells [J]. Adv. Energy Mater., 2017, 7, 1602121.
[19] Hossain MI, et al. Perovskite/perovskite planar tandem solar cells: A comprehensive guideline for reaching energy conversion efficiency beyond 30% [J]. Nano Energy, 2021, 79, 105400.
[20] Sahli F, et al. Fully textured monolithic perovskite/silicon tandem solar cells with 252% power conversion efficiency [J]. Nat. Mater. 2018, 17, 820–826.
[21] Nogay G, et al. 25.1%-Efficient monolithic perovskite/silicon tandem solar cell based on a p-type monocrystalline textured silicon wafer and high-temperature passivating contacts [J]. ACS Energy Lett., 2019, 4, 844–845.

[22] Chen B, et al. Blade-coated perovskites on textured silicon for 26%-efficient monolithic perovskite/silicon tandem solar cells [J]. Joule, 2020, 4, 850–864.
[23] Subbiah AS, et al. High-performance perovskite single-junction and textured perovskite/silicon tandem solar cells via slot-die-coating [J]. ACS Energy Lett., 2020, 5, 3034–3040.
[24] Gharibzadeh S, et al. 2D/3D heterostructure for semitransparent perovskite solar cells with engineered bandgap enables efficiencies exceeding 25% in four-terminal tandems with silicon and CIGS [J]. Adv. Func. Mater., 2020, 30, 1909919.
[25] Jacobsson T J, et al. 2-Terminal CIGS-perovskite tandem cells: A layer by layer exploration [J]. Solar Energy., 2020, 207, 270–288.
[26] Boukortt N E I, Patanè S, Loureiro, A G, AlAmri A M, Abdulraheem Y M, Lenka T R, Paul R and Abushattal A. Electrical and Optical Investigation of 2T–Perovskite/u-CIGS Tandem Solar Cells With~ 30% Efficiency [J]. IEEE Transactions on Electron Devices, 2022, *69*(7), pp.3798-3806.
[27] Eperon G E, Stranks S D, Menelaou C, Johnston M B, Herz L M, and Snaith H J. Formamidinium lead trihalide: a broadly tunable perovskite for efficient planar heterojunction solar cells [J]. Energy & Environmental Science, 2014, 7(3), pp.982-988.
[28] Wang F, Cao Y, Chen C, Chen Q, Wu X, Li X, Qin T, and Huang W. Materials toward the upscaling of perovskite solar cells: progress, challenges, and strategies [J]. Advanced Functional Materials, 2018, 28(52), p.1803753.
[29] Zhang L, Cui W, Zang Z, Tian F, Li X, and Qin G. Quantitative phase analysis on Cs-and Rb-doped FAPbI3 and corresponding solar cell efficiency simulations [J]. Solar Energy, 2019, 188, pp.224-229.
[30] Akin S. Hysteresis-free planar perovskite solar cells with a breakthrough efficiency of 22% and superior operational stability over 2000 h [J]. ACS applied materials & interfaces, 2019, 11(43), pp.39998-40005.
[31] Boukortt N E I, Patanè S, Hadri B, and Crupi G. Graded Bandgap Ultrathin CIGS Solar Cells [J]. Electronics, 2023, *12*(2), p.393.
[32] Lontchi J, Zhukova M, Kovacic M, Krc J, Chen W C, Edoff M, Bose S, Salomé P M, Goffard J, Cattoni A, et al. Optimization of Back Contact Grid Size in Al2O3-Rear-Passivated Ultrathin CIGS PV Cells by 2-D Simulations [J]. IEEE J. Photovolt., 2020, 10, 1908–1917.
[33] Boukortt N E I, Patanè S, and Abdulraheem Y M. Numerical investigation of CIGS thin-film solar cells [J]. Solar Energy, 2020, 204, pp.440-447.
[34] Boukortt N E I, AlAmri A M, Bouhjar F, and Bouhadiba K. Investigation and optimization of ultrathin Cu (In, Ga) Se 2 solar cells by using silvaco-TCAD tools [J]. Journal of Materials Science: Materials in Electronics, 2021, 32, pp.21525-21538.
[35] Boukortt N E I, Patanè S, Adouane M, and AlHammadi R. Numerical optimization of ultrathin CIGS solar cells with rear surface passivation [J]. Solar Energy, 2021, 220, pp.590-597.
[36] Minbashi M, and Yazdani E. Comprehensive study of anomalous hysteresis behavior in perovskite-based solar cells [J]. Scientific Reports, 2022, 12(1), p.14916.
[37] Bu T, Li J, Li H, Tian C, Su J, Tong G, Ono L K, Wang C, Lin Z, Chai N, et al. Lead Halide-Templated Crystallization of Methylamine-Free Perovskite for Efficient Photovoltaic Modules [J]. Science, 2021, 378, 1327–1332.
[38] Bi H, Liu B, He D, Bai L, Wang W, Zang Z, Chen J. Interfacial Defect Passivation and Stress Release by Multifunctional KPF6 Modification for Planar Perovskite Solar Cells with Enhanced Efficiency and Stability [J]. Chem. Eng. J., 2021, 418, 129375.
[39] Shao S and Loi M A. The role of the interfaces in perovskite solar cells [J]. Advanced Materials Interfaces, 2020, 7(1), p.1901469.
[40] Sun X, Li L, Shen S, Wang F. TiO2/SnO2 Bilayer Electron Transport Layer for High Efficiency Perovskite Solar Cells [J]. Nanomaterials, 2023, 13, 249.
[41] Zhu P, Gu S, Luo X, Gao Y, Li S, Zhu J, Tan H. Simultaneous Contact and Grain-Boundary Passivation in Planar Perovskite Solar Cells Using SnO2 -KCl Composite Electron Transport

Layer [J]. Adv. Energy Mater., 2019, 10, 1903083.
[42] Green M, Dunlop E, Hohl Ebinger J, Yoshita M, Kopidakis N, et al. Solar cell efficiency tables (version 57) [J]. Progress in photovoltaics: research and applications, 2021;29(1):3–15.
[43] Boukortt N E I, Alquennah A N, AlAmri A M, Patanè S, Lenka T R, and Paul R, Photogeneration losses from interface trap density in Passivated Ultrathin CIGS Solar Cell [C]. In IECON 2022–48th Annual Conference of the IEEE Industrial Electronics Society, 2022, pp. 1-6.
[44] ATLAS User's Manual Device Simulation Software. Silvaco International Ltd., Santa Clara, USA, 2020.

# Investigation of gamma induced effects on the properties of gamma irradiated Ce2S3 thin films

**S. Aldawood**[1]

[1]Department of Physics and Astronomy, College of Science, P.O. BOX 2455, King Saud University, Riyadh 11451, Saudi Arabia
Correspondence: sdawood@ksu.edu.sa

**Abstract**. Nanostructure cerium sulphide ($Ce_2S_3$) thin films were prepared using successive ionic layer adsorption and reaction (SILAR). The properties of the prepared samples were investigated as a function of gamma rays' doses of 0, 250, 500, and 1000Gy, respectively. The X-ray diffraction (XRD) results suggest an orthorhombic phase structure for $Ce_2S_3$ thin films and the crystallinity is enhanced with increasing gamma-ray doses. The irradiated thin films exhibit a variation in the energy band gap associated with the quantum confinement effect with larger grain size. This simple strategy of modifying properties of $Ce_2S_3$ thin films by the incident gamma rays can be an attractive way to investigate this material for dosimetry and radiation detection.

## 1. Introduction

Cerium is a most abundant element in the lanthanides series of elements and also called the rare earth elements. These elements have unique metal color pigments and unique luminescence characteristics [1-3]. Furthermore, the rare earth metal sulfide has attracted extraordinary attention due to the arrangement of sulfur with metal, which is predictable to generate surprising properties due to numerous oxidation states of sulfur [4]. Cerium sulfide ($Ce_2S_3$), is featured unusual properties due to remarkable thermal stability, electronic, and mechanical properties accompanied by superior electrical conductivity and optical characteristics [5-7]. It presents three type of structural phases, namely the orthorhombic α-$Ce_2S_3$, the metabolic β- $Ce_2S_3$ and the hexagonal γ- $Ce_2S_3$ with different structural, optical and electrical properties [8-10].

Several investigations have been directed to study the induced modification of high energy $\boldsymbol{\gamma}$ rays on the material's nanostructure, examine the phenomena of interaction of radiation with nanomaterials and explore the role of radiation dose. Not long ago, nanostructure thin films have been used in devices for detecting gamma rays. Irradiating thin films by high ionization energy radiation generates the structural imperfection or lattice defects in the form of deficiencies, as well as defect clusters forming due to the heat transformation and dislocation atoms near the interface. A frequent report has been published to detailly study the impacts of gamma ray irradiation on the physical characteristics of thin films [11-14]. In this study, the characterization and analysis of gamma induced modification of SILAR deposited $Ce_2S_3$ thin films have been comprehensively investigated. The main motivation of this research is to assess the degree of the radiation induced modification of the prepared $Ce_2S_3$ thin films properties in order to promote this material for radiation sensing and dosimatric applications. In order to do that, the structural, optical and electrical properties of $Ce_2S_3$ thin films have been experimentally measured at different gamma ray's doses.

## 2. Material and methods

### 2.1. Preparation of Cerium sulfide thin film

For the first time, glass substrates were used in a SILAR deposition process to deposit Ce2S3 thin films. The solutions concentration, PH values, immersion and rinsing times were optimized to deposited the homogeneous $Ce_2S_3$ thin film. All the chemicals were used of analytical grade (Sigma Aldrich). For the preparation of $Ce_2S_3$ thin films, the 0.025 M cesium nitrate Ce(NO3)2 cationic solution with pH~8 and

0.025 M and sodium sulfide (Na2S) anionic solution of pH~11 were used. A glass substrate was rinsed and then dipped into a Ce(NO3)2 solution for 2 minutes to produce cerium ions, which were then adsorbed to form Ce2S3 on the substrate. The cerium ion adsorbed substrate was washed by deionized water for 30 s to prevent unattached ions and impurities. In next step, same substrate was dipped in Na2S solution for 2 min, sulfide ions countered with the adsorbed cerium ions and washed again by deionized water for 30 s. For this case, one SILAR cycle of $Ce_2S_3$ thin film has been completed. To get the appropriate thickness of thin films 35 times SILAR cycles were repeated.

*2.2.* Gamma irradiation

The Cerium sulfide thin films were irradiated by a gamma ray source (Co-60), which has half-life of 5.2714 years. The source irradiation rate, which was used to monitor the received radiation dose, was found to be 927.299 Gy/h (Nordion, model GC-220). Three thin films of Cerium sulfide were exposed to different level of radiation doses: 250, 500, and 1000 Gy. The exposure time was used to determine the required dose.

*2.3.* Characterization

An X-ray diffractometer (Panalytical X'Pert3) equipped with a Cu-K of wavelength 1.5406 as the X-ray source was used to perform a structural analysis of both the prestine and gamma irradiated thin films. Gravimetric analysis was used to quantify the thickness of the samples. The samples' transmission and photoluminescence spectra were measured using a JASCO-V 670 UV-Vis spectrophotometer for the optical studies.

## 3. Experimental results and discussion

The XRD experimental results of the prestine and $\gamma$ rays exposed $Ce_2S_3$ samples with different gamma doses, ranged between (0-1000 Gy), is exhibited in Figure 1. Non-irradiated and gamma irradiated $Ce_2S_3$ samples are featured a crystalline structure with an orthorhombic phase, which presents a space group Pnam-62 [JCPDS No. JCPDS-21-0189]. The obtained 2θ values, which are 25.45, 35.97, 36.97, and 39.93° correspond to the (111), (051), (231) and (330), respectively. The intensity of XRD peaks reduced of all samples as the gamma rays' doses increased. This may be due to the reduction in the crystallinity of $Ce_2S_3$ structures in (111) direction. Moreover, slightly broadening of XRD main peak of irradiated samples have also been noticed as shown in the inset of Figure 1. The presented results show a good agreement with the reported effects obtained by Aldawood et al. [15].

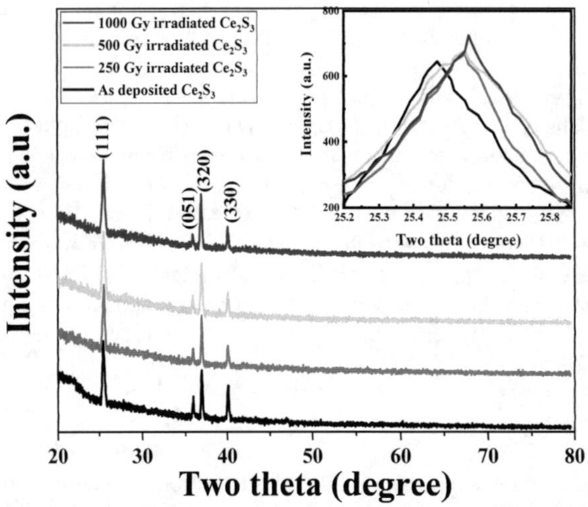


**Figure 1: (a) XRD patterns of un radiated and gamma exposed $Ce_2S_3$ samples.**

The structural parameters of $Ce_2S_3$ samples pre and post $\gamma$ rays exposed with variation of gamma ray doses value can be computed by (111) orientation using the Debye Scherrer formula (Eq. (1)) [16],

$$D = \frac{0.94\lambda}{\beta\cos\theta} \quad (1)$$

$$\varepsilon = \frac{\beta\cos\theta}{4} \quad \textbf{(2)}$$

The crystallite size reduced with increasing the gamma dose, as given in table 1. This may be due to the deep gamma rays penetration into the $Ce_2S_3$ leading to host thin films alteration by pushing the basis from their regular lattice and excruciating the molecules into small particles. In this phenomena, the gamma photons lose their energy that may generate or heal the defects in the structure [17]. This distortion in the crystal structure by gamma irradiation can be endorsed due to either the changes in lattice vacancies in host materials or the augmented in the reduction of the micro-strain in gamma exposed samples, which also be appeared in the peak broadening [18, 19].

The lattice specification values of the orthorhombic phase structure are measured by XRD diffraction planes using the following formula:

$$\frac{1}{d^2} = \frac{h^2}{a^2} + \frac{k^2}{b^2} + \frac{l^2}{c^2} \quad (3)$$

where the inter-planar spacing is represented by '*d*', the lattice parameters are represented by '*a, b* and *c*' and the Millar indices of the plane are '(*h, k, l*)'. The slightly deviation in lattice parameters of $Ce_2S_3$ thin films by means of gamma dose can be attributed to the modifications in lattice parameters. These changes can be due to the lattice vacancies alteration in the thin films, induced by gamma rays [20].

**Table 1: Nano-structural specification values of as prepared and $\gamma$ rays exposed $Ce_2S_3$ thin films.**

| Gamma –ray dose (Gy) | Lattice parameters (Å) | | | Crystallite size 'D' (nm) | Micro-strain 'ε' $10^{-3}$ |
|---|---|---|---|---|---|
| | '*a*' | '*b*' | '*c*' | | |
| 0 | 7. 6596 | 15.9838 | 4.2863 | 84.38 | 18.62 |
| 250 | 7. 6734 | 15.9637 | 4.2863 | 73.67 | 21.15 |
| 500 | 7.6388 | 15.9883 | 4.2863 | 57.54 | 29.68 |
| 1000 | 7.6835 | 15.9729 | 4.2863 | 53.36 | 35.75 |

Figure 2 (a) demonstrates the alteration in the proportion optical transmission in the range of 250–800 nm wavelength of the as prepared and $\gamma$ rays exposed $Ce_2S_3$ samples with numerous gamma dose (250-1000 Gy). It is experiential that the transmission in visible region has been reduced considerably with gamma ray exposure dose values. This could be due to the increases in the defects density with increasing the high power transformation to the host material structure, which provides the additional states of photon absorption in the thin films [14]. In the region of wavelength near the range up to 700 nm, the blue shift in the band edge is observed with increasing the gamma dose suggesting the decreased in the optical band gap.

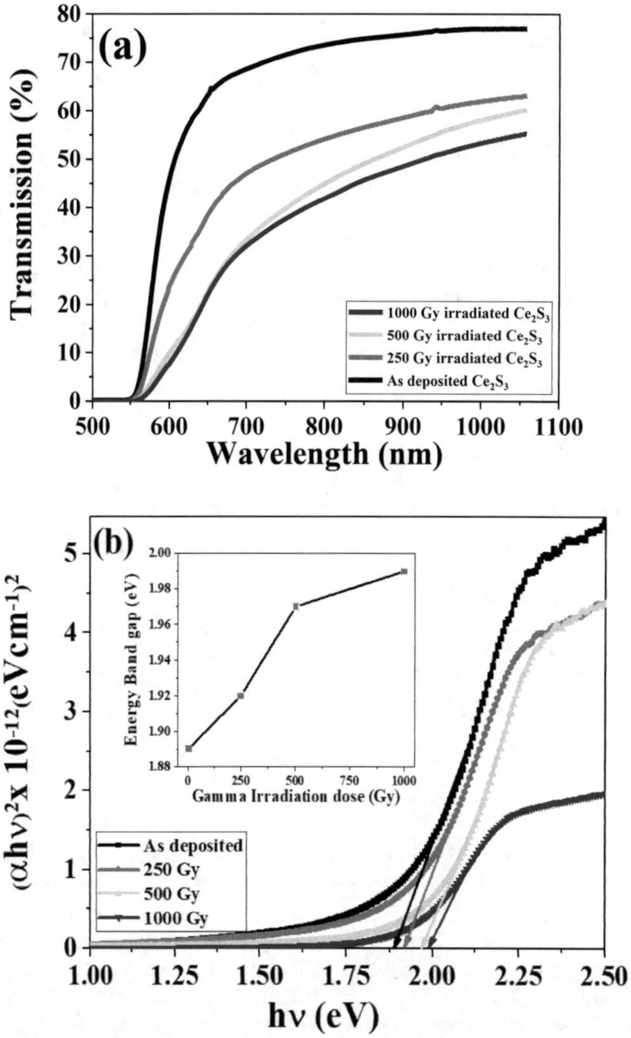


***Figure 2 (a) Transmittance spectra, (b) Tauc's plot for band gap calculation $Ce_2S_3$ thin films at different gamma irradiation doses and (inset) variation of band gap with gamma irradiation of $Ce_2S_3$ thin films.***

The optical energy band gap ($E_g$) is an imperious factor for different advanced applications of nano structural scale thin films. The UV–Vis transmission spectra has been used to compute the band gap of as prepared and gamma ray exposed $Ce_2S_3$ thin films. To analyze the band gap value ($E_g$), the optical absorption coefficient (α) values is obtained using formula 4 and 5, respectively [21]:

$$\alpha = \frac{1}{t}\ln\left(\frac{T}{100}\right) \tag{4}$$

$$(\alpha h\nu) = A(h\nu - E_g)^n \tag{5}$$

where 'T' represents the optical transmission, 't' is the width of the $Ce_2S_3$ film, 'ν' is the electromagnetic waves frequency, 'h' is the plank's constant, 'A' is an arbitrary constant, and '$n$ = ½ for direct electron

transition [22]. The expected optical band gap of the $Ce_2S_3$ films was calculated from the extrapolation of linear portion of the $(\alpha h\nu)^2$ vs. $(h\nu)$ graph to the horizontal axis as indicated in Figure 2(b).
The estimated optical band gap of $Ce_2S_3$ thin films before and after gamma exposure are calculated using the above equations. The increase in the optical band gap with increasing gamma ray dose from 0 to 1000 Gy is shown in a sub-figure inside Figure 2 (b). The deviations in the optical band gap were indorsed to radiation induced imperfections and induced traps concentration, that could change the absorption edge [22]. After gamma exposure, the traps concentration in conduction and valence bands of the thin films are dependent on the irradiation dose level. Hence, we can control that gamma ray modifications meaningfully and change the electrical properties of the interacting thin films [22]. Aldawood et al. described similar consequences for gamma exposed PbI thin films [15].

## 4. Conclusion

The method of sequential ionic layer adsorption and reaction (SILAR) was used to create Ce2S3 thin films. Structure and optical characteristics of the produced thin films were investigated as a function of gamma radiation with dosage variation. $Ce_2S_3$ crystallite size decreased from 84.39 to 53.36 nm as gamma dosage was increased, as seen by the XRD pattern. UV transmission spectra were used to determine the energy band gaps of both pre- and post-irradiated samples. As gamma doses were raised, a slight blue shift occur in the optical band gap, from 1.88 eV to 1.99 eV.

## References

[1] I. Kariper A 2014 *Prog. Nat. Sci. Mater. Int.*, **24 (6)** 663–670.
[2] Maqbool M, Ahmad I, Richardson H H, and Kordesch M E 2007 *Appl. Phys. Lett.* **91 (19)** 193511
[3] Maqbool M, Ahmad I, Richardson H H, and Kordesch M E 2007 *J. Mater. Sci. Technol.(Sofia),* **42** 5657–5660.
[4] Hui K, Fu J, Liu J, Chen Y, Gao X, Gao T, Wei Q, Zhang H and Tang M 2021 *Carbon Energy* **3** 709–720.
[5] Takeshita T, GschneidnerJr K A and Beaudry B J 1985 *J. Appl. Phys.* **57** 4633.
[6] Forster C M and White W B 1997 *J. Am. Ceram. Soc*, **80** 273.
[7] Forster C M and White W B 2006 *Mater. Res. Bull.* **41** 448
[8] Chopin T, Guichon H and Touret O. 1991 *E.P. Patent No. 545746.*
[9] Macaudière P 1995 *E.P. Patent No. 680930.*
[10] Babonas G, Dagys R, Pukinskas G and Vasil'eva I 1991 *Lituan. Phys. Coll.*, **31** 73.
[11] Aldawood S, Bannoob W M, AlGarawi M S, Alkhuraiji T S, Alashban Y, Shubayr N and Ali S M 2021 *J. Mater. Res. Technol.* **14** 108-120.
[12] Ali S M, AlGarawi M S, Aldawood S, Al Salman S A and AlGamdi S S 2020 *Rad. Phy. Chem.* **171** 108732.
[13] Ali S M, Ramay S M, Rehman N U, ALKhuraiji T S, Shar M A, Mahmood A, Hassan A and Riaz M 2019 *Mat. Sci. Semicon. Proc.* **93** 44-49.
[14] Ali S M 2017 *J. Mater. Sci: Mater. Electron.* **28** 16314
[15] Aldawood S, AlGarawi M S, Shar M A and Ali S M 2020 *J. of King Saud Uni. Sci.* **32** 2629-2634.
[16] Zinzuvadiya S, Pandya N C and Joshi U S 2019 *Thin Solid Films* **669** 525-530.
[17] Raneesh B, Saha A and Kalarikkal N 2013 *Radiat. Phys. Chem.* **89** 28–32.
[18] Dogra A, Singh M, Kumar R, 2003 *Nucl. Instrum. Method B,* **207** 296.
[19] B. Angadi, V.M. Jali, M.T. Lagare, N.S. Kini, A.M. Umarji, R. Kumar, S.K. Arora, D. Kanjilal, Nucl. Instrum. Method B, 187 (2002) 87.
[20] Tawfik A, Hamada I M and Hemeda O M 2017 *J. Magn. Magn. Mater.* **250** 77-82.
[21] Ejderha K, Yildirim N and Turat A 2010 *Superlattice. Microstr.* **47** 241.
[22] Souli M, Nefzi C, Seboui Z, Mejri A, Vidu R and Turki N K 2018 Mat. Sci. Semicon.Processing **83** 50–57.

# The Scalability of Third Generation Photovoltaics: Deposition Techniques and Modularity

**Mohamad Ayoub[1,*], Adnan Alashkar[2], Ayman Mdallal[1], Taleb Ibrahim[3], Mustafa Khamis[4] , Abdul Hai Alami[1,5]**

[1]Sustainable and Renewable Energy Engineering Department, University of Sharjah, 27272 Sharjah, United Arab Emirates
[2]PhD Program in Materials Science and Engineering, American University of Sharjah, Sharjah, 26666, United Arab Emirates
[3]Department of Chemical Engineering, American University of Sharjah, Sharjah, 26666, United Arab Emirates
[4]Department of Biology, Chemistry and Environmental Sciences, College of Arts and Sciences, American University of Sharjah, Sharjah, 26666, United Arab Emirates
[5]Center for Advanced Materials Research, Research Institute of Science and Engineering, University of Sharjah, 27272 Sharjah, United Arab Emirates

[*] Corresponding author's e-mail: mohamad.ayoub@sharjah.ac.ae

**Abstract**. The development of third-generation photovoltaic (TGPV) technologies promises to address some of the limitations of conventional silicon-based solar cells. However, the scalability of these technologies is a critical factor in their practical application. TGPV include Perovskite Solar Cells (PSC), Organic Solar Cells (OSC), and Dye-Sensitized Solar Cell (DSSC). This paper reviews and summarizes the recent trends and research on the deposition techniques and modularity of TGPV. Various deposition techniques such as slot-die coating, thermal evaporation, and spray pyrolysis are discussed, along with their advantages and limitations. Modularity, which allows for the integration of TGPV cells into different structures, is also examined as a critical factor in scalability. The paper concludes that the scalability of TGPV technologies depends on the development of efficient and cost-effective deposition techniques and modularity, which will facilitate the integration of the TGPV cells into various structures and enable the widespread use of these promising technologies.

**Keywords**: Third Generation Photovoltaics, Perovskite Solar Cells, Dye-Sensitized Solar Cells, Scalability, Deposition techniques.

## 1. Introduction

Third generation photovoltaics (TGPV) encompasses PV technologies that are manufactured through solution processing methods[1,2]. The fabrication of first (crystalline silicon) [3] and second (thin film) generation solar cells [4] is considered to energy intensive and bears adverse effects on the environment [5]. On the contrary, TGPV offer the advantages of the potential of very low capital and operation cost, ease of processing, the ability to scale up to large area fabrication and the use of flexible substrates.

Moreover, TGPV are electrochemical devices that operate and depend on the interactions between stacked layers, and are divided into three subcategories, which are perovskite solar cells (PSC) [6], organic solar cells (OSC) [7], and dye-sensitized solar cells (DSSC) [8].

DSSC (Grätzel cell) which was invented by O'Regan et al., [9] mainly consists of a bottom electrode, most commonly fluorine doped tin oxide (FTO), with a titanium dioxide ($TiO_2$) electron transport layer (ETL), a dye-sensitizer, a liquid or a solid electrolyte (as a hole transport layer HTL), and a counter electrode (CE), deposited in that order. OSC follow a similar pattern in terms of structure [10], however the ETL and the HTL are combined into a single blend to form a bulk-heterojunction (BHJ) and sandwiched between two electrodes. All while still having the option of utilizing passivating layers preceding or following the BHJ in order to further enhance performance metrics. Lastly, PSC, which were invented in 2009 by Kojima el al., [11], utilize a bottom electrode/ETL/Passivators/perovskite absorber/HTL/CE [12], with highly achieved efficiencies within the research and development scene and a very high potential for large scale applications.

There are plenty of techniques that are used for the fabrication of TGPV, some of which are spin coating (SC), slot-die coating (SDC), doctor blade coating (DBC), thermal evaporation (TE), electron beam evaporation (EBE), spray pyrolysis (SP), chemical bath deposition (CBD), etc. These techniques vary in defining limitations that affect the overall process of fabrication of TGPV such as cost, complexity, material wastage and scalability [13]. In this work, the different techniques that are used to fabricate TGPV, as well as their versatility in terms of scalability are comparatively presented.

## 2. Methodology

The data for this research was obtained from literature gathered from Scopus and Google Scholar databases, based on the information presented regarding the type of TGPV, the scalability of the modules/cells that are fabricated, and the performance parameters that were achieved. The search criteria used were based on the keywords and phrases that are commonly associated with third generation solar cells, including but not limited to “perovskite solar cells”, “organic solar cells”, “Dye-sensitized solar cells”, “spin coating”, “slot-die coating”, “doctor blade coating”. The search was mostly constrained to journal articles published between (2019-2023) due to the fact that the TGPV scene is rapidly developing, and recent literature provides a relevant insight on the current performance progress correlated to the utilized deposition technologies.

## 3. Results and Discussion

### 3.1. Perovskite Solar Cells (PSC)

PSC have been set on the track of rapid development since their induction into the TGPV scene in 2009, recently reaching a champion efficiency of 26.1% [14]. The scalability of TGPV is an aspect that should be tackled before the commercialization of this technology and its ascension to modularity (perovskite modules PM). Various deposition techniques have been utilized in that regard such as SDC, CBD, TE, SC (associated with moderate PSC areas), etc. Each fabrication technique defines the process of synthesizing the precursors of the different layers involved, as well as an optimization process for the fabrication parameters used. Throughout recent literature, scalability and fabrication of PSC has been put under the spotlight.

Kaleli et al. [15], fabricated a PM with an area of 36 $cm^2$, and an individual active cell area of 0.30 $cm^2$. The PM structure consisted of $FTO/TiO_2/MASnBr_3/P3HT/Ag$, electrode, ETL, perovskite absorber, HTL and CE layers, respectively. All layers were deposited via ultrasonic-spray pyrolysis, achieving a single cell power conversion efficiency (PCE) of 0.10%, an open circuit voltage ($V_{OC}$) of 0.37 V, a short circuit current density ($J_{SC}$) of 0.460 $mA/cm^2$ and a fill factor (FF) of 57.20%. Moreover, Fievez et al. [16] fabricated a 100 $cm^2$ PM with an active single cell area of 0.09 $cm^2$, which consisted of $FTO/SnO_2/TiO_2/CsFAPb(I_{0.88}Br_{0.12})_3$/Spiro OMeTAD/PTAA/Au, electrode, ETL, perovskite absorber, HTL and CE, respectively. The ETL and HTL were both deposited via SP/CBD and SP, respectively, whereas the perovskite absorber layer was deposited via slot-die/air knife coating and the gold CE via

TE. A PCE of 18% along with a 1.024 V $V_{OC}$, a 22.4 mA/cm$^2$ $J_{SC}$, and a 78% FF were achieved. Huang et al. [17] also fabricated a high performing PM via slot-die coating with an area of 16 cm$^2$ at a 10.34% PCE, a 5.73 V $V_{OC}$, a 2.85 mA/cm$^2$ $J_{SC}$ and a 63.34% FF. The fabricated PM consisted of FTO/NiO/$MAPbI_3$/PCBM/Ag electrode, HTL, perovskite absorber, ETL and CE, respectively. Both the HTL and the perovskite absorber were deposited via slot-die coating, the ETL via SC and the CE via TE. Bi et al. [18] achieved a 20.7% and a 16.5% PCE for PM with areas 0.2 cm$^2$ and 1 cm$^2$, respectively. The fabricated PM had a structure of FTO/$SnO_2$PbO/$SnO_2$/$MAPbI_3$/Spiro OMeTAD/Au, with the ETL and HTL being deposited via SP, the perovskite layer via DBC and the gold CE via thermal evaporation. EBE has also been utilized in upscaling PM, as Liu et al. [19] fabricated a 5 cm$^2$ PM that achieved 11.20% PCE along with a 1.14 V $V_{OC}$, a 13.54 mA/cm$^2$ $J_{SC}$, and a 72.52% FF, with a structure of ITO/NiO/$CsPbI_2Br$/$Nb_2O_5$/Ag, electrode, HTL, perovskite absorber, ETL and CE, respectively. All the layers were deposited via EBE, except for the perovskite absorber which was deposited via SC.
Adding to that SC is a prominent fabrication technique in the PSC and PM scene. For example, Qiu et al. [20] fabricated a 22.8 cm$^2$ PM with the structure of FTO/$SnO_2$/CsMAFAPb($I_{0.85}$ $Br_{0.15}$)/Spiro OMeTAD/Au, which achieved a 12.03% PCE, a 5.8 V $V_{OC}$, a 3.38 mA/cm$^{-2}$ and a 61.31% FF. Both the HTL and the perovskite absorber were deposited via SC, whereas the ETL was deposited via sputtering and the CE via thermal evaporation. Moreover, Liu et al. [21], was able to successfully fabricate a 10 cm$^2$ PM using spin coating, except for the CE which was deposited via TE. The fabricated structure consisted of FTO/$TiO_2$/$MAPbI_3$/Spiro OMeTAD/Ag, electrode, ETL, perovskite absorber/HTL and CE, and achieved a PCE of 15.14%, a 5.11 V $V_{OC}$, a 4.75 mA/cm$^2$ $J_{SC}$ and a 63% FF. Figure 1 shows a comparison between the PCE and module area for PSC based on the fabrication technique. Where slot-die coated modules maintain a relatively high efficiency with respect to the module area. Spin coated and slot-die coated modules show comparable performance with similar module areas, however scaling up limitations of spin coating quickly arise.

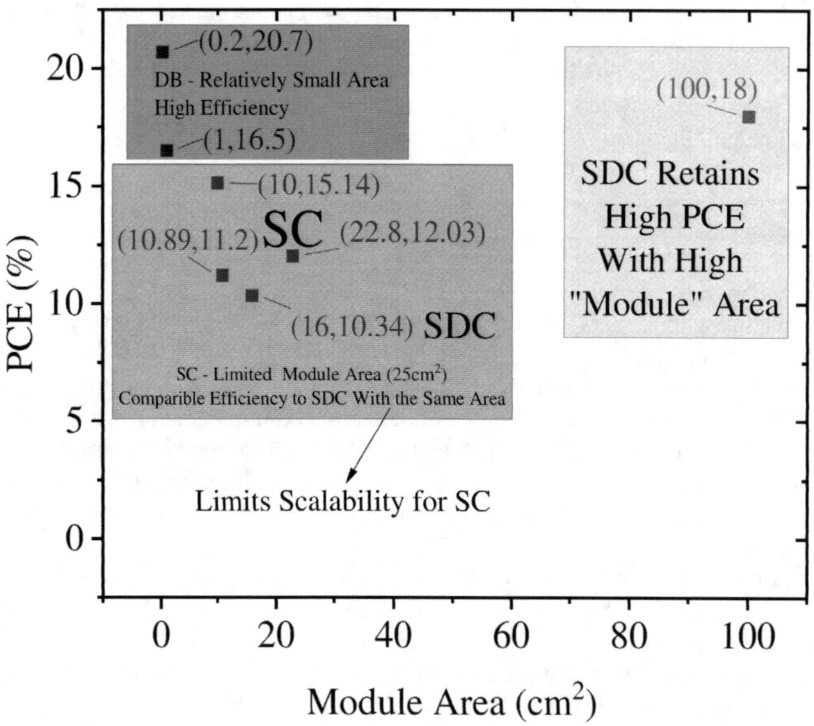


Figure 1. Perovskite Solar Cells – PCE vs Module Area for Different Fabrication Techniques

### 3.2. Organic Solar Cells (OSC)

OSC is a well-established TGPV technology that has been around since the 1980s [22]. Similar to succeeding technologies, such as PSC, OSC utilizes a structure consisting of a bottom electrode for charge collection, an electron and a hole transporting layer, and an active layer blend. An OSC blend (BHJ) has a unique structure of intermingling electron donor and acceptor materials, that create a network for efficient electron and hole transportation to adjacent ETL and/or HTL. Moreover, scalability has also been thoroughly studied for OSC by implementing flexible substrates, metal grid electrodes, as well as experimenting with various deposition techniques such as SDC, DBC, SC, transfer printing (TP), etc. The current champion OSC efficiency stands at 18.2% [23], and with large area OSC catching up, it has the potential to break through its commercialization stage if stability and degradation issues are overcome [24].

Although SC is limited to smaller areas, treatments that are facilitated by SC such as anti-solvent treatment compensate the devices with a high efficiency result due to a well-controlled crystallization of the layer that is being deposited. For instance, Jia et al., [25] fabricated an OSC with the structure ITO/ZnO(ETL)/PM6:Y6:ITIC:$PC_{71}BM$ (Active)/MoO(HTL)/Ag, where the active and ETL layers were deposited via SC, while the HTL and the silver CE were deposited via TE. The device achieved an efficiency of 13.25% with an area of 19.34 $cm^{-2}$, with a $V_{OC}$ of 6.06 V, a $J_{SC}$ of 3.07 mA/$cm^{-2}$ and a FF of 66.45%. Additionally, fabricating larger area modules is also viable using SC with the aid of TP. For example, Jahandar et al., [26] fabricated a 50 $cm^2$ OSC module using spin coating, with the structure ITO/CPEIE (ETL)/PBDB-T-2F:Y6:$PC_{70}BM$ (1:1:0.2)/MoO (HTL)/Ag, where the ETL (vacuum dried) and active layer were deposited via spin coating, and the HTL along with the silver CE via TE. They achieved an efficiency of 13.12%, a $V_{OC}$ of 8.74 V, a current density of 2.6 mA.$cm^{-2}$, and a FF of 58%. Additionally, and using the SC/TP method, Qin et al. [27] fabricated a 54 $cm^2$ OSC module with the structure of PET(Ag-grid)/AgNWs:PEI-Zn (ETL)/PBDB-T-2F:Y6:$PC_{71}BM$ (Active)/$MoO_3$ (HTL)/Ag, where the PET(Ag-grid) was deposited via DBC, the ETL and active layer via SC and both the HTL and Ag via TE. The fabricated module had a relatively good performance as the previous example reaching a PCE of 13.20%, a $V_{OC}$ of 7.34V, a $J_{SC}$ 2.57 mA.$cm^{-2}$, and a FF 70%. However, upscaling conventional active layer blends solely using SC resulted in relatively poor performances. As reported by Kang et al. [28], a 1.6 $cm^2$-OSC of the structure ITO/ZnO (ETL)/P3HT:PCBM (conventional OSC active blend)/$MoO_3$ (HTL)/Al, where the ETL and the active layer were solely deposited via SC and the HTL along with the Al CE were deposited via TE, had a PCE of 1.40%, a $V_{OC}$ of 0.597 V, a $J_{SC}$ of 7.6 mA.$cm^2$ and a FF of 30.3%. Figure 2 shows a comparison between the PCE and module area for OSC based on the fabrication technique. It can be seen that utilizing spin coating solely along with conventional OSC materials shows rather poor results on both performance and area spectrums. However, coupling spin coating with techniques such as transfer printing can elevate the modularity as well as performance drastically.

Roll-to-roll technologies have also been investigated for the deposition of OSC. Han et al. [29] reported a device of the structure ITO/ZnO (ETL)/AMD2:ITI-Th (Active)/HTL Solar/Ag, where all the layers were deposited via SDC, except for the Ag CE, which was deposited via BC. The 30 $cm^2$ module had a PCE of 8%, a $V_{OC}$ of 0.789 V, a $J_{SC}$ of 14.8 mA.$cm^{-2}$ and a FF of 66.4%. Moreover, DBC was also investigated by Fo et al. [30], through its utilization in a 1.2 $cm^2$ module implementing a conventional active layer blend with the structure of ITO/A-ZO (ETL)/PM6:Y6 (Active)/$MoO_3$ (HTL)/Al. The ETL was deposited via SC, the active layer via DBC and both the HTL and the CE via TE. The device achieved a relatively good efficiency at 11.69%, a $V_{OC}$ of 0.82 V, a high $J_{SC}$ of 21.68 mA.$cm^{-2}$ and a FF of 65.78%.

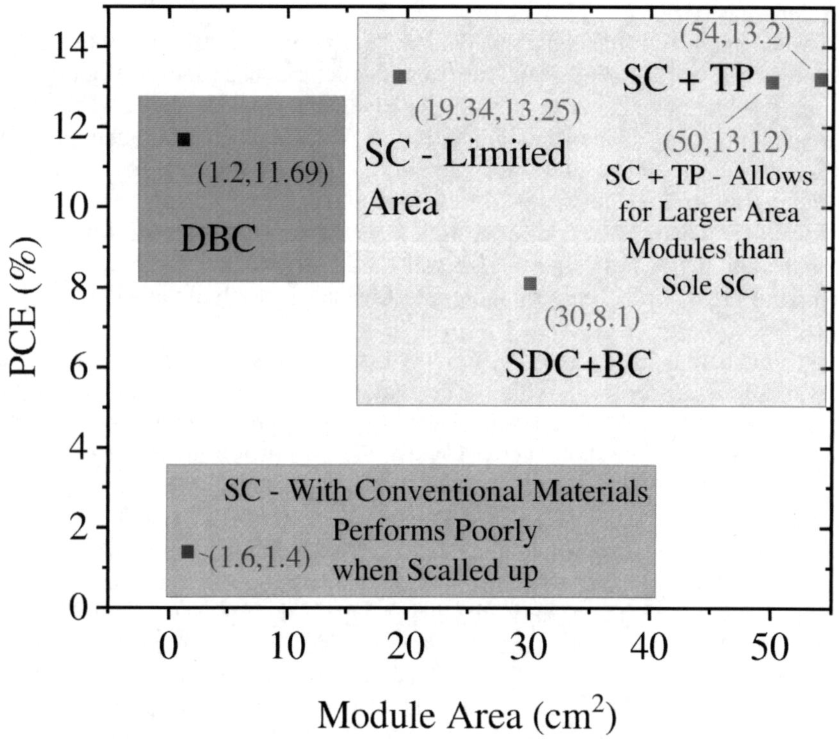


Figure 2. Organic Solar Cells – PCE vs Module Area for Different Fabrication Techniques

### 3.3. Dye-Sensitized Solar Cells (DSSC)

DSSC have been extensively studied as their performances depend greatly on each component of the cell. Hence, work has been carried to optimize each element of DSSC from the photoanode to the counter electrode including the electrolyte and dye [31]. The electrolyte utilized in the DSSC plays a vital role in its performance as it is responsible for the charge carrier transport between the photoanode and the counter electrode, in addition to the regeneration of the dye. Generally, three categories of electrolytes are present: liquid electrolytes, quasi-solid-state electrolytes, and solid-state electrolytes. In terms of performance, DSSC based on liquid electrolytes have attained the highest efficiencies, with a record efficiency of 14.3%. [32]. Hence, large modules of DSSC mainly consist of cells with liquid electrolytes [33]. A summary of performance vs module area which accounts for various deposition techniques and electrolyte injection considerations can be seen in Figure 3.

Novel natural dyes such as Calotropis and electrolyte composition are the center of recent research, especially electrolytes consisting of copper or cobalt redox shuttles and iodine-free electrolytes [34,35]. For instance, a DSSC with a conversion efficiency of 13% was achieved by using Co+2/+3 redox shuttle and porphyrin dye [36]. With the aid of co-sensitization, a combination of alkoxysilyl-anchor dye and carboxy-anchor based organic dye integrated with [Co(phen)2]2+/3+ redox couple resulted in a DSSC with a record efficiency of 14.3% [37]. Similarly, Cu-based electrolytes have shown great characteristics as potential electrolytes for DSSC especially in the terms of open voltage circuit, where voltages around 1.0 V are obtained [33]. A DSSC with an efficiency of 11.3% was achieved by employing a Cu-based redox mediator and a co-sensitized dye [38]. The main advantage of Co-based and Cu-based electrolytes is potential for scalability, as these electrolytes do not undergo severe leakage as compared to organic and iodine-based electrolytes.

Electrolytes in lab scale DSSC are traditionally introduced through the vacuum back filling method or the electrolyte injection filling method [39,40]. In both methods, a drop of electrolyte is dropped through a hole drilled on the counter electrode and photoanode. The electrodes are attached together through the use of thermoplastic, and a cell channel is created. The advantages of both methods are the potential for scalability, where they have been utilized for the production of large modulus. For instance, DSSC modules with an area of 5x5 $cm^2$ was assembled using the vacuum back filling method [41]. The modules with liquid electrolyte showed a PCE of 2.08%, while modules with polymer electrolytes recorded efficiencies around 1.6%. Through the use of electrolyte injection filling method, a DSSC module of size 30x30 $cm^2$ achieved an efficiency of 3.5% [42]. Another module of the same size also recorded an efficiency of 3.1%. Moreover, a smaller DSSC module of 10x10 $cm^2$ attained a PCE of 4.3%. Although extremely popular, traditional electrolyte filling methods suffer from certain disadvantages such as leaking, glass breaking during the drilling process, and uneven distribution of the electrolyte. In order to mitigate these issues, methods such as screen printing or inkjet printing are utilized as they contribute to the homogenous electrolyte composition in the device. Printing methods aid to eliminate the leakage of electrolyte, as the viscosity is optimized to successfully apply the methods. For example, a DSSC module achieved a PCE of 4.39% by employing screen printing [43]. Printing methods also show a great advantage in terms of the cell resistance, as the elimination of the drilled holes reduced the overall cell resistances. In addition, these methods allow for the homogenous distribution of the electrolyte and motivate the investigation various methods such as slot die and blade coating. Flexible DSSC can also be produced by the utilization of roll-to-roll methods.

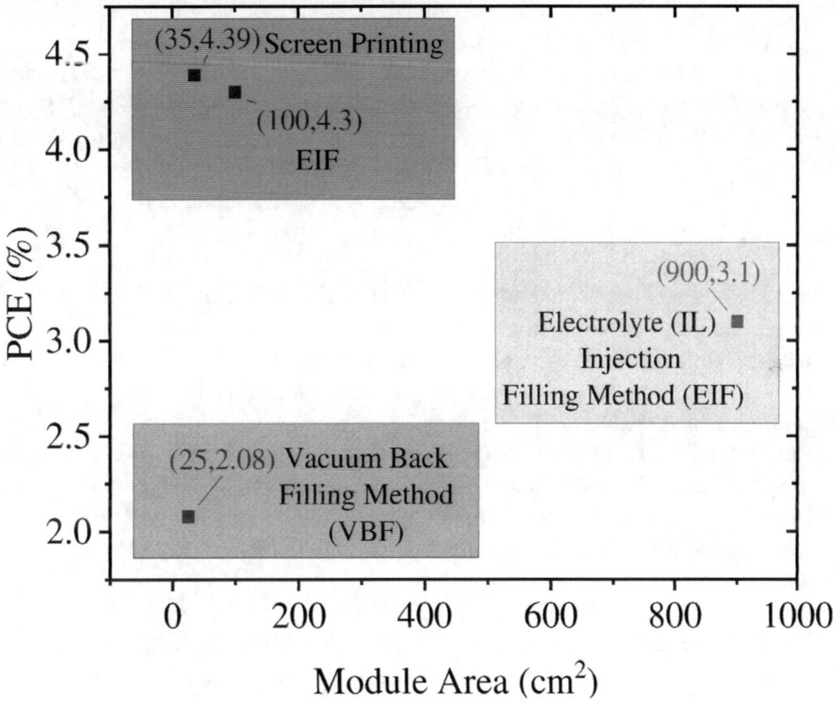


Figure 3. Dye-Sensitized Solar Cells – PCE vs Module Area for Different Fabrication Techniques

## 4. Conclusion

In summary, the scalability of third generation solar cells (TGPV) has been the subject of extensive research and optimization efforts. Various techniques, including slot-die coating (SDC), spin-coating (SC), and doctor blade coating (DBC), have been employed to achieve a balance between large module area and acceptable efficiency levels. Efficiencies reaching 18% for 100 $cm^2$ cells, 13.2% for 50 $cm^2$ cells, and 3.1% for 900 $cm^2$ cells, were achieved for perovskite (PSC), organic (OSC) and dye-sensitized (DSSC) solar cells, respectively. However, scaling up different TGPV technologies presents numerous challenges, including cost and toxicity issues that hinder the commercialization of PSC, degradation problems affecting OSC, and the incompatibility of high-performance liquid electrolyte DSSC with roll-to-roll techniques. As a result, further research and development are required to address these challenges and ensure a seamless transition towards commercialization. Only then can the full potential of third generation solar cells be realized, bringing us one step closer to a more sustainable future.

**References**

[1] Yan J and Saunders B R 2014 Third-generation solar cells: a review and comparison of polymer:fullerene, hybrid polymer and perovskite solar cells *RSC Adv.* **4** 43286–314
[2] Alashkar A, Ibrahim T, Khamis M and Hai Alami A 2022 Passivation of Perovskite Films Using Ionic Liquids *2022 Advances in Science and Engineering Technology International Conferences (ASET)* (IEEE) pp 1–4
[3] Alami A H, Alasad S, Aljaghoub H, Ayoub M, Alashkar A, Mdallal A and Hasan R 2023 Manufacturing of Silicon Solar Cells and Modules pp 45–63
[4] Alami A H, Alasad S, Aljaghoub H, Ayoub M, Alashkar A, Mdallal A and Hasan R 2023 Second-Generation Photovoltaics: Thin-Film Technologies pp 65–75
[5] Alami A H, Alasad S, Aljaghoub H, Ayoub M, Alashkar A, Mdallal A and Hasan R 2023 Silicon Feedstock and Ultra-Refinement pp 21–32
[6] Alami A H, Alasad S, Aljaghoub H, Ayoub M, Alashkar A, Mdallal A and Hasan R 2023 Third-generation Photovoltaics: Perovskite Solar Cells (PSC) pp 105–13
[7] Alami A H, Alasad S, Aljaghoub H, Ayoub M, Alashkar A, Mdallal A and Hasan R 2023 Third-Generation Photovoltaics: Organic Photovoltaics (OPV) pp 93–104
[8] Alami A H, Alasad S, Aljaghoub H, Ayoub M, Alashkar A, Mdallal A and Hasan R 2023 Third-Generation Photovoltaics: Dye-Sensitized Solar Cells (DSSC) pp 77–91
[9] O'Regan B and Grätzel M 1991 A low-cost, high-efficiency solar cell based on dye-sensitized colloidal TiO2 films *Nature* **353** 737–40
[10] Hoppe H and Sariciftci N S 2004 Organic solar cells: An overview *J Mater Res* **19** 1924–45
[11] Kojima A, Teshima K, Shirai Y and Miyasaka T 2009 Organometal Halide Perovskites as Visible-Light Sensitizers for Photovoltaic Cells *J Am Chem Soc* **131** 6050–1
[12] Alashkar A, Ayoub M, Ibrahim T, Khamis M, Nancarrow P, Alami A H and Tabet N 2023 Bulk passivation of perovskite films utilizing halide anion ionic liquids *International Journal of Thermofluids* **20** 100404
[13] Anon Thin Film Coating Solution-Processing Techniques Compared | Ossila
[14] Anon Best Research-Cell Efficiency Chart
[15] Kaleli M, Şen E, Lapa H E and Aldemir D A 2022 The production route of the MASnBr3 based perovskite solar cells with fully ultrasonic spray pyrolysis method *Physica B Condens Matter* **645** 414293
[16] Fievez M, Singh Rana P J, Koh T M, Manceau M, Lew J H, Jamaludin N F, Ghosh B, Bruno A, Cros S, Berson S, Mhaisalkar S G and Leong W L 2021 Slot-die coated methylammonium-free perovskite solar cells with 18% efficiency *Solar Energy Materials and Solar Cells* **230** 111189
[17] Huang S, Guan C, Lee P, Huang H, Li C, Huang Y and Su W 2020 Toward All Slot-Die Fabricated High Efficiency Large Area Perovskite Solar Cell Using Rapid Near Infrared Heating in Ambient Air *Adv Energy Mater* **10** 2001567
[18] Bi Z, Xu X, Chen X, Zhu Y, Liu C, Yu H, Zheng Y, Troshin P A, Guerrero A and Xu G 2022 High-performance large-area blade-coated perovskite solar cells with low ohmic loss for low lighting indoor applications *Chemical Engineering Journal* **446** 137164
[19] Liu X, Xiao Y, Zeng Q, Jiang J and Li Y 2019 Large-Area Organic-Free Perovskite Solar Cells with High Thermal Stability *J Phys Chem Lett* **10** 6382–8
[20] Qiu L, Liu Z, Ono L K, Jiang Y, Son D, Hawash Z, He S and Qi Y 2019 Scalable Fabrication of Stable High Efficiency Perovskite Solar Cells and Modules Utilizing Room Temperature Sputtered $SnO_2$ Electron Transport Layer *Adv Funct Mater* **29** 1806779

[21] Liu Q, Zhao Y, Ma Y, Sun X, Ge W, Fang Z, Bai H, Tian Q, Fan B and Zhang T 2019 A mixed solvent for rapid fabrication of large-area methylammonium lead iodide layers by one-step coating at room temperature *J Mater Chem A Mater* **7** 18275–84
[22] BERNÈDE J C 2008 ORGANIC PHOTOVOLTAIC CELLS: HISTORY, PRINCIPLE AND TECHNIQUES *Journal of the Chilean Chemical Society* **53**
[23] Anon Best Research-Cell Efficiency Chart | Photovoltaic Research | NREL
[24] Xu X, Li D, Yuan J, Zhou Y and Zou Y 2021 Recent advances in stability of organic solar cells *EnergyChem* **3** 100046
[25] Jia Z, Chen Z, Chen X, Yao J, Yan B, Sheng R, Zhu H and Yang Y (Michael) 2021 19.34 $cm^2$ large-area quaternary organic photovoltaic module with 12.36% certified efficiency *Photonics Res* **9** 324
[26] Jahandar M, Prasetio A, Lee C, Kim H, Kim A R, Heo J, Kim Y, Kim S and Lim D C 2022 Highly efficient flexible organic photovoltaic modules for sustainable energy harvesting under low-light condition via suppressing voltage-drop by metal-mediated cross-linkable polymer interfacial layer *Chemical Engineering Journal* **448** 137555
[27] Qin F, Sun L, Chen H, Liu Y, Lu X, Wang W, Liu T, Dong X, Jiang P, Jiang Y, Wang L and Zhou Y 2021 54 $cm^2$ Large-Area Flexible Organic Solar Modules with Efficiency Above 13% *Advanced Materials* **33** 2103017
[28] Kang M H, Cheon T and Kim H 2021 Fully vacuum-free large-area organic solar cell fabrication from polymer top electrode *Solid State Electron* **186** 108192
[29] Han Y W, Lee H S and Moon D K 2021 Printable and Semitransparent Nonfullerene Organic Solar Modules over 30 $cm^2$ Introducing an Energy-Level Controllable Hole Transport Layer *ACS Appl Mater Interfaces* **13** 19085–98
[30] Fo W, Xu G Y, Dong H, Liu L, Li Y W and Ding L 2021 Highly Efficient Binary Solvent Additive-Processed Organic Solar Cells by the Blade-Coating Method *Macromol Chem Phys* **222** 2100062
[31] Alashkar A, Ibrahim T, Khamis M and Alami A H 2022 Electrolytes, Dyes, and Perovskite Materials in Third Generation Photovoltaic Cells *Encyclopedia of Smart Materials* (Elsevier) pp 621–34
[32] Kokkonen M, Talebi P, Zhou J, Asgari S, Soomro S A, Elsehrawy F, Halme J, Ahmad S, Hagfeldt A and Hashmi S G 2021 Advanced research trends in dye-sensitized solar cells *J Mater Chem A Mater* **9** 10527–45
[33] Iftikhar H, Sonai G G, Hashmi S G, Nogueira A F and Lund P D 2019 Progress on Electrolytes Development in Dye-Sensitized Solar Cells *Materials* **12** 1998
[34] Alami A, Alachkar A, Alasad S, Alawadhi M and Zhang D 2021 Investigating Calotropis Procera natural dye extracts and PEDOT : PSS hole transport material for dye-sensitized solar cells *Agronomy Research* **19** 333–43
[35] Alashkar A, Alasad S, Aljaghoub H, Ayoub M, Francois T, El-Keblawy A, Zhang D and Alami A H 2022 Comprehensive assessment of the Calotropis procera natural dye extracts with weather effects for photovoltaic solar cell manufacturing *Int J Energy Res* **46** 17295–307
[36] Mathew S, Yella A, Gao P, Humphry-Baker R, Curchod B F E, Ashari-Astani N, Tavernelli I, Rothlisberger U, Nazeeruddin Md K and Grätzel M 2014 Dye-sensitized solar cells with 13% efficiency achieved through the molecular engineering of porphyrin sensitizers *Nat Chem* **6** 242–7
[37] Kakiage K, Aoyama Y, Yano T, Oya K, Fujisawa J and Hanaya M 2015 Highly-efficient dye-sensitized solar cells with collaborative sensitization by silyl-anchor and carboxy-anchor dyes *Chemical Communications* **51** 15894–7
[38] Freitag M, Teuscher J, Saygili Y, Zhang X, Giordano F, Liska P, Hua J, Zakeeruddin S M, Moser J-E, Grätzel M and Hagfeldt A 2017 Dye-sensitized solar cells for efficient power generation under ambient lighting *Nat Photonics* **11** 372–8

Journal of Physics: Conference Series **2751** (2024) 012018 doi:10.1088/1742-6596/2751/1/012018

[39] Ito S, Murakami T N, Comte P, Liska P, Grätzel C, Nazeeruddin M K and Grätzel M 2008 Fabrication of thin film dye sensitized solar cells with solar to electric power conversion efficiency over 10% *Thin Solid Films* **516** 4613–9

[40] Tsekouras G, Mozer A J and Wallace G G 2008 Enhanced Performance of Dye Sensitized Solar Cells Utilizing Platinum Electrodeposit Counter Electrodes *J Electrochem Soc* **155** K124

[41] Hirata M K, Freitas J N, Santos T E A, Mammana V P and Nogueira A F 2016 Assembly Considerations for Dye-Sensitized Solar Modules with Polymer Gel Electrolyte *Ind Eng Chem Res* **55** 10278–85

[42] Sastrawan R, Beier J, Belledin U, Hemming S, Hinsch A, Kern R, Vetter C, Petrat F M, Prodi-Schwab A, Lechner P and Hoffmann W 2006 A glass frit-sealed dye solar cell module with integrated series connections *Solar Energy Materials and Solar Cells* **90** 1680–91

[43] Wang C, Wang L, Shi Y, Zhang H and Ma T 2013 Printable electrolytes for highly efficient quasi-solid-state dye-sensitized solar cells *Electrochim Acta* **91** 302–6

# Fabrication of Surface Enhanced Raman Spectroscopy based sensor using Ag NPs on copper tape flexible substrate

**Mohamed Shameer. P[a,b], K. Vijai Anand[b]*, Soumya Columbus[a], Hussain Alawadhi[a,c], Mounir Gaidi[a,c,d].**

[a] Center for Advanced Materials Research, Research Institute of Sciences and Engineering, University of Sharjah, PO Box 27272, Sharjah, United Arab Emirates.
[b] Department of Physics, Sathyabama Institute of Science & Technology, Chennai 600 119, Tamil Nadu, India.
[c] Department of Applied Physics and Astronomy, University of Sharjah, PO Box 27272, Sharjah, United Arab Emirates
d Laboratoire de Photovoltaïque, Centre de Recherches et des Technologies de l′ Energie, Technopole de Borj C′edria, BP 95, 2050 Hammam-Lif, Tunisia
Email: *anandkvijai78@gmail.com

**Abstract**
In this work, the authors have investigated the properties of thermally evaporated silver nanoarrays on copper tape (Ag/Cu) as flexible and reusable sensors for organic contaminants. Uniformly deposited silver nanoparticles (Ag NPs) with well-controlled sizes have formed by thermal evaporation. The deposition time is varied to obtain different morphology of the Ag nanoparticles. The Surface-enhanced Raman spectroscopy (SERS) efficiency has been optimized using Rhodamine 6G (R6G) as a test molecule. Morphological studies have also been carried out using scanning electron and atomic force microscopes. The Ag NPs on the copper tape-based sensor show a substantial enhancement in the order of $8.3 \times 10^7$ in the SERS signal for a deposition time of 10 s. Furthermore, we observe that as-prepared Ag/Cu tape can detect micro-molar alizarine red S (ARS), which exhibited excellent reproducibility as well. Hence, the flexible copper substrate proves to be a promising material for detecting environmental hazards at a meager cost.

## 1. Introduction
Surface-enhanced Raman spectroscopy (SERS) is an essential tool for identifying molecules in different areas like food quality assessment, environmental monitoring, pharmaceutical research, and clinical diagnosis [1–4]. In SERS, the intensity of scattered Raman signals can be enhanced when analyte molecules are deposited on metal nanoparticle surfaces. Local surface Plasmon Resonance (LSPR) of metal nanostructures is responsible for electromagnetic enhancement (EM), whereas the charge transfer between the metal nanoparticles to analyte molecules leads to chemical enhancement (CM) [5–7]. In the case of EM, the intensification in the Raman signal is observed by $10^6$ to $10^8$ folds of its magnitude when the analyte has adsorbed on SERS active 'hotspot'. Also, the SERS effect can be greatly affected by interparticle distance, shape, and size [8]. Therefore, it has been considered that correlating the LSPR of the SERS substrate with the energy of incident radiation is highly crucial. Many research strategies have been employed for developing highly efficient and reproducible substrates, whereas ultra-low-level detection limits are reported for many molecules [9]. It has already been reported the detection of para amino benzoic acid (PABA) using silver nanoparticles (Ag NPs) deposited on a silicon substrate [10].

Due to the uniqueness of analysis, SERS has been widely used to detect environmental pollutants [11–14].

Different organic pollutants have been used in industries, usually discharged into the environment, which can cause serious water contamination issues [15]. Since many organic dyes are carcinogenic, it causes severe health issues to human and living organisms in the water [16–18]. Alizarin Red S (ARS) is a water-soluble organic dye used in fabric industries, and it is thermally and optically stable [19,20]. A complex metal form for ARS is also used as mordant cotton, wool, and silk dyes. In addition, ARS has been employed for various laboratory uses such as pH indicators, spot test reagents, etc. Also, it has a very high capacity to bind to metal ions [21]. ARS has been recognized as the most stable dye that can enter humans, which can lead to cancer and severe allergic reactions [22]. Hence, decomposing this dye is very important before it gets discharged to nature from any industry. Traditional water treatments like adsorption and desalination have many limitations in degrading ARS due to their high stability.

In this study, we are reporting the detection of hydroxy anthraquinone, also known as alizarin red S (ARS, 1,2-dihydroxy-3-sulfonate-9,10-anthraquinone), over silver nanoarrays deposited on a copper flexible substrate. Highly flexible copper tape decorated with silver nanoparticles was fabricated by a thermal evaporation process, and the efficacy of ARS detection was investigated.

## 2. Experimental

Silver nanoparticles were deposited on a copper flexible substrate (Ag/Cu) by a thermal evaporation method. A current of 250 A was applied for 5, 10, 15, 20, 25, and 30 seconds to a high-purity silver target bought from a local jewellery shop. The instrument used for this experiment was from Denton benchtop thermal evaporator, and a tungsten boat was used to hold the silver target. A vacuum of $2\times10^{-5}$ torr was created inside the chamber, and the substrates were allowed to rotate at a speed of 15 RPM.

The morphological studies were carried out on the prepared samples using Thermo scientific Apreo 2C field emission scanning electron microscope (FESEM) and Nanosurf Flex axiom 3000 atomic force microscope (AFM). The presence of silver deposition was confirmed using the Oxford Instruments UltimMax 100 energy dispersive spectrometer (EDS) detector attached to the FESEM. The surface-enhanced Raman spectroscopy studies were carried out using a Renishaw Invia Raman microscope with an objective of 50x. An excitation wavelength of 514 nm was used with a power of 100 μwatt for an integration time of 10 seconds.

## 3. Results and Discussion

FESEM micrographs show a uniform deposition of spherical silver nanoparticles on the surfaces of flexible copper tape (Figure 1a) and the corresponding EDS spectra, which confirms the presence of the silver (Figure 1b). Figure 1c represents the morphology obtained from the AFM image. Both the FESEM and AFM show a uniform distribution of the silver nanoparticles with uniform sizes. The deposition process was done by applying a very high current for a short duration. Hence the particles were deposited instead of continuous film which is proved from the SEM image.

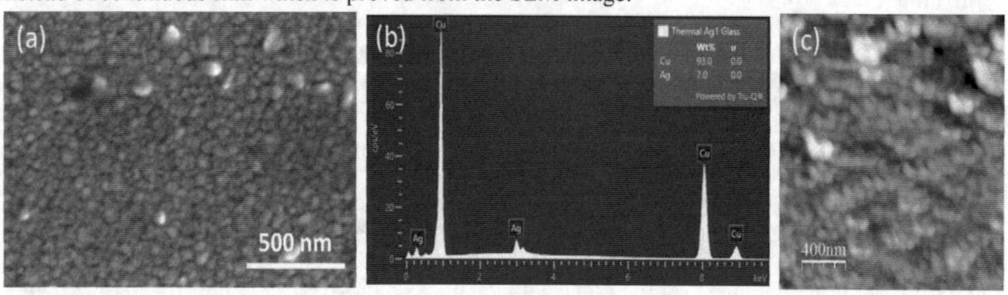


**Figure 1**: (a) FESEM image of Ag NPs on copper substrate, (b)corresponding EDS spectra that show the presence of silver and copper, and (c) represents the AFM image of the sensor.

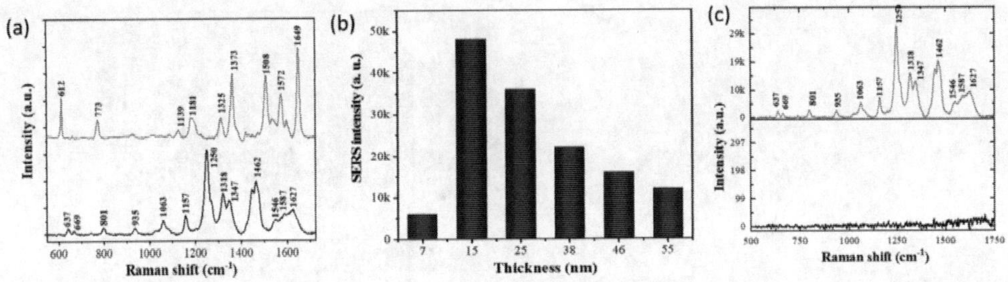

**Figure 2**: (a) SERS spectra depicting vibrational peaks of Rhodamine 6G (R6G) (in red) and ARS (in black), (b) variation of SERS intensity of R6G with respect to the thickness of Ag NPs, (c) Raman spectra of ARS on a pure copper substrate (in Black) and ARS on Ag NPs (in Red) [23].

The SERS activities of fabricated Ag/Cu tape sensors have been analysed using both R6G and ARS (Figure 2a). One can see the selectivity of SERS analysis as both dyes show distinct fingerprint bands [24,25]. The main Raman modes such as C-C ring bending (612 $cm^{-1}$), C-H bending (773, 1128, and 1187 $cm^{-1}$), C-C stretching (1373 $cm^{-1}$), C-N stretching (1449 $cm^{-1}$), C-C stretching (1508, 1572 and 1649 $cm^{-1}$) are observed for R6G. In the case of ARS, the characteristics of vibrational modes such as $SO_3$ stretching (935 and 1063 $cm^{-1}$), 1248 (C=O stretching), C–C stretching (1318 $cm^{-1}$), C–C stretching (1318 cm-1) and C–OH bending (1420 $cm^{-1}$), C=O stretching (1635 $cm^{-1}$) are observed.

The SERS activities of Ag nanoarrays deposited on copper are investigated using R6G. It has been found that the intensity of the Raman signal is significantly affected by the deposition time. In the thermal evaporator process, the time has been varied to obtain and tune Ag nanoparticles' morphological characteristics. The Ag-NPs sizes are estimated and plotted against SERS intensity, as shown in Figure 2b. It is observed that the highest SERS intensity is obtained for the evaporation duration of 10 seconds. The superior SERS enhancement can correlate to the confined nanoparticle morphology, which can provide better LSPR effects [26]. In comparison with copper tape, we have evaluated Ag NPs on glass substrates as well which we have already demonstrated on our earlier work [10].

The enhancement factor (EF) of Ag/Cu can be calculated from the concentrations of Raman and the SERS spectra of the R6G by taking the counts for the peak position at 612 $cm^{-1}$. This is estimated about $8.3x10^7$ using the equation [27]

$$EF = \frac{I_{SERS}}{I_{Ref}} \times \frac{C_{Ref}}{C_{SERS}}$$

The optimized Ag/Cu tape sensor has been evaluated for reproducibility of Raman signals by performing Raman 2D mapping. From the 2D map, we have selected 15 random spectra of ARS and plotted them, as shown in Figure 3a. Moreover, the intensities at 1250 $cm^{-1}$ are plotted, and the relative standard deviation (RSD) is estimated (Figure 3b). It has been observed that an RSD value of 11% is obtained for the as-fabricated sensor. Thus, Ag/Cu tape demonstrates excellent spot-to-spot reproducibility for being used as chemo-sensors for environmental applications.

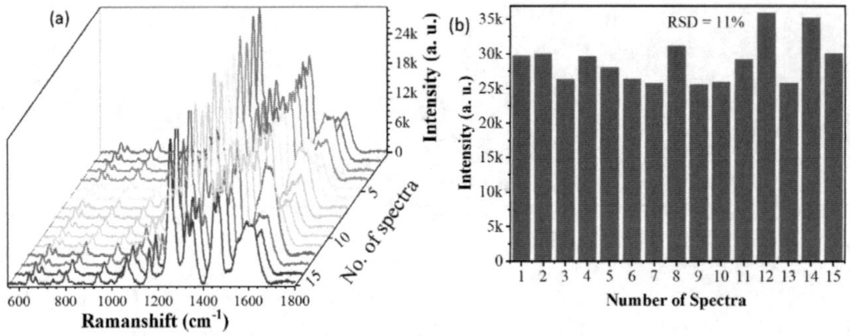


**Figure 3**: (a) Reproducibility analysis of Ag/Cu tape sensor (b) corresponding intensities at 612 cm$^{-1}$ peak of R6G for 15 spectra collected from 2D Raman mapping.

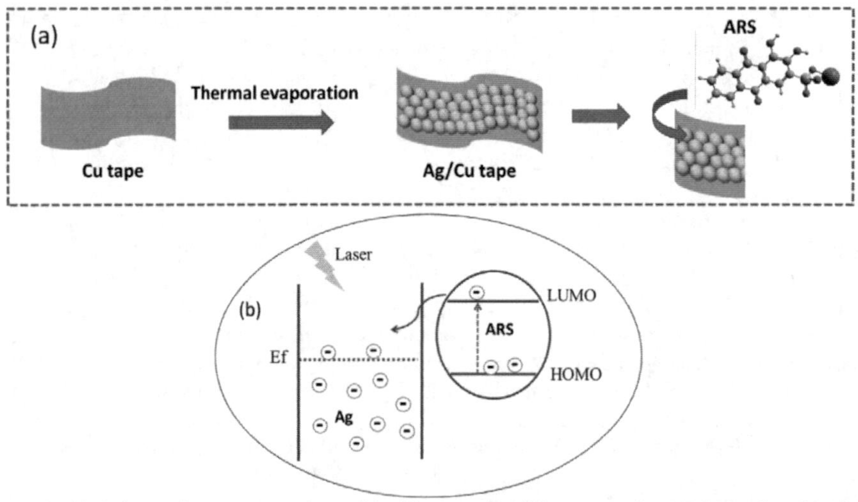


**Figure 4**: (a) Schematic representation of Ag/Cu tape flexible sensor for ARS sensing (b) charge transfer mechanism of ARS sensing.

In the present work, we have attempted to investigate the feasibility of SERS sensing of stable dye like ARS using Ag/Cu tape as a flexible and cost-effective sensor (Figure 4a). The electromagnetic mechanism can mainly account for the SERS enhancement observed for Ag/Cu tape sensor. In addition, the charge transfer mechanism can also occur between silver nanoparticles and ARS, as represented in Figure 4b. Upon laser exposure, the free electrons present in silver can be excited to the fermi level (Ef), while the electrons present in the HOMO of ARS can also be excited to LUMO. SPR electrons of silver can be further transferred to LUMO of ARS, resulting in a charge transfer mechanism. EM and CM resulted in superior SERS activities of Ag/Cu tape flexible substrates, which can be employed to efficiently detect organic pollutants.

## 4. Conclusions

Ag nanoparticles/Cu tape has been successfully fabricated by thermal evaporation. The highest SERS activity is observed for 10 s deposition time. SERS sensing of micromolar solutions of R6G and ARS are demonstrated. The mechanism of sensing has been illustrated in detail. Reproducibility analysis has shown that as-fabricated sensor has high repeatability for being used as excellent sensors for environmental applications.

**References**

[1] Sun Y, Zhai X, Xu Y, Liu C, Zou X, Li Z, Shi J and Huang X 2021 Facile fabrication of three-dimensional gold nanodendrites decorated by silver nanoparticles as hybrid SERS-active substrate for the detection of food contaminants *Food Control* **122** 107772

[2] Godad S P and Kamal S 2021 Multifunctional Ag(2)CO(3) microrods as SERS-active substrate for the detection and degradation of Rhodamine B dye. *Spectrochim. Acta. A. Mol. Biomol. Spectrosc.* **263** 120176

[3] Cailletaud J, De Bleye C, Dumont E, Sacré P-Y, Gut Y, Leblanc N, Letellier P, Ginot Y-M, Hubert P and Ziemons E 2020 Detection of low dose of piroxicam polymorph in pharmaceutical tablets by surface-enhanced Raman chemical imaging (SER-CI) and multivariate analysis *Int. J. Pharm.* **574** 118913

[4] Haroon M, Tahir M, Nawaz H, Majeed M I and Al-Saadi A A 2022 Surface-enhanced Raman scattering (SERS) spectroscopy for prostate cancer diagnosis: A review *Photodiagnosis Photodyn. Ther.* **37** 102690

[5] Liu C, Song Q, Chen J, Li X, Cai J, Lu Z, Li W, Fang N X and Feng S-P 2019 Electromagnetic and Chemical Enhancements of Surface-Enhanced Raman Scattering Spectra from Cu2O Hexagonal Nanoplates *Adv. Mater. Interfaces* **6** 1900534

[6] Trabelsi K, Hajjaji A, Gaidi M, Bessais B and El Khakani M A 2017 Enhancing the photoelectrochemical response of TiO2 nanotubes through their nanodecoration by pulsed-laser-deposited Ag nanoparticles *J. Appl. Phys.* **122** 64503

[7] Gaidi M, Trabelsi K, Hajjaji A, Chourou M L, Alhazaa A N, Bessais B and El Khakani M A 2018 Optimizing the photochemical conversion of UV–vis light of silver-nanoparticles decorated TiO2 nanotubes based photoanodes *Nanotechnology* **29** 15703

[8] Hao E and Schatz G C 2004 Electromagnetic fields around silver nanoparticles and dimers. *J. Chem. Phys.* **120** 357–66

[9] Xu D, Zhang Y, Zhang S, Yang W and Chen J 2021 Ultrasensitive SERS detection of crystal violet and malachite green based on high surface roughness copper nanocorns prepared via solid-state ionics method *Sensors Actuators A Phys.* **331** 113042

[10] Shameer P M, Vijai Anand K, Columbus S, Alawadhi H, Daoudi K, Gaidi M and Govindaraju K 2023 Highly efficient, multiplexed SERS sensing of para-aminobenzoic acid using reusable silver nanoarrays for environmental monitoring *Mater. Sci. Eng. B* **295** 116576

[11] Hering K, Cialla D, Ackermann K, Dörfer T, Möller R, Schneidewind H, Mattheis R, Fritzsche W, Rösch P and Popp J 2008 SERS: a versatile tool in chemical and biochemical diagnostics. *Anal. Bioanal. Chem.* **390** 113–24

[12] Quang L X, Lim C, Seong G H, Choo J, Do K J and Yoo S-K 2008 A portable surface-enhanced Raman scattering sensor integrated with a lab-on-a-chip for field analysis *Lab Chip* **8** 2214–9

[13] Alvarez-Puebla R A, dos Santos David S. J and Aroca R F 2007 SERS detection of environmental pollutants in humic acid–gold nanoparticle composite materials *Analyst* **132** 1210–4

[14] Bao L, Mahurin S M, Haire R G and Dai S 2003 Silver-Doped Sol−Gel Film as a Surface-Enhanced Raman Scattering Substrate for Detection of Uranyl and Neptunyl Ions *Anal. Chem.* **75** 6614–20

[15] Al-Musharafi S K, Mahmoud I Y and Al-Bahry S N 2013 Heavy Metal Pollution from Treated Sewage Effluent *APCBEE Procedia* **5** 344–8

[16] Tang H, Zhu C, Meng G and Wu N 2018 Review—Surface-Enhanced Raman Scattering Sensors for Food Safety and Environmental Monitoring *J. Electrochem. Soc.* **165** B3098

[17] Nilsson R, Nordlinder R, Wass U, Meding B and Belin L 1993 Asthma, rhinitis, and dermatitis in workers exposed to reactive dyes. *Br. J. Ind. Med.* **50** 65 LP – 70

[18] Padhi R A-R A-B S 2012 Pollution due to synthetic dyes toxicity & carcinogenicity studies and remediation *Int. J. Environ. Sci.* **v. 3** 940-955–2012 v.3 no.3

[19] Fu F, Gao Z, Gao L and Li D 2011 Effective Adsorption of Anionic Dye, Alizarin Red S, from

Aqueous Solutions on Activated Clay Modified by Iron Oxide *Ind. Eng. Chem. Res.* **50** 9712–7

[20] Ding F, Liu W, Diao J-X and Sun Y 2011 Characterization of Alizarin Red S binding sites and structural changes on human serum albumin: A biophysical study *J. Hazard. Mater.* **186** 352–9

[21] Swarnkar R K, Singh S C and Gopal R 2009 SERS Study of Alizarin Red Dye on Colloidal Silver Nanoparticles *MRS Online Proc. Libr.* **1241** 608

[22] Rajeshwar K, Osugi M E, Chanmanee W, Chenthamarakshan C R, Zanoni M V B, Kajitvichyanukul P and Krishnan-Ayer R 2008 Heterogeneous photocatalytic treatment of organic dyes in air and aqueous media *J. Photochem. Photobiol. C Photochem. Rev.* **9** 171–92

[23] Columbus S, Hammouche J, Ramachandran K, Daoudi K and Gaidi M 2022 Assessing the efficiency of photocatalytic removal of alizarin red using copper doped zinc oxide nanostructures by combining SERS optical detection *J. Photochem. Photobiol. A Chem.* **432** 114123

[24] de Souza M L and Corio P 2010 Surface-enhanced Raman scattering study of alizarin red S *Vib. Spectrosc.* **54** 137–41

[25] Ferreira de Sousa Junior J, Columbus S, Hammouche J, Ramachandran K, Daoudi K and Gaidi M 2023 Engineered micro-pyramids functionalized with silver nanoarrays as excellent cost-effective SERS chemosensors for multi-hazardous pollutants detection *Appl. Surf. Sci.* **613** 156092

[26] Israelsen N D, Hanson C and Vargis E 2015 Nanoparticle Properties and Synthesis Effects on Surface-Enhanced Raman Scattering Enhancement Factor: An Introduction ed X-F Zhao *Sci. World J.* **2015** 124582

[27] Daoudi K, Gaidi M, Columbus S, Shameer M and Alawadhi H 2022 Hierarchically assembled silver nanoprism-graphene oxide-silicon nanowire arrays for ultrasensitive surface enhanced Raman spectroscopy sensing of atrazine *Mater. Sci. Semicond. Process.* **138** 106288

Journal of Physics: Conference Series **2751** (2024) 012020 doi:10.1088/1742-6596/2751/1/012020

# Fast and Facile Synthesis of Passivated Perovskite Single Crystals with Zn-Porphyrin Derivatives

**Abdul Kareem K. Soopy, Bhaskar Parida and Adel Najar***

Department of Physics, United Arab Emirates University, Al Ain, United Arab Emirates.

*adel.najar@uaeu.ac.ae

**Abstract**. Organic-inorganic trihalide, $ABX_3$ perovskite are excellent candidates for advanced optoelectronic applications beyond photovoltaics. Perovskite single crystals offer several advantageous special features, such as a long diffusion length, low trapped carrier density, and outstanding environmental stability to use in optoelectronics. In this work, Zn has been incorporated into $MAPbBr_3$ single crystals as Zn- Porphyrin to passivate the bulk crystals and enhance their optical and electrical properties. The inverse temperature crystallization phenomenon was used to crystallize $MAPbBr_3$ rapidly in hot solutions. The structural and optical characterization confirms that the passivation enhanced the optoelectronic properties of the single crystals.

## 1. Introduction

Recently, there has been significant interest in the use of organic and inorganic lead halide perovskites (referred to as $MAPbX_3$) in various optoelectronic applications, including solar cells, light-emitting diodes, photodetectors, and other devices [1-3], light-emitting diodes [4], photodetectors [5], and other devices. These materials possess advantageous characteristics such as a precisely controllable band gap, a high light-absorption coefficient, an extensive carrier transport distance, and a minimal number of defects [6-8]. Additionally, they can be deposited using low-cost and uncomplicated techniques. To enhance the performance of these devices, numerous endeavors have been undertaken to produce high-quality perovskite materials, such as polycrystalline films or single crystals [9, 10].

Single-crystal (SC) perovskite materials possess numerous desirable properties, including a long diffusion length, a low trapped carrier density, and exceptional environmental stability [11]. Their minimal defect densities and absence of grain boundaries make them suitable candidates for optoelectronic devices. Among the various members of the $MAPbX_3$ family, $MAPbBr_3$ is well-known for its stability compared to $MAPbI_3$ and its wider range of visible light absorption compared to $MAPbCl_3$. Furthermore, $MAPbBr_3$ exhibits cubic crystal symmetry, which facilitates the formation of high-quality single crystals [12]. Bulk $MAPbBr_3$ single crystals are grown using various techniques such as laser-induced localized growth, antisolvent vapor-assisted crystallization, and inverse temperature crystallization (ITC) [13]. However, these reported methods for perovskite single crystallization have certain drawbacks, including slow growth rates for large-scale applications or unsatisfactory quality for demanding optoelectronic applications [14]. The development of large-scale perovskite single crystals with rapid growth and excellent quality would enable broader applications. The ITC method, which relies on the inverse solubility of perovskite halides in specific solvents, has emerged as a popular

Journal of Physics: Conference Series **2751** (2024) 012020 doi:10.1088/1742-6596/2751/1/012020

approach for achieving rapid growth of large-scale, high-quality perovskite single crystals. This process takes advantage of the temperature-dependent solubility of perovskite molecules in the solvent. Unlike conventional dissolution processes, where perovskite molecules dissolve freely, most of the perovskite molecules form complexes with the solvent, reducing the number of unbound molecules. At lower temperatures, the complexes remain stable, and the solution is not saturated with unbound molecules. However, as the temperature increases, the binding energy of the complexes decreases. Eventually, the solution becomes supersaturated, causing the unbound perovskite molecules to rise and initiate crystallization.

The presence of defects and trap states in perovskite materials can hinder charge carrier mobility and lead to charge carrier recombination, thereby reducing the efficiency of solar cells. Researchers have sought to address this issue by employing chemical passivation techniques to fill in surface dangling bonds and reduce the density of defects and trap states [15-17]. Recently, porphyrin derivatives have emerged as promising additives in perovskite optoelectronics due to their high light absorption, favorable charge transport properties, and excellent thermal stability [18, 19]. To achieve high-performance optoelectronic devices, the advantages of multifunctional porphyrins in defect passivation need to be fully exploited.

Bulk passivation, which involves passivating the entire crystal during the growth process, offers advantages over surface passivation, particularly in terms of controlling growth defects and improving trap density. In this study, zinc (Zn) was incorporated into $MAPbBr_3$ single crystals as Zn-Porphyrin (Zn-PP) to passivate the bulk crystal and enhance its optical and electrical properties. Large-scale and high-quality single crystals of $MAPbBr_3$ were successfully produced using the inverse temperature crystallization (ITC) method, taking advantage of the high solubility of hybrid perovskites at low temperatures. The structural, optical, and electrical properties of the single crystals were extensively investigated as crucial parameters influencing device performance. Moreover, through analysis of charge transport characteristics, the synthesized perovskite single crystal demonstrated potential for application in photodetectors. The low-temperature crystallization process offers a scalable approach for generating numerous high-quality perovskite single crystals, thus facilitating their use in various perovskite-based applications.

## 2. Experimental Method

The inverse temperature crystallization method was employed to grow single crystals of methylammonium lead bromide ($MAPbBr_3$) [20]. To prepare the growth solution, methylammonium bromide (MABr) and lead bromide ($PbBr_2$) were dissolved in N,N-dimethylformamide (DMF) with a 1:1 molar ratio. The solution was continuously stirred at room temperature until complete dissolution. Subsequently, the solution was filtered using a 0.2 μm PTFE filter to remove any impurities. After filtration, the transparent solution was divided into 2 mL portions and placed in vials, which were then immersed in an oil bath. The oil bath was maintained at a temperature of 80°C throughout the process. For the preparation of $MAPbBr_3$ single crystals passivated with Zn-PP, 0.1 wt% of Zn-PP was added to the solution described above. The crystals used for measurements were grown by transferring seed crystals into freshly prepared solutions. The crystallization procedure was repeated every 3 hours to facilitate the growth of single crystals with a size of approximately 5 mm.

All the procedures were conducted under ambient conditions with a humidity level of 55-60%. The optical and structural properties of the single crystals were analyzed using various techniques. Photoluminescence spectroscopy (PL) was employed to investigate the emission properties of the crystals upon excitation. UV-Vis-NIR transmittance spectroscopy was used to study the absorption properties of the crystals across a wide range of wavelengths. X-ray diffraction (XRD) measurements were conducted to determine the crystal structure, lattice parameters, and overall crystal quality. These techniques provided valuable insights into the optical and structural characteristics of the $MAPbBr_3$ single crystals.

To fabricate planar photodetectors, interdigital Au electrodes were deposited on the top surface of the $MAPbBr_3$ single crystals. The deposition was carried out using the vacuum evaporation method,

Journal of Physics: Conference Series **2751** (2024) 012020 doi:10.1088/1742-6596/2751/1/012020

resulting in Au electrodes with a thickness of approximately 100 nm. Each electrode was composed of a group of fine Au wires, which allowed for further investigation of the device performance. The effective illuminated area of each photodetector was approximately $0.84\times10^{-3}$ $cm^2$. This area represents the region on the surface of the photodetector that receives light and contributes to the device's photo response.

## 3. Results and Discussion

At elevated temperatures (80°C) and concentrated precursor solutions (1M) containing equal amounts of Methylammonium bromide (MABr) and lead bromide ($PbBr_2$), we observed the rapid formation of small $MAPbBr_3$ perovskite precipitates. However, by carefully controlling the temperature and concentration of the precursor in N,N-dimethylformamide (DMF) at 80°C and 1M, respectively, we were able to produce only a few crystals. To synthesize $MAPbBr_3$ single crystals passivated with Zn porphyrin (Zn-PP), we combined the perovskite solution with Zn-PP using the same method. The concentration of Zn-PP added to the precursor solutions was expressed as a weight percentage relative to the $PbBr_2$ content.

As a result of the crystallization reaction at 80°C, we successfully obtained 5mm-size $MAPbBr_3$ single crystal cubes, as shown in Figure 1(a). Each individual $MAPbBr_3$ crystal was found to grow at a rate of approximately 40 $mm^3/h$, which is significantly faster than the growth rates reported in previous methods [21]. Notably, we observed that the crystallization process is reversible for both $MAPbBr_3$ single crystals and Zn-PP passivated $MAPbBr_3$ single crystals. When the crystals were cooled back to room temperature, they dissolved, indicating the reversible nature of the crystallization process.

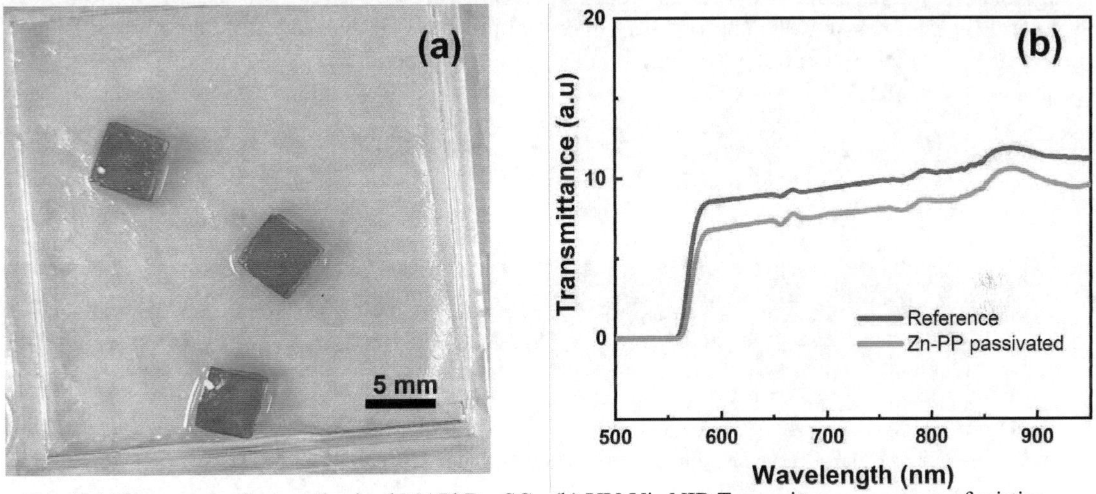


Fig. 1(a) Photograph of as-synthesized $MAPbBr_3$ SCs. (b) UV-Vis-NIR Transmittance spectrum of pristine and Zn-PP passivated SCs

The UV-Vis-NIR transmission spectrum of the unprocessed and bulk-passivated Zn-PP $MAPbBr_3$ SCs is shown in Fig. 1(b). After Zn-PP passivation, a substantial increase in transmittance was seen. This proves that Zn-PP has an impact on the optical characteristics of $MAPbBr_3$ crystals. Using the Tauc-plot method, the band gap of the crystal was calculated and was found to be 2.216 eV for the untreated samples and 2.203 eV for the Zn-PP treated samples, respectively. The slight reduction in bandgap in Zn-passivated perovskite single crystals suggests improved crystallinity [15] with Zn-PP, which is further supported by the XRD high-intensity peaks for the passivated perovskite single crystals relative to the nontreated perovskite single crystals as seen in Fig. 2(a).

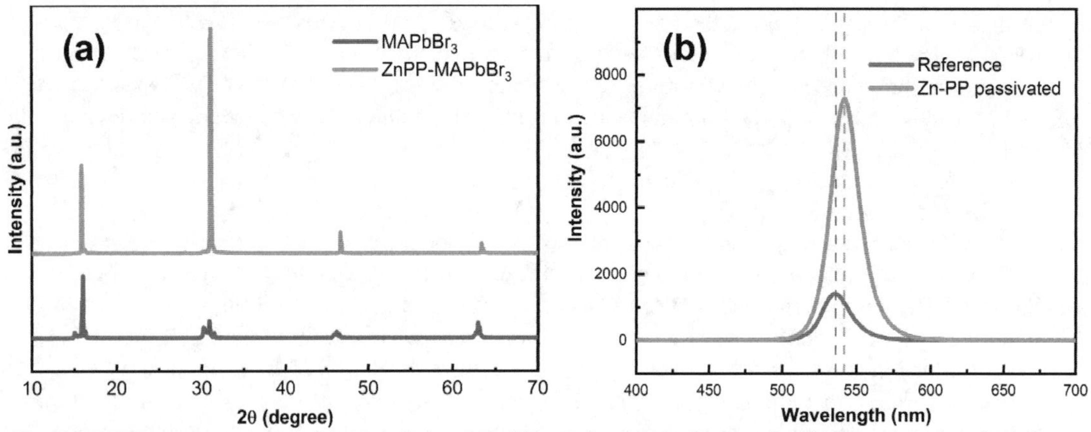


Fig. 2(a) XRD measurements of the pristine and Zn-PP passivated single crystals. (b) PL spectrum of pristine and Zn-PP passivated SCs

Using photoluminescence spectroscopy, the optical characteristics of the single crystals were further assessed. The PL spectrum of the SCs with and without Zn-PP passivation is depicted in Fig. 2(b). It is noticed that the PL emission peak position of Zn-PP passivated $MAPbBr_3$ SCs exhibits a 7 nm red-shift under similar conditions. This observation may be explained by the larger ion radii of Zn-PP than Pb, which causes the change in the lattice. An important point to remember is that the addition of Zn-PP to $MAPbBr_3$ single crystals passivates the bulk perovskite and decreases defects, leading to a notably higher PL peak intensity. The improvement in crystallinity caused by the Zn-PP derivative is due to this red shifting, which is supported by the XRD measurement in Fig.2(a).

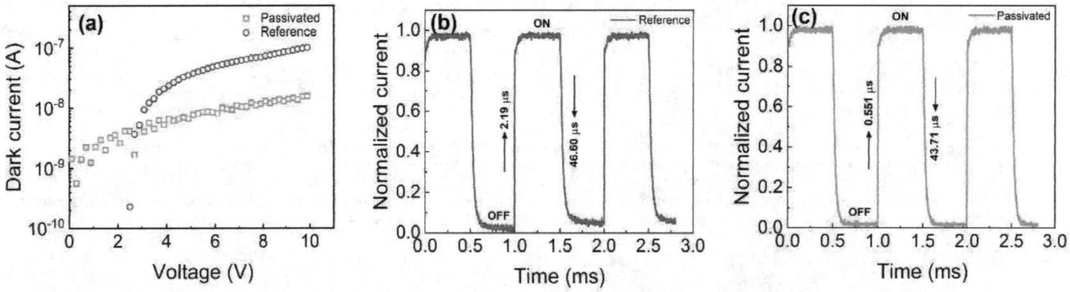


Fig. 3(a) Photocurrents of devices at different voltages. Photocurrent response of MAPbBr3 Photodetector measured under 532 nm switching light *of (b) pristine and (c) passivated Photodetector device.*

Fig. 3(a) shows the dark current of the $MAPbBr_3$ and Zn-PP passivated $MAPbBr_3$ photodetector. It was observed that the dark current of the Zn-PP passivated sample was significantly lower than the un-passivated $MAPbBr_3$ sample. The decrease in dark current is due to the effective defect passivation of $MAPbBr_3$ single crystal. Even at lower light intensities, this detector offers a high on-off ratio and an excellent linear response over a wide range of light intensities. This suggests that the detector has improved in terms of photo response performance. The rate of response of a photosensor is strongly influenced by charge collection and transport. The present photosensor's transient photocurrent responsiveness was measured under 532 nm chopped laser light. The rising time ($t_{rise}$) and fall time ($t_{decay}$) of unpassivated $MAPbBr_3$ SC photodetector was found to be 2.1 μs and 46.60 μs respectively as shown in Fig 3(b). The $t_{rise}$ and $t_{decay}$ of Zn-PP passivated $MAPbBr_3$ SC photodetector was found to be 0.55 μs and 43.71 μs respectively as depicted in Fig 3(c). It is obvious that the Zn-PP passivation improved the Photo response characteristics of $MAPbBr_3$. The devices show a certain amount of reversibility and stability after going through multiple cycles of "on" (under light) and "off" (under dark) states.

## 4. Conclusion

In this work, we successfully synthesized quickly crystalizing, a few millimeters in size pristine $MAPbBr_3$ single crystals and Zn Porphyrin passivated single crystals. In this inverse temperature crystallization method, the precursor solution concentration and crystallization temperature play a vital role in determining the rate and size of a single crystal in a hot solution. A few mm size $MAPbBr_3$ single crystals cubes were obtained after crystallization reaction at 80°C. A significant increase in the PL peak intensity after the incorporation of Zn-PP in the $MAPbBr_3$ single crystals passivates the bulk perovskite and reduces the defects. It is confirmed by the optical and structural characterization that the optical properties of the $MAPbBr_3$ SCs can be modified by introducing Zn-PP, without destroying the perovskite single crystal structure. By introducing bulk passivation on the $MAPbBr_3$ single crystal, the photo response performance of the photodetector has improved significantly. This study demonstrates the passivation of $MAPbBr_3$ by Zn-PP SCs using solution-processed growth is a viable method for improving the optoelectronic characteristics of photo-detector performance in the future.

## References

[1] Park N-G 2020 Research Direction toward Scalable, Stable, and High Efficiency Perovskite Solar Cells *Advanced Energy Materials* **10** 1903106

[2] Parida B, Singh A, Kalathil Soopy A K, Sangaraju S, Sundaray M, Mishra S, Liu S and Najar A 2022 Recent Developments in Upscalable Printing Techniques for Perovskite Solar Cells *Advanced Science* **9** 2200308

[3] Qamhieh N, Najar A, Qamhieh Z N, Abdel Aziz B, Mansour A and Alghoul I 2018 Synthesis and characterization of a perovskite film for solar cells applications *Optik* **171** 648-51

[4] Lu M, Zhang Y, Wang S, Guo J, Yu W W and Rogach A L 2019 Metal Halide Perovskite Light-Emitting Devices: Promising Technology for Next-Generation Displays *Advanced Functional Materials* **29** 1902008

[5] Miao J and Zhang F 2019 Recent progress on highly sensitive perovskite photodetectors *Journal of Materials Chemistry C* **7** 1741-91

[6] Kojima A, Teshima K, Shirai Y and Miyasaka T 2009 Organometal Halide Perovskites as Visible-Light Sensitizers for Photovoltaic Cells *Journal of the American Chemical Society* **131** 6050-1

[7] Stranks S D and Snaith H J 2015 Metal-halide perovskites for photovoltaic and light-emitting devices *Nature nanotechnology* **10** 391-402

[8] Yang W S, Noh J H, Jeon N J, Kim Y C, Ryu S, Seo J and Seok S I 2015 SOLAR CELLS. High-performance photovoltaic perovskite layers fabricated through intramolecular exchange *Science (New York, N.Y.)* **348** 1234-7

[9] Rao H S, Li W G, Chen B X, Kuang D B and Su C Y 2017 In Situ Growth of 120 cm(2) CH(3) NH(3) PbBr(3) Perovskite Crystal Film on FTO Glass for Narrowband-Photodetectors *Advanced materials (Deerfield Beach, Fla.)* **29**

[10] Snaith H J 2013 Perovskites: The Emergence of a New Era for Low-Cost, High-Efficiency Solar Cells *The Journal of Physical Chemistry Letters* **4** 3623-30

[11] Gao L and Yan Q 2020 Recent Advances in Lead Halide Perovskites for Radiation Detectors *Solar RRL* **4** 1900210

[12] Lozhkina O A, Yudin V I, Murashkina A A, Shilovskikh V V, Davydov V G, Kevorkyants R, Emeline A V, Kapitonov Y V and Bahnemann D W 2018 Low Inhomogeneous Broadening of Excitonic Resonance in MAPbBr3 Single Crystals *The Journal of Physical Chemistry Letters* **9** 302-5

[13] Zuo Z, Ding J, Zhao Y, Du S, Li Y, Zhan X and Cui H 2017 Enhanced Optoelectronic Performance on the (110) Lattice Plane of an MAPbBr3 Single Crystal *The Journal of Physical Chemistry Letters* **8** 684-9

[14] Zhang Y, Liu Y, Xu Z, Ye H, Li Q, Hu M, Yang Z and Liu S 2019 Two-dimensional (PEA)2PbBr4 perovskite single crystals for a high performance UV-detector *Journal of Materials Chemistry C* **7** 1584-91

[15] Zhang H, Tian Q, Xiang W, Du Y, Wang Z, Liu Y, Liu L, Yang T, Wu H, Nie T, Huang W, Najar A and Liu S 2023 Tailored Cysteine-Derived Molecular Structures towards Efficient and Stable Inorganic Perovskite Solar Cells *Advanced Materials* **n/a** 2301140

[16] Wang K, Ma S, Xue X, Li T, Sha S, Ren X, Zhang J, Lu H, Ma J, Guo S, Liu Y, Feng J, Najar A and Liu S 2022 Highly Efficient and Stable CsPbTh3 (Th = I, Br, Cl) Perovskite Solar Cells by Combinational Passivation Strategy *Advanced Science* **9** 2105103

[17] Zhou Y, Najar A, Zhang J, Feng J, Cao Y, Li Z, Zhu X, Yang D and Liu S F 2022 Effect of Solvent Residue in the Thin-Film Fabrication on Perovskite Solar Cell Performance *ACS Applied Materials & Interfaces* **14** 28729-37

[18] Gkini K, Balis N, Papadakis M, Verykios A, Skoulikidou M-C, Drivas C, Kennou S, Golomb M, Walsh A, Coutsolelos A G, Vasilopoulou M and Falaras P 2020 Manganese Porphyrin Interface Engineering in Perovskite Solar Cells *ACS Applied Energy Materials* **3** 7353-63

[19] Fang Z, Wang L, Mu X, Chen B, Xiong Q, Wang W D, Ding J, Gao P, Wu Y and Cao J 2021 Grain Boundary Engineering with Self-Assembled Porphyrin Supramolecules for Highly Efficient Large-Area Perovskite Photovoltaics *Journal of the American Chemical Society* **143** 18989-96

[20] Liu Y, Zhang Y, Yang Z, Feng J, Xu Z, Li Q, Hu M, Ye H, Zhang X, Liu M, Zhao K and Liu S 2019 Low-temperature-gradient crystallization for multi-inch high-quality perovskite single crystals for record performance photodetectors *Materials Today* **22** 67-75

[21] Bai D, Wang H, Bai Y, Najar A, Saleh N i, Wang L and Liu S F 2021 ASnX3—Better than Pb-based Perovskite *Nano Select* **2** 159-86

# Parametric study of bio-char production from the co-pyrolysis of tire and coffee wastes via fixed bed reactor

**Martina Žabčić[1,2], Lina Hilal[1,3], Lisandra Rocha-Meneses[4], Abrar Inayat[1,5,*], Chaouki Ghenai[1,5], Abdallah Shanableh[1,6]**

[1]Biomass and Bioenergy Research Group, Center for Sustainable Energy and Power Systems Research, Research Institute of Sciences and Engineering, University of Sharjah, 27272 Sharjah, United Arab Emirates
[2]Faculty of Technology and Metallurgy, University of Belgrade, Belgrade, Serbia
[3]Chemical Engineering Department, Kocaeli University, Kocaeli, Türkiye
[4]Technology Innovation Institute, Abu Dhabi, UAE
[5]Department of Sustainable and Renewable Energy Engineering, University of Sharjah, 27272 Sharjah, United Arab Emirates
[6]Department of Civil and Environmental Engineering, University of Sharjah, 27272 Sharjah, United Arab Emirates
*Corresponding author email: ainayat@sharjah.ac.ae

**Abstract**
Co-pyrolysis has become an eye-catching process for obtaining fuel and biochar and is attracting attention from both the scientific and industrial communities. One of the reasons for this is the wide variety of raw materials for the process itself and the possibility of obtaining attractive and high-quality products. The influence of three distinctive parameters has been studied on the co-pyrolysis reaction. These variables were temperature, reaction time, and plastic ratio. The results were analysed using Response Surface Methodology (RSM), and it was found that the optimum conditions to produce biochar from the co-pyrolysis of coffee and tire wastes were 359.441 ˚C, 10 minutes of reaction duration, and a plastic ratio of 97.8528 wt.%.

Keywords: Co-pyrolysis; tire waste; coffee waste; biochar; RSM

## 1. Introduction

Due to the increase in fuel demands and prices, dwindling fossil fuel supplies, and negative environmental impact, there is an intention to replace these fuels with those obtained from renewable energy sources [1, 2]. There are various alternative energy sources: solar, wind, and hydro-energy. However, bioenergy has attracted much attention [3, 4]. The conversion of biomass into biofuels and other value-added renewable commodities is the subject of extensive research [5, 6]. Several types and processes of biomass-to-energy conversion are known, which can be biochemical, thermochemical, or physicochemical. Drying, torrefaction, combustion, gasification, and pyrolysis were utilized to convert biomass into usable products [7]. Pyrolysis is the name given to the thermal decomposition of biomass in the absence of oxygen, and the outputs of this process are biochar, bio-oil, and syngas [8]. The higher the heating rate, the higher the amount of biofuel would reach at moderate temperatures. In contrast, the low heating rate, besides normal temperature conditions, would increase biochar production [9]. Finally, enhancing syngas production requires high temperatures and low heating rates. When biomass roasts in a closed container under inert conditions, biochar, a carbon-rich substance- forms. Technically, biochar is produced by what is known as the thermal decomposition of organic material at relatively low temperatures and with limited oxygen [10]. Biochar usually contains multiple components: carbon, hydrogen, oxygen, nitrogen, sulfur, phosphorus, potassium, calcium, magnesium, and sodium. The largest proportion, usually more than 60%, is carbon, followed by hydrogen and oxygen [11]. In the production of biochar, co-pyrolysis is particularly interesting. The concept of co-pyrolysis is that two

feedstocks are used at the beginning, and then the process is continued like conventional pyrolysis. The co-pyrolysis of different feedstocks has already been carried out and optimized many times [12]. In the literature, there are a lot of articles on this topic, although the process of obtaining energy from biomass in this way is still being studied. Zanella et al. [13] studied the low-temperature co-pyrolysis of polypropylene and coffee waste. In their study, they investigated the effects of the composition of a sample and the pyrolysis temperatures. The process was monitored with a constant heating rate of 5 °C/min and a pyrolysis temperature between 360 and 420 °C. A higher polypropylene content leads to a decrease in the number of liquid products and an increase in the number of solid products. Sajdak et al. [14] performed co-pyrolysis of polypropylene waste and two different types of biomass: alder and pine wood. The variable parameters in this study were the temperature and the polymer concentration, both for five different values, and the analysis of variance (ANOVA) investigated the influence of the varied parameters on the properties of the obtained products. The addition of polypropylene can improve the quality of biochar and has a good effect on the properties of bio-oil and syngas. The temperature and the type of material used for co-pyrolysis also have a noticeable influence on the properties of the coal and oil fractions. Inayat et al. [15] studied the production of biochar from coffee waste by slow pyrolysis at different reaction temperatures (200 °C–300 °C) and residence times (20–40 minutes). More biochar is obtained at low reaction temperatures and long reaction times, while bio-oil production is highest at moderate time and temperature: 30 minutes and 250 °C. Hossain et al. [16] performed the co-pyrolysis process of tire waste and rice husk to obtain biofuels and chemicals. The main parameter was the tire content in the feedstock, while the temperature was constant at 450°C. As the percentage of tire waste in the blends decreased, the biochar and gas products increased, while the bio-oil yield decreased.
A different study by Inayat et al. [17] showed the possibility of obtaining bio-oil, biochar, and syngas from the co-pyrolysis of plastic waste and date seeds. The bio-oil's highest yield occurred at 500 °C, and the highest biochar's yield was detected at 300 °C. The biochar amount increases while the plastic ratio increases and the temperature decreases. Raza et al. [18] simulated the effect of pyrolysis operating temperature on the products of the process in a temperature range from 300 °C to 500 °C. Mixtures of date seeds and coffee waste were used as feedstock. The simulation shows that biochar production decreases with increasing temperature, while gas and bio-oil production increases at 450 °C. Therefore, the bio-oil production rate diminished, while gases continued to enhance with temperature. Since there is a large amount of tire and coffee waste globally, it is a readily available and suitable feedstock for co-pyrolysis. In this work, the co-pyrolysis of tire and coffee waste was carried out to obtain biochar and to analyze and optimize biochar production using Response Surface Methodology (RSM).

## 2. Materials and methods

### 2.1 Feedstocks

The co-pyrolysis reaction was carried out in a pyrolysis reactor (Weihai Borui Chemical Machinery Mfg. Co., Ltd., China, model number BFK-1L/12.5). Tyre waste was supplied from a nearby auto repair shop. Coffee waste (Americano and espresso) was collected from the cafeteria of the Faculty of Engineering, University of Sharjah, UAE.

### 2.2 Experimental section

The tire wastes have been chopped into small pieces to fit into the reactor, and the coffee wastes have been in the oven at 110 °C. Each coffee layer took 8–12 minutes to dry; the duration depends on its thickness. The reactor should be well closed after putting the waste inside. The reason for this is mass losses. Nitrogen was used as the inert media. After the reactor had sealed well, the temperature had settled to the required level. Response Surface Methodology (RSM) was the method to optimize the reaction sequences (Table 1).

**Table 1:** Co-pyrolysis reaction conditions.

| Run | Reaction temperature (°C) | Plastic ratio (wt%) | Reaction time (mins) |
|---|---|---|---|
| 1. | 400 | 50% | 10 |
| 2. | 400 | 50% | 20 |
| 3. | 400 | 50% | 30 |
| 4. | 350 | 50% | 20 |
| 5. | 450 | 50% | 20 |
| 6. | 400 | 0% | 20 |
| 7. | 400 | 100% | 20 |

The variables examined in each run were as follows:

**Table 2:** The changing parameters in pyrolysis.

| Experimental variables | -1 | 0 | 1 |
|---|---|---|---|
| Reaction temperature (°C) | 350 | 400 | 450 |
| Reaction time (min) | 10 | 20 | 30 |
| Plastic ratio (%) | 0 | 50 | 100 |

## 3. Results and discussion

The effect of the three parameters (reaction time, reaction temperature, and plastic ratio) on biochar production is shown in this section. Based on the literature data mentioned in the introduction of this paper, we saw that this reaction is usually tested at temperatures between 300°C and 500°C, and that at higher temperatures a larger amount of bio-oil is obtained, and at lower temperatures a larger proportion of bio-char is obtained. In order to obtain the largest possible share of both biochar and biooil, we decided to test our reaction at three temperatures (350, 400, and 450°C). Also considering that we wanted to check how the share of plastic affects the amount of biochar obtained, we decided on three values: 0%, 50%, and 100% share of plastic. Fig. 1a shows different amounts of biochar obtained as a function of diverse reaction times (10, 20, and 30 minutes) at a constant temperature of 400 °C and a plastic ratio of 50%. The highest amount of biochar is obtained at a reaction time of 20 minutes, while the lowest amount is obtained after 10 minutes. In conclusion, the amount of biochar increases with the duration of the reaction, but up to a specific time, it decreases.

Fig. 1b shows the effects of different reaction temperatures on biochar production at a constant reaction time of 20 minutes and a plastic content of 50%. The amount of biochar recorded the highest value at a temperature of 400°C. The latter starts decreasing after the temperature crosses 400 °C.

The last parameter that has been studied is the plastic ratio. All reactions were performed at a reaction time of 20 minutes and a temperature of 400°C but with a different plastic ratio (0, 50, and 100 wt.%). Fig. 1c shows that the biochar yield increases with the plastic ratio and that the most biochar (48.18%) was obtained from the experiment with 100 wt.% plastic and the least (26.18%) with no plastic waste at all.
3D surface plots (Fig. 2a-c) show the combined effect of two process parameters on biochar production. The impact of reaction temperature and time is shown in Figure 2a. As can be seen from the green-yellow area on the graph, the amount of biochar produced is consistently close even when the temperature and time conditions are changed. The best values for time and temperature are 20 minutes and 370°C, respectively.

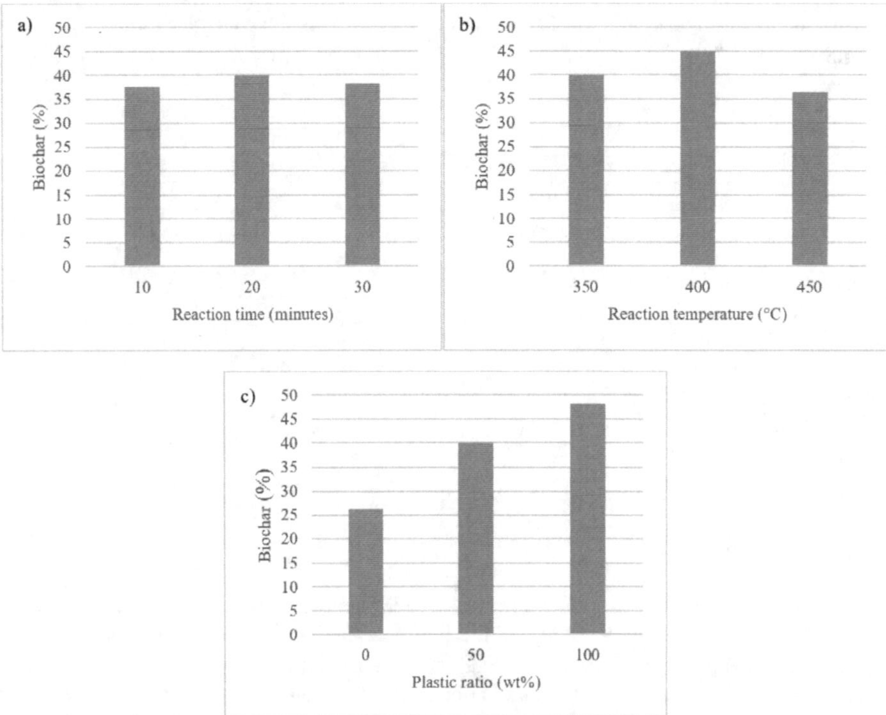


*Figure 1. Effect of different: a) reaction times in biochar production, b) reaction temperatures, and c) plastic to coffee waste ratio in biochar production.*

The significance of the plastic ratio in the process is shown in Figs. 2b and 2c. From the red area of the graphs shown, a higher plastic ratio results in higher biochar production. Fig. 2c shows that the optimum reaction time is 20 minutes since most biochar is produced at this time when the plastic content is highest.

Perturbation diagrams are prepared for biochar. Fig. 3 shows the influence of reaction time on the result. The temperature in Figs. 3a and 3b is 375 °C, the plastic content is 50%, and the reaction times are 20 and 30 minutes. The letters A, B, and C represent the reaction temperature, reaction time, and plastic ratio, respectively. The plastic ratio impacts the amount of biochar the most, as noticed.

The optimized values (Fig. 5) to produce biochar from tire and coffee waste obtained by the RSM study were a temperature of 359.441 °C, a reaction time of 10 minutes, and 97.8528 wt% plastic. Under these conditions, 48.5293 wt% biochar and 24.4238 wt% bio-oil can be produced.

From all the obtained results, it can be seen that all the examined parameters affect the amount of biochar obtained to different extents. The biggest influence is the share of plastic, when during pyrolysis the largest part of biochar is obtained from plastic while the greater part of bio-oil is obtained primarily from coffee waste.

Journal of Physics: Conference Series **2751** (2024) 012021 doi:10.1088/1742-6596/2751/1/012021

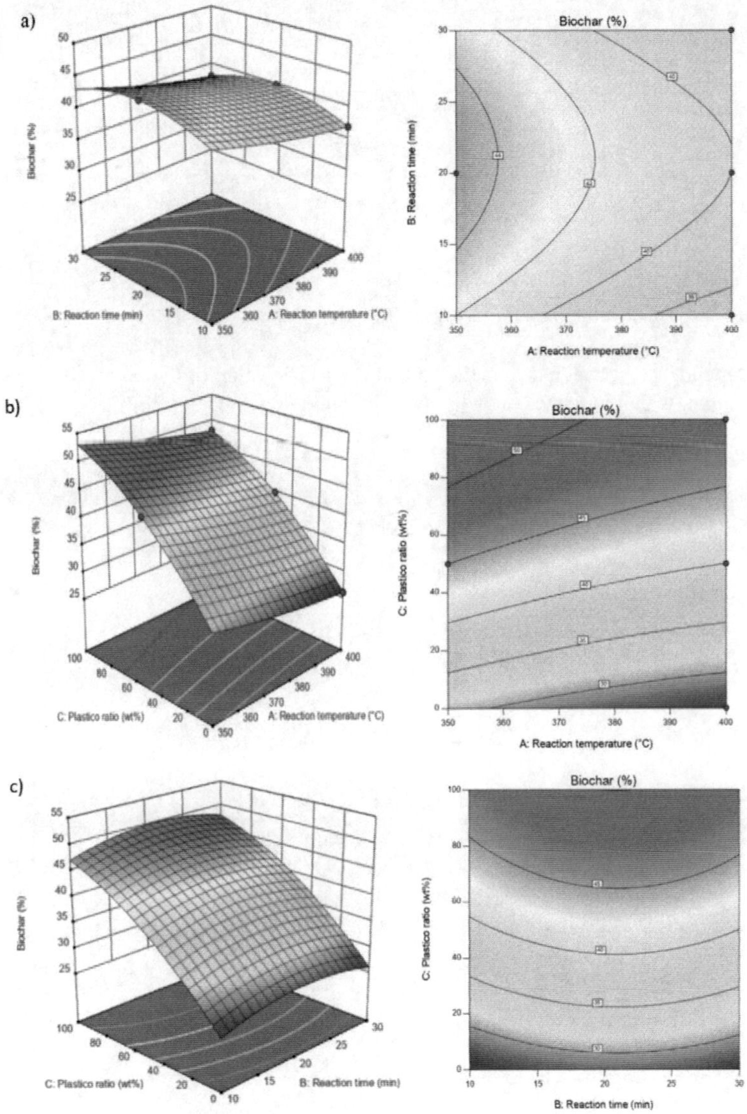


*Figure 2. Influence of different a) reaction times, b) temperatures, and c) plastic ratio in biochar production.*

Journal of Physics: Conference Series **2751** (2024) 012021 doi:10.1088/1742-6596/2751/1/012021

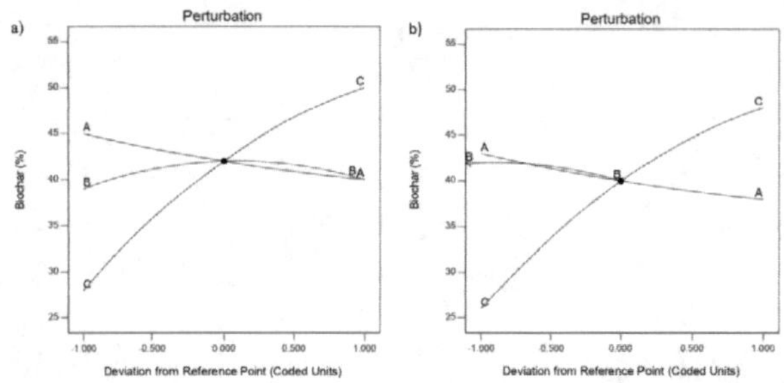


*Figure 3. Default perturbation graph for biochar results based on actual factors for: a) 375 °C, 50% plastic ratio and 20 minutes reaction time, and b) 375 °C, 50% plastic ratio and 30 minutes reaction time.*

Fig. 4 shows the dependence between the predicted and the actual value of the obtained biochar. Only two values do not agree, which indicates the validity of the model created.

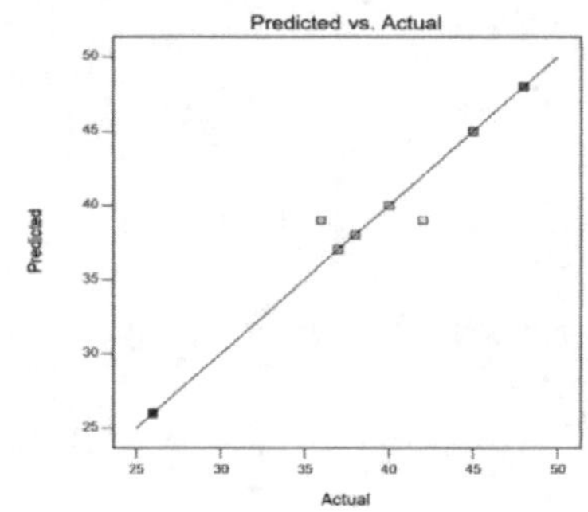


*Figure 4. Predicted and actual biochar production.*

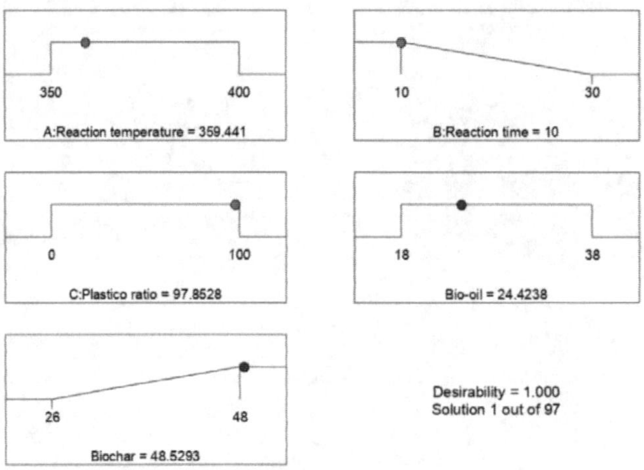


*Figure 5. Optimized values of bio-char production from the co-pyrolysis*

## 4. Conclusion

Tire and coffee wastes were used as feedstock for the co-pyrolysis reaction, and the results were shown and optimized by RSM. In the experiments, the variations of three parameters were monitored - temperature, plastic content, and reaction length. The highest biochar yield is exhibited at reaction conditions of 20 minutes, 400 °C, and 100% plastic content. In addition, RSM results show that the plastic ratio has the highest influence on biochar production. The optimum conditions for biochar production are 359.441 °C, 10 minutes reaction time, and 97.8528 wt.% plastic.

## Acknowledgment

The authors would like to thank the University of Sharjah for supporting this study.

## References

[1] Cavalcanti C J S, Ravagnani M A S S, Stragevitch L, Carvalho F R, and Pimentel M F 2022 Simulation of the soybean oil hydrotreating process for green diesel production. *Cleaner Chemical Engineering* **1** 100004.

[2] Esteves V P P, Esteves E M M, Bungenstab D J, Feijó G L D, Araújo O d Q F, and Morgado C d R V 2017 Assessment of greenhouse gases (GHG) emissions from the tallow biodiesel production chain including land use change (LUC) *Journal of Cleaner Production* **151** 578-591.

[3] Al-Dawody M F, Rajak U, Jazie A A, Al-Farhany K, Saini G, Verma T N, and Nashine P 2022 Production and performance of biodiesel from Cladophora and Fucus green diesel *Sustainable Energy Technologies and Assessments* **53** 102761.

[4] Yani F T, Husin H, Darmadi, Muhammad S, Abnisa F, Nurhazanah, Nasution F, and Erdiwansyah 2022 Palm oil hydrodeoxygenation into green diesel over NiO/NbOPO4 catalyst: A novel approach of synthesizing NbOPO4 from NbCl5 *Journal of Cleaner Production* **354** 131704.

[5] Safa-Gamal M, Asikin-Mijan N, Arumugam M, Khalit W N A W, Nur Azreena I, Hafez F S, and Taufiq-Yap Y H 2021 Catalytic deoxygenation by H2-free single-step conversion of free fatty acid feedstock over a Co-Ag carbon-based catalyst for green diesel production *Journal of Analytical and Applied Pyrolysis* **160** 105334.

[6] Jeon K-W, Park H-R, Lee Y-L, Kim J-E, Jang W-J, Shim J-O, and Roh H-S 2022 Deoxygenation of non-edible fatty acid for green diesel production: Effect of metal loading amount over Ni/MgO–Al2O3 on the catalytic performance and reaction pathway *Fuel* **311** 122488.

[7] Abdul Razak N A, Mijan N-A, Taufiq-Yap Y H, and Derawi D 2022 Production of green diesel via hydrogen-free and solventless deoxygenation reaction of waste cooking oil *Journal of Cleaner Production* **366** 132971.

[8] Li X, Chen G, Liu C, Ma W, Yan B, and Zhang J 2017 Hydrodeoxygenation of lignin-derived bio-oil using molecular sieves supported metal catalysts: A critical review *Renewable and Sustainable Energy Reviews* **71** 296-308.

[9] Arun N, Sharma R V, and Dalai A K 2015 Green diesel synthesis by hydrodeoxygenation of bio-based feedstocks: Strategies for catalyst design and development *Renewable and Sustainable Energy Reviews* **48** 240-255.

[10] Chen S, Zhou G, and Miao C 2019 Green and renewable bio-diesel produce from oil hydrodeoxygenation: Strategies for catalyst development and mechanism *Renewable and Sustainable Energy Reviews* **101** 568-589.

[11] Xu H, Ju J, and Li H 2022 Toward efficient heterogeneous catalysts for in-situ hydrodeoxygenation of biomass *Fuel* **320** 123891.

[12] Bu Q, Lei H, Zacher A H, Wang L, Ren S, Liang J, Wei Y, Liu Y, Tang J, Zhang Q, and Ruan R 2012 A review of catalytic hydrodeoxygenation of lignin-derived phenols from biomass pyrolysis *Bioresource Technology* **124** 470-477.

[13] Zanella E, Zassa M D, Navarini L and Canu P 2013 Low-temperature co-pyrolysis of polypropylene and coffee wastes to fuels *Energy & Fuels* **27** 1357-1364.
[14] Sajdak M and Słowik K 2014 Use of plastic waste as a fuel in the co-pyrolysis of biomass: Part II. Variance analysis of the co-pyrolysis process *Journal of Analytical and Applied Pyrolysis* **109** 152-158.
[15] Inayat A, Rocha-Meneses L, Said Z, Ghenai C, Ahmad F F, Al-Ali A M, Mahmood F and Abdallah N 2022 Activated Carbon Production from Coffee Waste via Slow Pyrolysis Using a Fixed Bed Reactor *Environmental and Climate Technologies* **26** 720-729.
[16] Hossain M, Islam M, Rahman M, Kader M and Haniu H 2017 Biofuel from co-pyrolysis of solid tire waste and rice husk *Energy Procedia* **110** 453-458.
[17] Inayat A, Rocha-Meneses L, Ghenai C, Abdallah M, Shanableh A, Al-Ali K, Alghfeli A and Alsuwaidi R 2022 Co-pyrolysis for bio-oil production via fixed bed reactor using date seeds and plastic waste as biomass *Case Studies in Thermal Engineering* **31** 101841.
[18] Raza M, Inayat A, Al Jaberi B, Said Z and Ghenai C 2020 Simulation of the pyrolysis process using blend of date seeds and coffee waste as biomass Advances in Science and Engineering Technology *International Conferences (ASET)* 1-5.

# Highly ordered $TiO_2$ Nanotubes decorated with plasmonic silver nanoparticles ($TiO_2$/Ag) for ultra-sensitive SERS detection of textile dye contaminant Rhodamine 6 G

Krithikadevi Ramachandran[a*], Ameni Rebhi[c,d], Khaled Trabelsi[c], Anouar Hajjaji[c], Bernabé Mari Soucase[d], Brahim Bessais[c], Mounir Gaidi[a,b**]

[a] Centre for Advanced Materials Research, Research Institute of Sciences and Engineering, University of Sharjah, P. O. Box 27272, Sharjah, United Arab Emirates

[b] Department of Applied Physics and Astronomy, College of Sciences, University of Sharjah, Sharjah 27272, United Arab Emirates.

[c] Laboratoire de Photovoltaïque, Centre de Recherches et des Technologies de l'Energie, Technopole de Borj-Cédria, BP 95, 2050 Hammam-Lif, Tunisia.

[d] School of Design Engineering, Departamento de Fisica Aplicada, Universitat Politecnica de Valencia, Cami de Vera, Spain.

*Corresponding author: kramachandran@sharjah.ac.ae

** Co-corresponding author: mkaidi@sharjah.ac.ae

## Abstract

We report a susceptible, ultrafast, molecular fingerprint-specific detection tool of analytes at an extreme trace concentration using surface-enhanced Raman spectroscopy (SERS). Highly ordered $TiO_2$ Nanotubes (NT) modified with plasmonic Ag nanoparticles (Ag) are fabricated by wet chemical strategy. $TiO_2$/Ag nanostructures could benefit the SERS mechanism arising from the effects of both the charge transfer process within $TiO_2$/Ag/analyte and the enhancement in the electromagnetic field by the twining of plasmonic Ag nanoparticles with the construction of semiconductor/plasmonic bi-component system ($TiO_2$/Ag). To achieve a better density of Ag NPs on $TiO_2$ nanotubes, the photoinduction process irradiation time was varied from 10 min to 40 min. The prepared bicomponent SERS system was validated for its structural and morphological properties. SERS optimization studies were analyzed with the probe molecule Rhodamine 6 G (R6G) with an initial concentration of $10^{-6}$M. The highest SERS enhancement was notified with $TiO_2/Ag_{20\ min}$ for R6G sensing. The fabricated optical sensor reproducibility in signal detection was scrutinized with Raman mapping studies and statistically validated by low RSD values of 9.8%. Furthermore, their detectability was gauged with low concentration detection of R6G down to Pico-molar levels ($10^{-12}$M). Synergistic effects of both CT and EM contributed to SERS enhancement to trace and detected pico-levels of R6G by the fabricated bicomponent semiconductor/ plasmonic heterostructure system ($TiO_2/Ag_{20\ min}$).

Keywords: Semiconductor, plasmonic, charge transfer, electromagnetic enhancement, Rhodamine 6 G

## 1. Introduction

The rising population is forcing textile industries to discharge more harmful chemicals into the environment [1]. Therefore, it is vital to monitor the environment for hazardous dyes, particularly in water, with sensors for public health safety [2]. Although traditional techniques like gas chromatography and HPLC have been standard, surface-enhanced Raman spectroscopy (SERS) is emerging as a diagnostic tool to screen various categories of contaminants present in the environment, such as dyes, food additives, bacteria, and pesticides [3]. SERS enhancement is achieved with localized surface plasmon resonance (LSPR) originating from plasmonic nanomaterials. Hotspots promoting signal sensitivity are achieved in the LSPR state when the incoming laser frequency matches the plasmons' oscillating frequency. Hotspots trap as much electric field from the incident laser photons within the stipulated volume, reading out the analyte fingerprint characteristics [4, 5].

The origination of SERS relies on substrate design but still has drawbacks in providing stability, rescale, and uniform signal sensitivity [6]. So far, SERS exploration has been achieved dramatically with plasmonic nanomaterials with various dimensionality-based substrates. Nowadays, there is great interest in the cost-effective design of SERS substrates by conjugating plasmonic/semiconductor materials to promote signal sensitivity by multiple orders through the synergistic effect of both electromagnetic enhancement (EM) and charge transfer enhancement (CT) [7, 8]. $TiO_2$ is an attractive semiconducting material with excellent optical, electrical, and chemical properties that can be easily conjugated with any other nanomaterials for its realization [9]. Amongst plasmonic nanomaterials, silver, being relatively low-cost than gold, is chosen to function with EM enhancement in the proposed SERS architecture [10].

The proposed research focuses on fabricating 1D $TiO_2$ nanotube arrays (NT) decorated with Ag nanoparticles ($TiO_2$/Ag) to achieve the coactive SERS mechanisms of EM and CT. An electrochemical anodization process followed by a photo reduction phenomenon to decorate AgNPs on the $TiO_2$ NT surface at various reduction time intervals was adopted. The fabricated SERS substrates are explored in SERS sensing of organic environmental pollutants such as Rhodamine 6G (R6G).

## 2. Experimental Details

An electrochemical anodization process [11, 12] was carried out to obtain $TiO_2$ NT arrays. Firstly, Titanium foils (TF-99% purity) were polished with abrasive paper and then cleaned in deionized water (DI), ethanol, and acetone under ultrasonication for 15-20 minutes each, respectively, and finally rinsed with DI water. TF was used as an anode, and the counter electrode cathode was chosen as a platinum electrode. The electrolyte consisted of a mixture of solvents such as ethylene glycol (100 mL), DI water (2%), and ammonium fluoride (1% vol/0.065 M concentration). The anode and cathode were completely immersed in the electrolytic solution for 100 minutes at a stable temperature of 25°C. The $TiO_2$ NT was finally heat-treated at 400°C for 3 hours to ensure the formation of the anatase crystalline structure. AgNPs decoration was achieved by a facile photoreduction process, within which the $TiO_2$ NT was soaked in a 0.1M silver nitrate solution for 24 hours. A longer immersion time ensured the adsorption of silver ions within the $TiO_2$ NT, followed by rinsing with DI water. Finally, the $TiO_2$ NT with silver ions was fully immersed in methanol and exposed to UV light ($\lambda = 254$ nm) for 10 minutes, 20 minutes, and 40 minutes, respectively. Lastly, the samples were washed with DI water and dried in a vacuum oven at 75°C for an hour. The samples were labeled throughout the paper as $TiO_2$/Ag 10 min (TA-10), $TiO_2$/Ag 20 min (TA-20), and $TiO_2$/Ag 40 min (TA-40), respectively.

## 3. Results and Discussion

### 3.1 Morphological features of SERS substrates

The morphology properties of $TiO_2$/Ag SERS substrates at different photoreduction times (Fig. 1) exhibit a 1D NT shape with small spherical-shaped Ag nanoparticles decoration. It is evident that porous 1D nanotubes in a vertical order were formed, and the diameter of the $TiO_2$ NT is 100 nm. For the TA-10 SERS substrate, the average size of the AgNPs is 60-80 nm and appears to cluster in some places. With the TA-20 substrate, the AgNPs form flower-like architectures, increasing the average diameter of their size to almost (~120 nm), which is higher than the TA-10 substrate. Similarly, with the TA-40 substrate design, it is clearly noticed that branched clustering of AgNPs resulted in a larger particle size upon longer UV irradiation time (~180 nm). Furthermore, their elemental composition was proven with EDS analysis (Fig. 1, second row). The elemental analysis provides the phase purity of the samples prepared by the adopted method with the key elements of Ti, O, and Ag, respectively

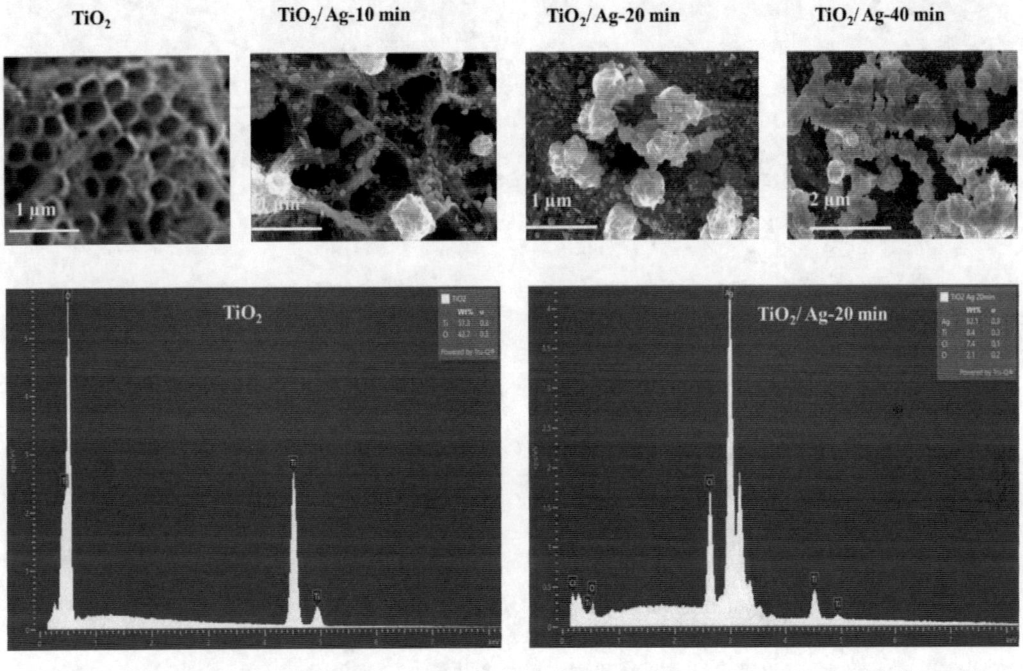


Figure 1. FESEM analysis on the morphology of the prepared $TiO_2$/ Ag SERS substrates (First row); Elemental analysis of $TiO_2$ NT and TA-20 (Second row).

### 3.2 SERS sensing of environmental contaminant dye R6G:

Environmental pollution monitoring with fabricated SERS substrates is the prime focus of the research. Rhodamine 6 G (R6G) is a synthetic dye extensively used as a colorant for tracer applications, printing, laser probes, etc.,[13]. Even though it is useful, it belongs to the class of xanthene derivatives, which causes potential risks like corrosion process in metals, carcinogen to the public and other organisms if not adequately treated before its discharge into the environment [14]. Studies revealed its toxicity in rats causing severe mutilation to retinal ganglion cells and severe damage to DNA in ovarian cells of Chinese hamsters [15]. Being highly hazardous, R6G needs to be adequately treated and monitored to reduce its harmful risks to the environment [16]. The developed SERS substrates will benefit the quality of the treated water to trace and detect the R6G with their unique EM and CT synergistic effects.

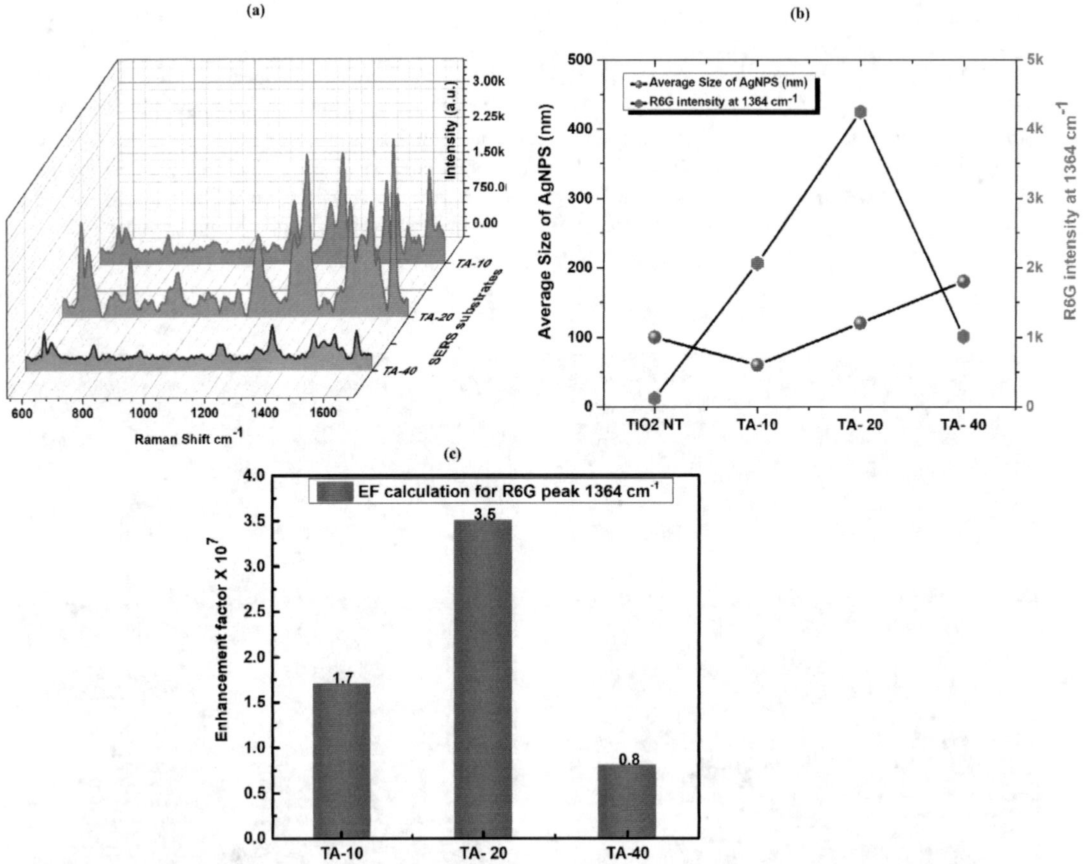


Figure 2: (a) SERS sensing of organic environmental contaminant R6G with TA-10; TA-20; TA-40 SERS substrates (b) Size of AgNPs Vs R6G intensity of detection with prepared SERS substrates (c) Enhancement Factor evaluation with fabricated $TiO_2$/AgNPs SERS substrates.

Different concentrations of R6G solutions were prepared by serial dilutions (ranging from $2x10^{-3}$ M to $2x10^{-6}$ M) and 7 μL of the solution was dropped onto the fabricated substrates [1-3]. After drying, the substrate was analyzed using SERS spectroscopy for detection. It was observed that the unique chemical signature of R6G was clearly detected with the fabricated substrates (Fig 1a). The SERS fingerprint of R6G was identified at 614 cm-1 (in-plane bending of C-C-C ring), 1186 $cm^{-1}$ (in-plane bending of C-H), 770 $cm^{-1}$ (out-of-plane bending of C-H), 1307 $cm^{-1}$, 1364 $cm^{-1}$, 1508 $cm^{-1}$, 1582 $cm^{-1}$, and 1654 $cm^{-1}$ (stretching modes of C-C) [1, 3]. The SERS intensity was found to vary with the size of the AgNPs, and the highest sensitivity was achieved with the TA-20 SERS substrate (Fig 1b). Therefore, the optimum size of AgNPs for achieving high sensitivity in this study was determined to be around 210 nm. The enhancement factors were also calculated (Fig 1c), and the TA-20 substrate had the highest value at 3.5 X $10^7$.

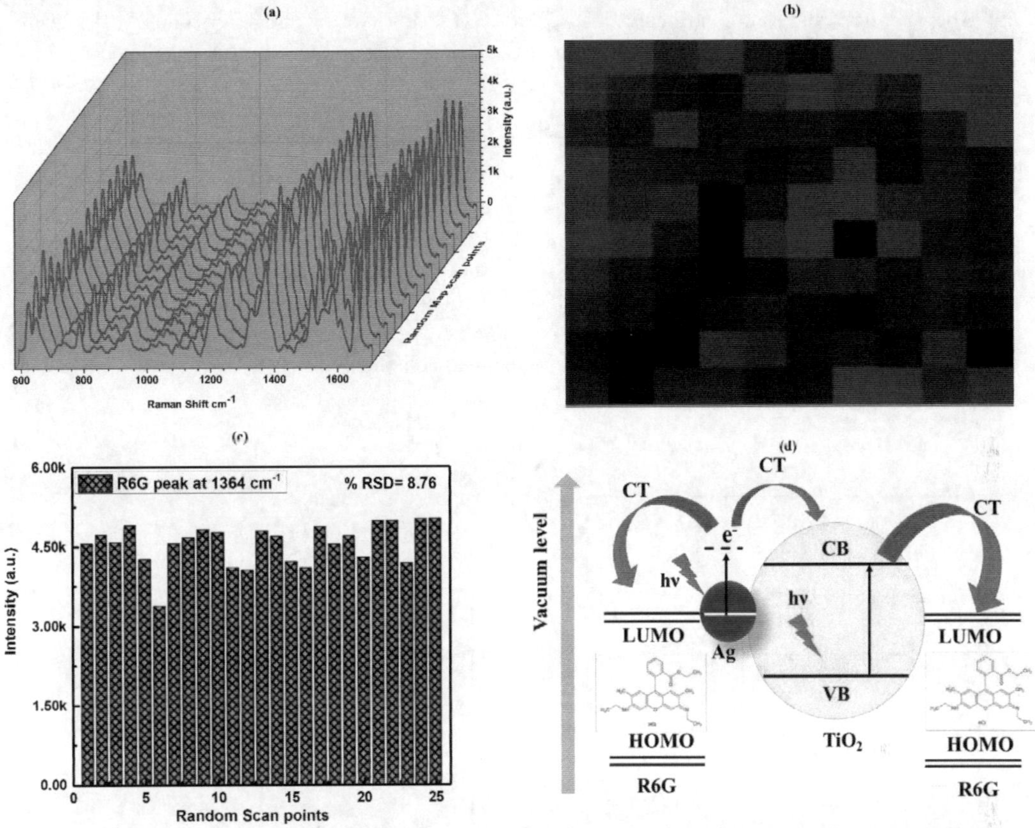


Figure 3: (a) Mapping spectrum for random scan points (b) 2D view of Map scan for R6G at peak 1364 $cm^{-1}$ (c) RSD analysis (d) CT mechanism with fabricated $TiO_2$/Ag SERS substrate

The mapping spectrum analysis (Fig 3a) demonstrated the uniformity of the SERS substrate in signal detection, with the optimized substrate TA-20 exhibiting good uniformity in R6G signal sensing without any loss of its fingerprint characteristics. The 2D scan spectrum (Fig 3b) revealed areas of higher sensitivity denoted by dark blue squares and low sensitive regions represented by black squares. Additionally, the RSD analysis was carried out for the R6G peak intensity at 1364 $cm^{-1}$ to validate the sensor characteristics. Around 25 scan spectrums were randomly derived from the map data to calculate the % RSD, which was found to be 8.76, within the acceptable range for the sensor [3].

The SERS mechanism was found to be due to the synergistic process of both EM and CT effects arising from individual AgNPs and $TiO_2$/AgNPs/analyte. The TA-20 substrate, which exhibited dominant sensing behavior, was attributed to the equally spaced arrangement of AgNPs on the surface of the nanotube. SERS is affected by particle size, shape, and interparticle distance of the fabricated architecture [4, 6]. Hotspots originated from the spherical-shaped AgNPs with an identical interparticle distance in the TA-20 substrate, promoting the R6G sensing intensity more than the other substrates.

With TA-10, although the AgNPs are fairly located, the long-range distances fail to bring about the stronger EM effect, and with TA-40, the particle size enlarges, forming aggregated structures that reduce their surface area, and therefore, their ability to sense the analyte. On the other hand, the direct bond between $TiO_2$ and R6G contributes to the CT effect (Fig 3d). The total SERS effect of the designed SERS substrate is explained by the conjunction of EM and CT effects, which promotes their ability to trace and detect R6G.

### 3.3 Limit of detection studies with fabricated SERS substrate TA-20:

Detectability studies were conducted with TA-20 for the R6G molecule at varying concentrations ranging from $10^{-7}$M to $10^{-12}$M (see Fig 4a). A decrease in the intensity of R6G peaks was observed at low detection concentrations with TA-20. This decrease in signal sensitivity can be attributed to the lower number of R6G molecules present at lower concentrations. The TA-20 SERS substrate was able to detect the R6G signal at $10^{-12}$M, but could not detect the signal at concentrations lower than this. Therefore, the prominent peak of R6G at 1364 $cm^{-1}$ was chosen to plot the calibration curves (see Fig 4b). The regression statistics coefficient was computed to be $R^2$=0.99, indicating a linear correlation.

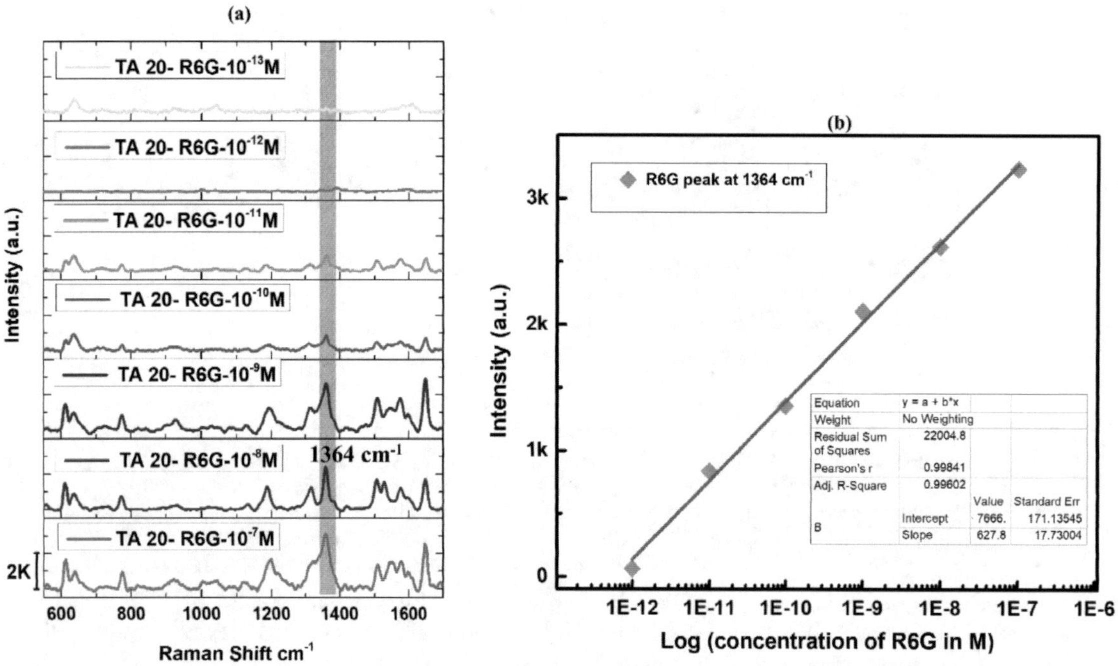


Figure 4: (a) LOD studies of R6G with TA-20 (b) Linear fit analysis of LOD studies for R6G peak at 1364 $cm^{-1}$.

### 4. Conclusion

The current research proposes a bicomponent architecture design ($TiO_2$/Ag) for ultra-sensitive SERS detection of the textile dye contaminant Rhodamine 6G. By varying the photoreduction time from 10 to 40 minutes, the size of the AgNPs varied accordingly as well as their dispersion over the $TiO_2$

surface. The optimum size of AgNPs providing the highest SERS sensitivity of R6G is 210 nm, which was attained with the TA-20 SERS substrate. A higher enhancement factor of $3.5 \times 10^7$ was observed for the TA-20 substrate. The fabricated TA-20 substrates exhibited good reproducibility in SERS signal sensing with an RSD of 8.76%. Further limit of detection studies revealed the TA-20 substrate's capability to sense picomolar levels ($10^{-12}$M) of the water contaminant dye R6G. The underlying SERS mechanism that enables exorbitant, ultra-low level detection of R6G with the best TA-20 SERS substrate is owing to the synergistic effects of EM and CT from the plasmonic AgNPs and semiconductor $TiO_2$ NT.

## 5. Acknowledgement

The authors would like to acknowledge the financial support from the University of Sharjah-United Arab Emirates (Grant No. 23021430136).

## References

[1] Ramachandran K, Daoudi K, Columbus S, Chidambaram S and Gaidi M 2022 Environ. Technol. Innov. 27 102409
[2] Ziad R, Columbus S, Elgamouz A, Daoudi K, Kawde A N, Ramachandran K and Gaidi M 2023 SAA. 297 122701
[3] Daoudi K, Ramachandran K, Alawadhi H, Boukherroub R, Dogheche E, Khakani M A El and Gaidi M 2021 Surf. Interfaces .27 101454
[4] Ferreira de Sousa Junior J, Columbus S, Hammouche J, Ramachandran K, Daoudi K, and Gaidi M 2023 Appl. Surf. Sci. 613
[5] Columbus S, Hammouche J, Ramachandran K, Daoudi K and Gaidi M 2022 *J.* Photochem. Photobiol. A Chem. 432 114123
[6] Yang W, Tang J, Ou Q, Yan X, Liu L and Liu Y 2021 ACS Omega 6 27271–8
[7] Columbus S, Hamdi A, Ramachandran K, Daoudi K, Dogheche E H, and Kaidi M 2022 Sensors Actuators A Phys. 347 113962
[8] Ramachandran K, Hamdi A, Columbus S, Meselmene N Al, Dogheche E, Daoudi K and Kaidi M 2022 Surf. Interfaces .35 102407
[9] Ramachandran K, Ziad R, Columbus S and Daoudi K 2022 J. Mater. Sci. Mater. Electron. 33 9194–9203
[10] Pereira R N and Peripolli S B 2023 SAA 290 122262
[11] Gaidi M, Trabelsi K, Hajjaji A, Chourou M L, Alhazaa A N, Bessais B and El Khakani M A 2018 Nanotechnology 29 015703
[12] Trabelsi K, Hajjaji A, Gaidi M, Bessais B and El khakani M.A 2017 Journal of Applied Physics 122, 064503
[13] Sabnis R W 2015 Handbook of Fluorescent Dyes and Probes
[14] Choyke P L, Alford R, Simpson H M, Duberman J, Craig Hill G, Ogawa M, Regino C, and Kobayashi H 2009 Mol. Imaging 8 341–54
[15] Nestmann E R, Douglas G R, Matula T I, Grant C E and Kowbel D J 1979 Cancer Res. 39 4412–7
[16] Yudasari N, Anugrahwidya R, Tahir D, Suliyanti M M, Herbani Y, Imawan C, Khalil M and Djuhana D 2021 J. Alloys Compd. 886 161291

# Development and characterization of a clay-graphite-$CeO_2$ composite electrode for simultaneous detection of heavy metals

**Sondos Nizar[1], Abderrazek El-Kordy[1], Abdelaziz Elgamouz*[1], Abdelnasser Kawde*[1], Tahar Laoui[2]**

[1] Pure and Applied chemistry Research Group, Department of Chemistry, College of Science, University of Sharjah, Sharjah P.O. Box 27272, United Arab Emirates

[2] Department of Mechanical and Nuclear Engineering, College of Engineering, University of Sharjah, Sharjah P.O. Box 27272, United Arab Emirates

Corresponding author: aelgamouz@sharjah.ac.ae, akawde@sharjah.ac.ae

**Abstract.** A carbon paste electrochemical sensor doped with clay impregnated with cerium oxide ($CeO_2$), CPE/Clay/$CeO_2$, was developed and used for the electrochemical sensing of four heavy metals, Cd(II) , Pb(II), Cu(II) and Hg(II). The optimum ratio of clay/$CeO_2$ and sintering temperature were found to be 10 wt.% Clay/$CeO_2$ and 1000°C respectively. Heavy metals were sensed simultaneously on the 10 wt.% Clay/$CeO_2$@1000°C sensor using square wave voltammetry (SWV), which was found to yield the highest current density, with optimum parameters: potential increment 0.001 V, amplitude 0.07 V, frequency 15 Hz, preconditioning potential -1.0 V, preconditioning time of 30 s. Linear range calibration curves were obtained in the ranges of 2.0-28 μM, 2.0-25 μM, 2.0-28 μM, 2.0-28 μM for Cd(II), Pb(II), Cu(II) and Hg(II), with respective limit of detection (LOD) and limit of quantitation (LOQ) of 0.42, 1.4 μM for Cd(II), 0.37, 1.2 μM for Pb(II), 0.28, 0.94 μM for Cu(II) and 0.31, 1.0 μM for Hg(II). The prepared CPE/Clay/$CeO_2$ sensor showed a very high selectivity toward the selected heavy metals.

## 1. Introduction

Heavy metal contamination is dangerous to both environment and human health due to their non-biodegradable nature [1]. Heavy metal contamination originates from different sources, including industrial activities such as producing chemicals, fertilizers, mining, foundries, smelters, oil refineries, and coal-burning stations [2]. Therefore, the wastewater decontamination from toxic heavy metals is becoming a serious problem because this contamination is dangerous and toxic to the environment, plants, and humans when it is available at contaminated levels [3]. The toxic metals can leach into the soil and bioaccumulate in the plants resulting in biomagnification in apex predators [1]. Due to their high soil-to-plant transfer rate, these contaminants are found in most human foodstuffs.

Lead, cadmium, copper, and mercury are four heavy metals that are known to be toxic, they are found in wastewater and groundwater, mainly originated from anthropogenic sources such as industrial activities. The scale of toxicity of the four studied heavy metals is Pb < Cu < Cd < Hg [4].
Studies reported that in cadmium-polluted areas, there is an overly high mortality risk among people with glomerular proteinuria. Cadmium was also found to cause osteoporosis [5] and implicated in kidney toxicity [1]. For instance, studies in Japan and the United States have reported that excess cancer mortality was related to environmental exposure to cadmium. An increase in endometrial cancer risk was observed in participants who consumed > 15 μg/day of cadmium, from a vegetable source [6].
Mercury is emitted to the environment in its inorganic form, which is converted to methylmercury, an organic form of mercury in soil and water. This toxic form of mercury enters the food chain and then the human body through consumption of fish and shellfish contaminated with methylmercury [7]. Low-level mercury exposure was linked to cardiovascular and renal diseases [5], it was found to affect the kidney, peripheral and central nervous systems, and gastrointestinal track [7].
Lead was also associated with vascular disease and mortality [5]. Among both adults and children, higher exposure to lead causes a delay in neurological development, decrease/inhibition of Delta-Aminolaevulinic Acid Dehydratase (ALAD) activity, anaemia, sperm count, and blood pressure [7].
Copper metal is also responsible for many human health problems, including Wilson disease, kidney failure, stomach and intestine problems, and inhibition of microbial activity such as cell metabolism [4].
Conventional optical techniques are used for the sensing of heavy metals like spectrophotometric analysis [10] and spectroscopic techniques like neutron activation analysis (NAA) [11], atomic absorption spectroscopy (AAS) [12], and X-ray fluorescence (XRF) [13]. The above-mentioned techniques are presented with low detection limit, very good sensitivity, and simultaneous detection of elements [2]. However, optical techniques are not as effective as electrochemical techniques due to their high energy consumption, multi-step sample preparation, expensive, the requirement for individual instrument operation training, and low volume concentration removal of metal ions [2, 3,8-14].
Electrochemical techniques are promising methods and are becoming interestingly important for their many advantages, such as applying a notable selectivity rate to the decontamination procedure [11], easy to use, low-cost, high-energy efficiency, rapidness, simplicity, lower sensitivity [2], easily combined with other technologies and the conditions needed are moderate [16].
Carbon-paste electrode (CPE) have been used widely for the sensing of heavy metals due to the broad range of potential, their versatility, and their high electrochemical surface and conductive properties that allow sensitive determination of the metals. The four most common carbon-based electrodes are carbon-paste, graphite, glassy carbon, and screen-printed electrodes [18].
Many materials have been used to modify the surface area of the working electrodes, due to their oxide composition, large specific area, and inherent availability, especially the aluminosilicate like zeolites, clays or their hybrid materials have gain special interest in the modification of various electrochemical platforms amongst various other electrode modifiers [23].
Clays are naturally occurring fine particles which get plasticized when mixed with an appropriate amount of water and harden when dried and calcinated [24], with particle diameter ranging from 2 to 5 μm [25]. Clays are physically and chemically active. The specific crystal shape and the small size give clays their physical activity, whereas the internal structure gives a specific characteristic of their chemical activity [25]. The interesting internal features brought about by the laminar structures created by the series of tetrahedral silicate $SiO_4$ and/or octahedral aluminates Al-O, Al-OH are the source of clay's chemical activity. A sheet is produced when oxygen atoms in tetrahedral or octahedral structures are shared [26]. By separating analytes from the matrix and prompting a certain shape, the interlayer restricted space act as an interior hollow to capture the analyte [2].
Clays have a complex porous structure with negative surface charge due to the substitution of tetrahedral and/or octahedral that attract cations more than anions and non-ionic species [27], the surface modification of the clay provide the opportunity to increase their suitability for use in electrochemical analysis [28]. This provides the ability of ion exchange on the surface of the clay that will help with detecting and attracting the cationic heavy metals. Modifiers of the clay can be added to the surface of

the electrode as a film or it can prepare and mixed with the carbon paste and clay [27] (as in our study). In this case, the detection of cationic heavy metals is based on ion exchange reaction on clay modifiers. In our study, the clay is modified with cerium oxide mainly due the crystallographic structure stability of the cerium in its solid form when converted from one form to the other ($Ce^{4+}$ to $Ce^{3+}$) during an electrochemical redox process. Furthermore, there are no published studies that implies the use of cerium oxide as a clay modifier in carbon paste electrodes. It is expected that the $CeO_2$ will improve the conductivity of the electrode paste and helps in electron transfer.
This present study attempts to prepare carbon paste electrochemical sensors doped with Clay-$CeO_2$ for the electrochemical sensing of heavy metals, various techniques namely: cyclic voltammetry (CV), square wave voltammetry (SWV), differential pulse voltammetry (DPV), normal pulse voltammetry (NPV), and linear scan anodic stripping voltammetry (LSASV), are examined in this study to find the suitable technique to sense heavy metal cations in aqueous solutions. After selecting the best electrochemical technique based on the current density, the parameters of the technique are optimized. Different clay compositions and different temperatures at which the clay composition is sintered are examined to find the best conditions at which our sensor works better.

## 2. Experimental section

### *2.1. Chemicals*

The chemicals listed in this study were used without further purification. Potassium ferro-cyanide ($K_4(Fe(CN)_6).3H_2O$, min assay 99%) obtained from Wardle chemicals, potassium ferricyanide ($K_3(Fe(CN)_6)$ 99%) was obtained from sigma Aldrich, mercury (II) chloride ($HgCl_2$) was obtained from AnalaR, lead (II) nitrate ($Pb(NO_3)_2$) was obtained from Riedel-de haen. Graphite powder (<20 μm, synthetic) and mineral oil were acquired from sigma Aldrich, potassium chloride KCl (assay, above 99.0%) obtained from Eurolab, Cadmium nitrate-4-hydrate ($Cd(NO_3)_2.4H_2O$, assay 99%) was obtained from Riedel-de haen, Cerous nitrate ($Ce(NO_3)_3.6H_2O$), Activated charcoal (≤8% loss on drying, 120 °C, pH: 6-9) were obtained from Supleco analytical. 0.1 M acetate buffer pH 5.0 was freshly prepared from acetic acid ($CH_3COOH$) and sodium acetate anhydrous ($CH_3COONa$) which were obtained from Breckland scientific supplies. Ultra-pure distilled water was obtained from Water Still Aquatron A 4000 D (UK), water purifier, and was used for blanks, buffers preparations and cleaning glassware.

### *2.2. Clay Composition Preparation*

The graphite paste electrode was prepared by grinding and mixing clay collected from Ras-Al-Khaimah-UAE and saturated with $Ce(NO_3)_3.6H_2O$ solution. The prepared solution was kept stirring at room temperature overnight. In the next day, the solvent layer was decanted, and the solid particles were dried to obtain clay powder dopped with cerium oxide. To determine how clay/$CeO_2$ structure will react to temperature change and to study the enhancement of electrochemical properties of the CPE, the clay loaded with $CeO_2$ was calcinated to final temperatures of 250 °C, 500 °C, 700 °C, 900 °C, 1000 °C and 1100 °C, 10 wt% Clay/$CeO_2$, then 10 wt% CPE electrodes were prepared from Clay/$CeO_2$ sintered at different temperatures by making 1.0 g preparation of paste composed of 0.1 g Clay/$CeO_2$, 0.6 g of graphite and 0.3 g of mineral oil as binder. In addition, different clay-$CeO_2$ compositions, containing 10 wt.%, 20 wt.%, 30 wt.%, 40 wt.%, and 50 wt.% clay, were prepared at sintering temperature of 1000 °C to obtain the best composition of the electrode. Clay-$CeO_2$ was mixed with mineral oil that serves as a binder to ensure the homogeneity of the mixture and graphite using mortar and pestle for 10 minutes. The paste mixture was put onto the electrode's sensing part. The working electrode is 3D printed cylindrical hollow tube with dimensions identical commercial GCE (7.0 cm long, and disc diameter of 2.0 mm and surface area of 0.0314 $cm^2$), a copper wire was used as an electric connector. The paste surface on the working electrode was smoothed on wax paper. The electrochemical cell consists of saturated silver/silver chloride (Ag/AgCl) electrode as reference electrode, the platinum (Pt) wire as a auxiliary electrode and the prepared clay-graphite-$CeO_2$ electrode as the working electrode. $K_4Fe(CN)_6/K_3Fe(CN)_6$ equimolar solution was used for CV and EIS analysis to find the active electrochemical surfaces and impedances of different prepared sensors. Buffer acetate (pH=5.0) was used as electrolyte for all analysis based on our previous studies [2].

### *2.3. Physicochemical characterization*

Physical characterization of the raw clay, different clay compositions ranging from 10 to 50 wt.%, and different sintering temperatures ranging from 250°C to 1100°C is carried out by XRD and Field Emission Scanning Microscopy (FESEM) to study the morphological structure of the modified materials used in the electrochemical sensor fabrication. FESEM micrographs are obtained at a 10 kV accelerating voltage with 2µm and 10 µm. X-ray diffraction (XRD) is used to provide us information about the material properties like crystal structure, crystalline size, strain, and phase transformation. Results are obtained with CuKα radiation (λ= 1.5418 Å) on Bruker D8 machine with Cu tube. Copper is used as an node and graphite monochromator is used as a cathode. The data is collected between 2θ= 5.000° 85.000° with a speed of 0.400 seconds/step.

### *2.4. CPE Electrodes' Characterization Using Electrochemical Techniques*

In an electrochemical cell with three electrodes setup, the clay-modified carbon paste electrode is used as the working electrode, the Ag/AgCl (saturated KCl) as the reference electrode, and a Pt wire as the counter electrode. CV measurements were recorded on a CHI-1232c electrochemical workstation. CV was scanned for all electrodes made from materials sintered at temperatures 250-1100°C, higher current response is associated with a higher electrochemical surface area. The CPE with higher response was selected then scanned again with different clay mass ratio. Electrochemical impedance spectroscopy (EIS) was recorded on a Zahner Ennium Pro 43,243 electrochemical workstation to study the conductivity of different CPE prepared from different clay's ratios and different sintering temperatures. LSASV, DPV, SWV and NPV were all recorded on CHI-1232c electrochemical workstation to study the selectivity toward heavy metals. Peak current signals of the five techniques were recorded and current densities calculated by dividing the current by the electrochemical surface area were compared. The technique with highest current density was selected and parameters were optimized for Cd (II), Pb (II), Cu (II), and Hg (II). The selected technique with optimum parameters was then exploited for simultaneous sensing of heavy metals using clay-modified carbon paste electrode with the selected clay mass ratio and sintering temperature to study the percentage of recovery. The limit of quantitation LOQ and limit of detection LOD were then determined using calibration curves.

### *2.5. Electrochemical analysis of heavy metals*

In a plastic electrochemical cell with our clay-modified carbon paste electrode as the working electrode, Ag/AgCl (Sat. KCl) as the reference electrode, and Pt wire as counter electrode, 2.00 mL of buffer acetate were added to the cell, the CV, LSV, DPV, NPV, and SWV were recorded separately as a blank signal using CHI-1232c electrochemical workstation. Two methods were used, the single metal analysis is carried out by spiking 100 µL of $10^{-2}$ M of the single metal Pb(II), Cd(II), Cu(II) or Hg(II) to the same electrochemical cell, the stirring is switched on for 30 seconds to ensure the homogeneity. For multi-metal analysis, the later step is repeated by spiking consecutively 100 µL $10^{-2}$ M Pb(II), Cd(II), Cu(II) and Hg(II). CV, LSV, DPV, NPV, and SWV signals were recorded separately, after the addition of each metal. The surface of the working electrode was refreshed with new paste between each technique. The electrodes were washed with diluted nitric acid between each run to remove any metal traces.

## 3. Results and Discussion

### *3.1. Clay/$CeO_2$ material's Characterization*

In Figure 1, is represented the crystallographic structure of raw clay (Figure 1a) and $CeO_2$ doped clay material (Figure 1b). The figure shows different phase mixtures that originally consist mainly of illite, kaolinite, quartz, and calcite. The miller indices, interreticular distances, and the 2θ position of the quartz reflexions were found to be (100), 4.2, 20.8°; (101), 3.3, 26.5°; (102), 2.3, 39.5°; and (112), 1.8, 48.5°

which were identified using quartz diffraction file No 00-046- 1045. The kaolinite phase was characterized with miller indices, interreticular distances, and the 2θ position as follow: (001), 7.2, 12.2°; (020), 4.5, 19.8°; and (002), 3.5, 24.2°, these parameters were identified through powder diffraction file No 00-014-0164. The illite phase was identified by miller indices, interreticular distances, and the 2θ as follow: (002), 10.1, 8.4; (006), 3.34, 26.5, which were identified using illite diffraction file No 00-026-091. Carbonate was found to be present in the form of calcite with miller indices, interreticular distances, and the 2θ as follow: (012), 3.9, 23.1°; (104), 3.0, 29.4°; (110), 2.5, 39.6°, and (018), 1.9, 47. 47.5°, which were identified using the calcite diffraction file No 00-005-0586. The XRD of 10% clay-graphite-$CeO_2$ sintered to 700 °C shows a significant change in the clay structure and appearance of strong peaks corresponding to Cerianite ($CeO_2$) identified with miller indices, interreticular distances, and the 2θ position as follow: (111), 3.37, 26.4°; (220), 2.1, 44.4°; (311), 1.7, 54.5°; (222), 1.5, 59.8; (400), 1.2, 77.3°, which were identified using $CeO_2$ diffraction file No R05037-9. This suggests that $CeO_2$ was successfully dopped with the clay. Therefore, the prepared clay-$CeO_2$ paste consists mainly of quartz and kaolinite minerals and $CeO_2$.

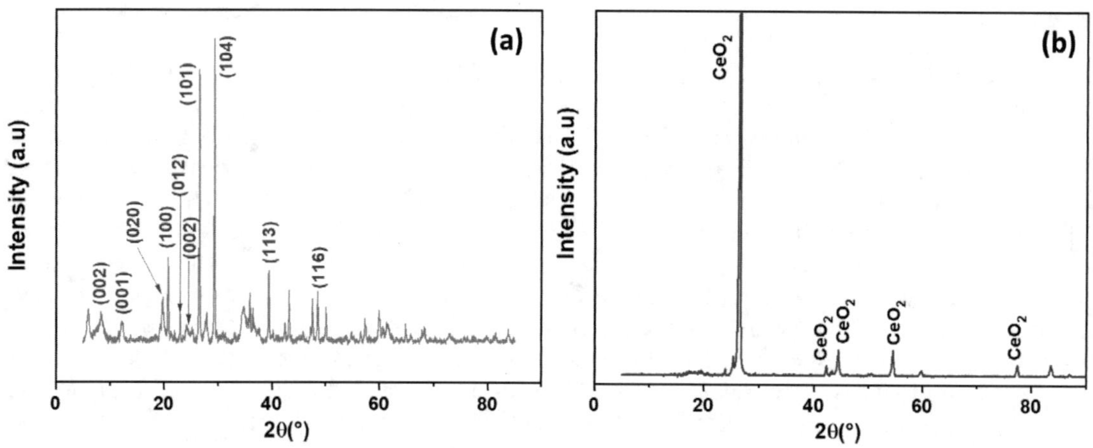


**Figure 1**. X-ray diffraction analysis of (a) Raw clay material and (b) 10% clay-graphite-$CeO_2$ @ 700°C, C: calcite, K: Kaolinite, I: illite, Q: quartz.

To learn about morphology and elemental composition of raw and sintered clays, Field Emission Scanning electron microscopy (FESEM) is used. FESEM analysis was carried out, in Figure 2, are presented micrographs of raw clay as well as different wt% doped $CeO_2$ at different sintering temperatures. The morphology of the clay surface shows that they are generally rough surfaces. The roughness and porosity of the clay sample is increasing as the temperature increases to 1000°C where the porosity and particle size decreased which is likely due to grains' fusion. If a comparison is to be made between the cerium oxide modified clays, it is evident that the cerium modified clay sintered at 1000°C displays a homogeneous layer as compared to the other cerium oxide pastes displayed.

### *3.2. Cyclic Voltammetry to Study Electrochemical Surface Area*

The electrodes electroactive surface areas were determined using cyclic voltammetry analysis of CPE@10wt.%Clay/$CeO_2$ sintered to various final temperatures (250°C, 500°C, 700°C, 900°C, 1000°C, and 1100°C) were performed at different scan rates (200mV, 100mV, 75mV, 50mV, 40mV, 30mV, 20mV, 10mV, 5 mV) in the presence of 5.0 mM equimolar solution of $K_3Fe(CN)_6$ /$K_4Fe(CN)_6$. The largest electrochemical surface area is demonstrated by the surface of CPE@10wt.%Clay/$CeO_2$ sintered to 1000 °C.

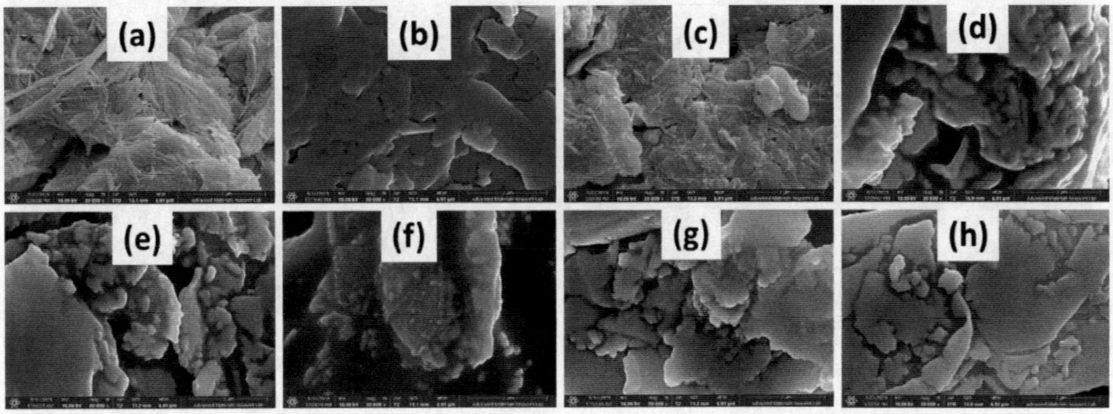


**Figure 2.** FESEM micrographs of raw clay from UAE (a), clay-$CeO_2$ (b), (10 wt.% clay sintered to 250°C (c), 500°C (d),700°C (e), 900°C (f), 1000°C (g)) and 40 wt.% clay sintered to 1000°C (h)

Therefore, the surface area development of various wt.% of the composite on the CPE electrode composition was further investigated using the Clay/$CeO_2$ sintered to 1000 °C composite material. The modified CPE/Clay/$CeO_2$ electrodes' electrochemical surface areas were calculated after the Randles-Sevick equation given in Eq. 1

$$I_p = 2.69 \times 10^5\, n^{3/2}\, A\, D^{1/2}\, C\, \gamma^{1/2} \quad \text{Eq. 1}$$

In Eq. 1, Ip is the oxidation peak current of the $K_4Fe(CN)_6$ given in ampere (A), A, is the electrochemical active surface in $cm^2$, D, is the diffusion coefficient of the $[Fe(CN)_6]^{4-}$ which is subjected to oxidation at the surface of the electrode taken as $7.19 \times 10^{-6}$ $cm^2s^{-1}$, C is the concentration of the same ion, taken as $5.0 \times 10^{-5}$ $mol.cm^{-3}$, n represent the number of electrons involved in the ferro/ferricyanide redox reaction, taken as 1 $e^-$, and $v$ is the specific slew at which the reaction is taking place is given in V/s.

All electrodes show that the $Fe(CN)_6^{4-/3-}$ system is reversible with the highest current density of $1.17\times10^{-2}$ $A/cm^2$ presented for the 10%wt./Clay/$CeO_2$@1000°C (Figure 3a, 3c). The calculated electroactive surface areas for different electrode compositions shows that 10%wt. $CeO_2$ at sintering temperature of 1000 °C, was presented with the highest electroactive surface area of 0.111 $cm^2$ (Figure 3b, 3e). Whereas at 700 °C the electrode shows the lowest electroactive surface area. This could be attributed to the conversion of $Ce(NO_3)_2$ to $CeO_2$ at high temperatures.

There is a trend in the surface area, and we can see that the surface area was a little bit higher because of higher surface area presented by the raw clay material, on the further increase of the temperature of sintering of the material to 500°C, the electrochemical surface decrease by 6%, this is mainly due to the loss of the clay surface because of grains consolidation. Interestingly the electrochemical surface at 700 °C is represented with the lowest value, with a 66% decrease, this may be linked with the transformations that happen in the clay, in fact, clay, according to the TGA analysis is subjected to a dramatic transformation due to the loss of carbonates and transformation of kaolinite to metakaolin [3]. On further increase of the temperature to 900 °C, we can see an increase in the electrochemical surface by 100% compared to 700 °C, this may be due to the start of the formation of Clay/$CeO_2$ composite, remarkably at 1000 °C, the electrochemical surface area increases by 700% which may be due to the formation of a highly conductive composite formed between the clay and $CeO_2$. While on continuous increase of temperature to final temperature to 1100°C, the electrochemical surface area drops by 18%, the 10%wt./Clay/$CeO_2$@1000°C composition was chosen as optimum preparation for further studies.

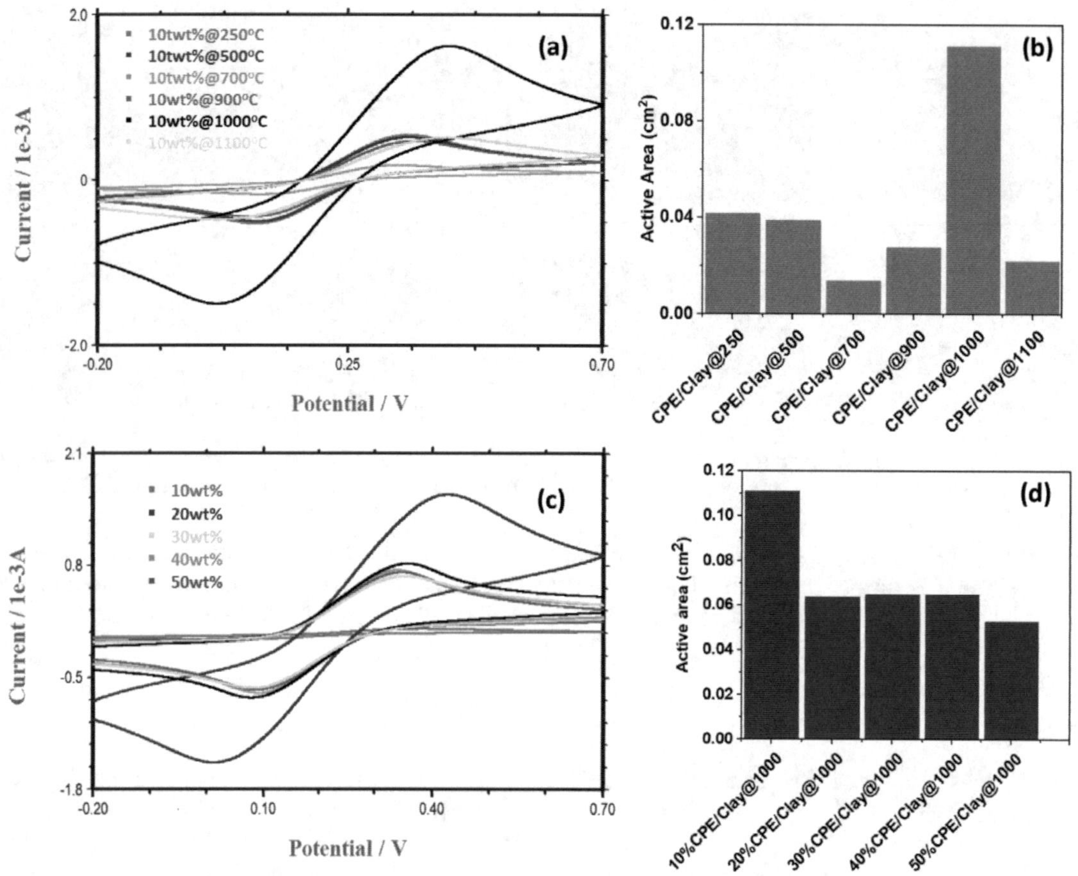


**Figure 3.** Cyclic voltammetry of 5.0 mM $K_3Fe(CN)_6$ / $K_4Fe(CN)_6$ system at a 100.0 $mV.s^{-1}$ of different CPEs (a) electrodes prepared from Clay/$CeO_2$ sintered from 250°C to 1000°C, (c) electrodes prepared from different wt.% of Clay/$CeO_2$ ranging from 10wt.% to 50wt.%, (b, d) their corresponding electrochemical surface areas.

*3.3. Impedance to study conductivity*

EIS analysis was carried out to study the conductivity of different electrode compositions. Figure 4 show an overlay of 10wt.% Clay/$CeO_2$ sintered at different temperatures of 250 °C, 500 °C, 700 °C, 900 °C, 1000 °C, and 1100 °C.

From the EIS data, the equivalent circuits (Figure 4c, d) and their parameters, solution resistance ($R_s$) and charge transfer resistance ($R_{CT}$) were determined using the Zahner data fitting. Values are plotted in Figure 4b. The electrodes prepared from Clay/$CeO_2$ sintered to temperature from 250°C to 1000°C follow the simplified Randles cell represented in Figure 4c, which include the solution resistance in series with a double layer capacitor ($C_{dL}$), charge transfer resistance ($R_{CT}$) and a Warburg impedance (W). While the CPE/10wt.%Clay/CeO2@1100°C was fitted with the Randels circuit represented in Figure 4d, this model is equivalent to two semicircles (Figure 4a) and represented a coating capacity $C_c$ and a pore resistance ($R_{po}$), in addition to components of a simplified circuit ($R_s$, $C_{dL}$, and $R_{CT}$). Comparable Rs and $R_{CT}$ were obtained for all electrodes prepared with Clay/$CeO_2$ sintered from 250 to 1000°C, values varying between 17.3 and 27 Ω for $R_S$ and 284 to 298 for $R_{CT}$. A slight increase in the $R_S$ and $R_{CT}$ to 87 and 373 Ω respectively, was observed for the CPE/10wt.%Clay/$CeO_2$@1000°C electrode, which may be due to transformations happening in the clay at higher temperatures. Surprisingly a second semicircle appear with a resistance of mass transfer (RMT) of the electrode made from 10wt% Clay/$CeO_2$ and sintered to 1100 °C was presented with the highest values of the charge transfer or polarization resistance $R_{ct}$ with comparison to other electrodes prepared from composite (Clay/$CeO_2$) sintered at different temperature varying between 250°C and 1000°C. Different electrochemical processes that are taking place in the system are commonly represented by the two semicircles. The charge transfer resistance at the electrode-electrolyte interface is linked to the high-frequency semi-circle, which sheds light on the kinetics of the electrochemical reaction. On the other hand, the low-frequency semi-circle, which reflects the electrode material's porosity and ion diffusion properties, is connected to the diffusion or transport processes within it.

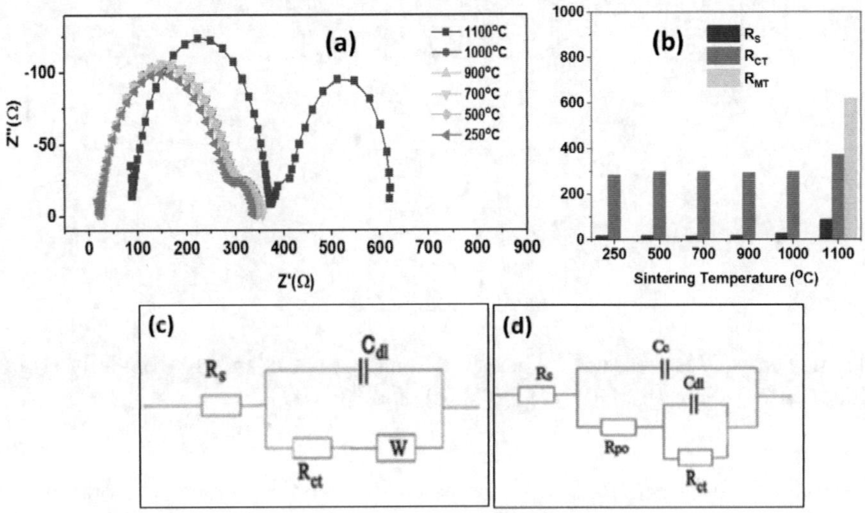


**Figure 4.** EIS analysis of the CPEs/fabricated from 10wt%Clay/$CeO_2$ composite sintered to the temperature varying from 250°C to 1100°C. (a) Fitted Nyquist plots, (b) Impedance analysis, (c) equivalent circuit for 250-1000°C and (d) 1100°C.

### *3.4. Heavy metals analysis*

CV, DPV, NPV, LSV, and SQWV were used to find the most suitable technique for the detection of heavy metals at the surface of the CPE/10wt%Clay/$CeO_2$ are performed on stock solution of $10^{-4}$ M of Cd (II), Pb (II), Cu (II), and Hg (II) for the determination of trace heavy metals. The oxidation peaks of the heavy metals were positioned at -0.703, -0.477, 0.004, and 0.379 V for Cd, Pb, Cu and Hg respectively in the CV analysis (Figure 5a). Nonreversible peaks were observed for Pb and Cu only. These potentials were found to change when using LSV (Figure 5b), DPV (Figure 5c), NPV (Figure 5d), and SWV (Figure 5e). The higher redox potential of mercury Hg compared to copper Cu could be explained by the different electrochemical behavior of the metals, $Hg^{2+}$ (aq) + $2e^-$ → Hg(s) half-cell reaction has an $E^\circ$=0.80V which is higher than that of copper $Cu^{2+}$ (aq) + $2e^-$ → Cu(s) $E^\circ$=0.34V, which means that the tendency of mercury to be reduced is higher than that if copper. Among the five used techniques, SWV was presented with the highest current densities for the four heavy metals investigated.

### *3.5. Optimization of the technique and calibration curves*

SWV parameters were optimized to obtain the best parameters with better signals. Potential increment was varied between 0.001, and 0.04 V, amplitude between 0.01, and 0.09 V, frequency between 10, and 35 Hz, pre-conditioning potential was fixed at -0.5 V, and accumulation time changed between 0, and 300 s. and pre-conditioning time fixed at 305 s, while accumulation potential was changed between – 0.5, and +1.0 V

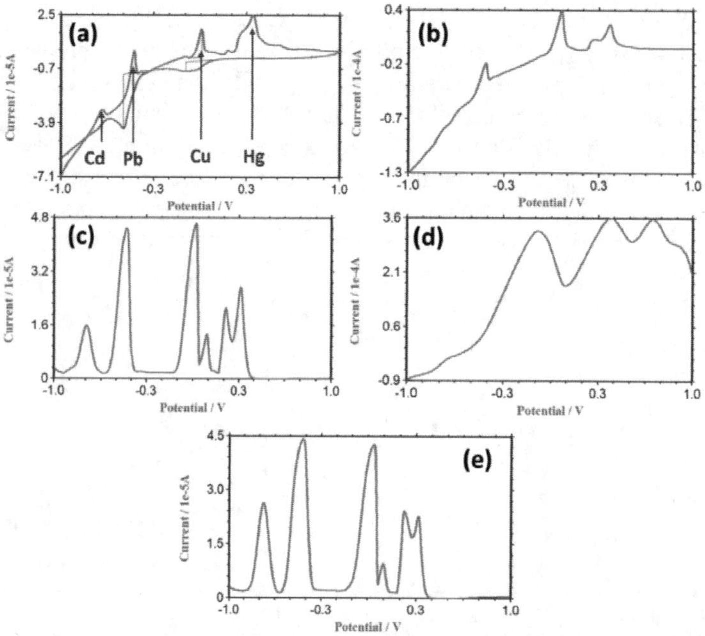


**Figure 5.** Electrochemical responses of the heavy metals Cd, Pb, Cu, and Hg from left to right measured obtained using: CV (a), LSVS (b), DPV (c), NPV (d), and SWV (e).

The signal response for each parameter was compared to the other signals. The optimum parameters were found to be: potential increment 0.001 V, frequency 15 Hz, amplitude 0.07 V, preconditioning time 30 s, preconditioning voltage -1.0 V. These parameters were used to build calibration curves for simultaneous detection of the four analyzed heavy metals.
Figure 6 shows different spike increments of the four heavy metals ranging from 2.0 μL to 50. μL of $10^{-4}$ M metal solution mixture. Peak currents for different metals were found to increase with the increase of the metal spikes. This study yields 4 calibration curves with linear ranges of 2.0-28, 2.0-25, 2.0-28 and 2.0-12 μM for Cd, Pb, Cu and Hg respectively, with respective regression coefficients $R^2$ = 0.9909, 0.9908, 0.9963 and 0.9910.

*3.6. Limit of Detection and Limit of Quantitation of the Test*
The limit of quantitation LOQ and limit of detection LOD were calculated using the calibration curves established previously for each metal separately. LOD was calculated using the expression LOD=3σ/S while LOQ calculated from the formula LOQ = 10σ/S, where σ, is the blank standard deviation and s is the slope of the calibration curve. The calculated values are presented in Table 1.

**Table 1.** Calibration range, limit of detection, limit of quantitation obtained using the proposed sensor.

| Metal | Calibration range | $R^2$ | LOD | LOQ |
|---|---|---|---|---|
| Cd | 2.0-28 μM | 0.9909 | 0.42 | 1.4 |
| Pb | 2.0-25 μM | 0.9908 | 0.37 | 1.2 |
| Cu | 2.0-28 μM | 0.9963 | 0.28 | 0.94 |
| Hg | 2.0-28 μM | 0.9910 | 0.31 | 1.0 |

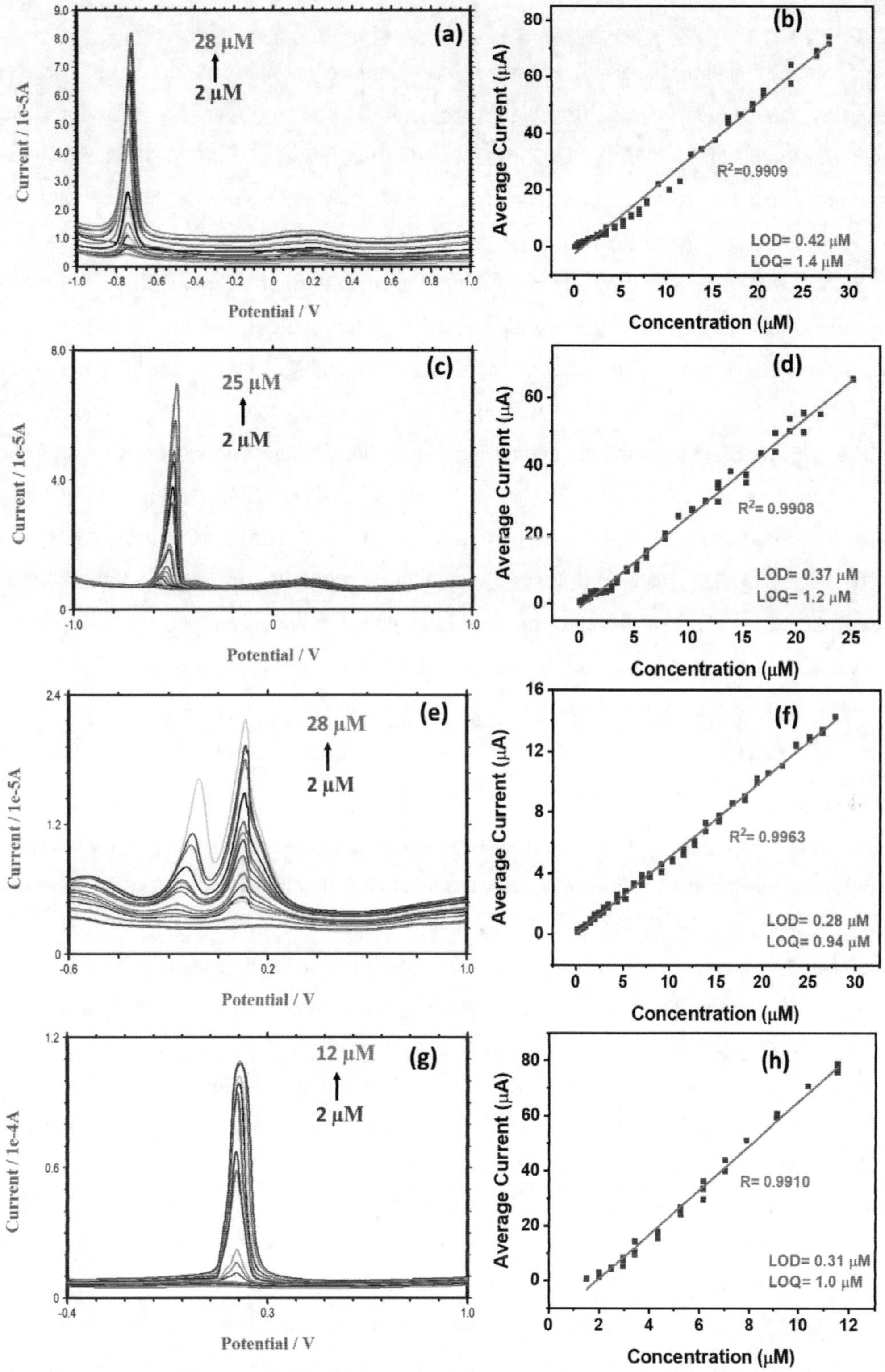


**Figure 6.** Overlay of SWV of different heavy metal concentration spikes (left) and it is corresponding calibration curve (right), Cd (a, b) Pb (c, d), Cu (e, f) and Hg (g, h).

Journal of Physics: Conference Series **2751** (2024) 012023 doi:10.1088/1742-6596/2751/1/012023

## 4. Conclusion

A clay-graphite-$CeO_2$ composite electrode for the sensing of heavy metals was successfully developed. Different clay contents of 10wt.%, 20 wt.%, 30 wt.%, 40 wt.%, and 50 wt.%. sintered at different sintering temperatures of 250 °C, 500 °C, 700 °C, 900°C, 1000°C, and 1100°C were examined to find the best working conditions for our electrode. The prepared clay-$CeO_2$ sensor was further studied for the sensing of toxic heavy metals, namely Cd(II), Pb(II), Cu(II) and Hg(II). Different analysis techniques including CV, LSV, DPV, NPV, and SWV, were used to find the right technique for sensing the four heavy metals simultaneously. SWV yielded the highest current densities for all metals and was adopted as a technique of further investigation. Later it was optimized, and sensing was carried out at potential increment of 0.001 V, amplitude of 0.07 V, frequency of 15 Hz, preconditioning potential of -1.0 V, preconditioning time of 30 s, as optimum conditions. Wide linear ranges for the considered four heavy metals were obtained between 2 μM to 30 μM, with limit of detection (LOD) and limit of quantitation (LOQ) of 0.42, 1.4 μM for Cd (II), 0.37, 1.2 μM for Pb (II), 0.28, 0.94 μM for Cu (II) and 0.31, 1.0 μM for Hg (II). The Clay/$CeO_2$ material displays a very high sensitivity towards accumulation of the four investigated heavy metals. This characteristic could be applied to understand the behavior of ceramic membranes made from such materials and could also help in understanding the mechanism by which the studied heavy metals are filtrated through ceramic membranes.

## 5. References

[1]. Lahnafi, A., Elgamouz, A., Tijani, N., Jaber, L., & Kawde, A. N. (2022). Hydrothermal synthesis and electrochemical characterization of novel zeolite membranes supported on flat porous clay-based microfiltration system and its application of heavy metals removal of synthetic wastewaters. *Microporous and Mesoporous Materials*, *334*, 111778.

[2]. Jaber, L., Elgamouz, A., & Kawde, A. N. (2022). An insight to the filtration mechanism of Pb (II) at the surface of a clay ceramic membrane through its preconcentration at the surface of a graphite/clay composite working electrode. *Arabian Journal of Chemistry*, *15*(12), 104303.

[3]. Lahnafi, A., Elgamouz, A., Tijani, N., & Shehadi, I. (2020). Hydrothermal synthesis of zeolite A and Y membrane layers on clay flat disc support and their potential use in the decontamination of water polluted with toxic heavy metals. *Desalin. Water Treat*, *182*, 175-186.

[4]. Mallik, A. K., Kabir, S. F., Rahman, F. B. A., Sakib, M. N., Efty, S. S., & Rahman, M. M. (2022). Cu (II) removal from wastewater using chitosan-based adsorbents: a review. Journal of Environmental Chemical Engineering, 108048.

[5]. Whitfield, J. B., Dy, V., McQuilty, R., Zhu, G., Heath, A. C., Montgomery, G. W., & Martin, N. G. (2010). Genetic effects on toxic and essential elements in humans: arsenic, cadmium, copper, lead, mercury, selenium, and zinc in erythrocytes. Environmental health perspectives, 118(6), 776-782.

[6]. Satarug, S., Garrett, S. H., Sens, M. A., & Sens, D. A. (2010). Cadmium, environmental exposure, and health outcomes. Environmental health perspectives, 118(2), 182-190.

[7]. Joint, W. H. O., & World Health Organization. (2007). Health risks of heavy metals from long-range transboundary air pollution (No. EUR/06/5067592). World Health Organization. Regional Office for Europe.

[8]. Kurniawan, T. A., Chan, G. Y., Lo, W. H., & Babel, S. (2006). Physico–chemical treatment techniques for wastewater laden with heavy metals. *Chemical engineering journal*, *118*(1-2), 83-98.

[9]. Gunatilake, S. K. (2015). Methods of removing heavy metals from industrial wastewater. *Methods*, *1*(1), 14.

[10]. Kocaoba, S. (2007). Comparison of Amberlite IR 120 and dolomite's performances for removal of heavy metals. *Journal of Hazardous Materials*, *147*(1-2), 488-496.

[11]. McBride, M. B., Richards, B. K., Steenhuis, T., Russo, J. J., & Sauvé, S. (1997). Mobility and solubility of toxic metals and nutrients in soil fifteen years after sludge application. *Soil Science*, *162*(7), 487-500.

[12]. Kunkel, R., & Manahan, S. E. (1973). Atomic absorption analysis of strong heavy metal chelating agents in water and waste water. *Analytical Chemistry*, *45*(8), 1465-1468.

[13]. Rezk, R. A., Galmed, A. H., Abdelkreem, M., Ghany, N. A., & Harith, M. A. (2018). Detachment of Cu (II) and Co (II) ions from synthetic wastewater via adsorption on Lates niloticus fish bones using LIBS and XRF. *Journal of advanced research*, *14*, 1-9.

[14]. Baby Shaikh, R., Saifullah, B., & Rehman, F. U. (2018). Greener method for the removal of toxic metal ions from the wastewater by application of agricultural waste as an adsorbent. *Water*, *10*(10), 1316.

[15]. Tran, T. K., Chiu, K. F., Lin, C. Y., & Leu, H. J. (2017). Electrochemical treatment of wastewater: Selectivity of the heavy metals removal process. *International Journal of hydrogen energy*, *42*(45), 27741-27748.

[16]. Chen, X., Huang, G., & Wang, J. (2013). Electrochemical reduction/oxidation in the treatment of heavy metal wastewater. *Journal of Metallurgical Engineering (ME) Volume*, *2*(4).

[17]. Tayim, H. A., & Al-Yazouri, A. H. (2005). Industrial wastewater treatment using local natural soil in Abu Dhabi, UAE. *American Journal of Environmental Sciences*, *1*(3), 190-193.

[18]. Ustabasi, G. S., Ozcan, M., & Yilmaz, I. (2021). Voltammetric Determination of Heavy Metals with Carbon-Based Electrodes. *Journal of The Electrochemical Society*, *168*(9), 097508.

[19]. Pu, Z., Tu, J., Han, R., Zhang, X., Wu, J., Fang, C., ... & Li, D. (2018). A flexible enzyme-electrode sensor with cylindrical working electrode modified with a 3D nanostructure for implantable continuous glucose monitoring. *Lab on a Chip*, *18*(23), 3570-3577.

[20]. Suhanto, R. N., Rahmawati, R., Setyorini, D. A., Noviandri, I., & Yuliarto, B. (2018, May). Modified working electrode by magnetite nanocomposite for electrochemical sensor application. In *IOP Conference Series: Materials Science and Engineering* (Vol. 367, No. 1, p. 012054). IOP Publishing.

[21]. Arduini, F., Di Nardo, F., Amine, A., Micheli, L., Palleschi, G., & Moscone, D. (2012). Carbon black-modified screen-printed electrodes as electroanalytical tools. *Electroanalysis*, *24*(4), 743-751.

[22]. Wang, Y., Li, Y., Tang, L., Lu, J., & Li, J. (2009). Application of graphene-modified electrode for selective detection of dopamine. *Electrochemistry communications*, *11*(4), 889-892.

[23]. Mousty, C. (2004). Sensors and biosensors based on clay-modified electrodes—new trends. *Applied Clay Science*, *27*(3-4), 159-177.

[24]. Elgamouz, A., Tijani, N., Shehadi, I., Hasan, K., & Kawam, M. A. F. (2019). Characterization of the firing behaviour of an illite-kaolinite clay mineral and its potential use as membrane support. *Heliyon*, *5*(8), e02281.

[25]. Issa, A. A. A., Al-Degs, Y. S., & Al-Rabady, N. A. (2009). Selective electrochemical detection of toxic heavy metals at ultra trace levels using natural clay-modified-electrode. *Eurasian Journal of Analytical Chemistry*, *4*(3), 245-256.

[26]. Velde, B., & Meunier, A. (2008). Fundamentals of clay mineral crystal structure and physicochemical properties. In *THE ORIGIN of clay minerals in soils and weathered rocks* (pp. 3-73). Springer, Berlin, Heidelberg.

[27]. Zhang, T., Wang, W., Zhao, Y., Bai, H., Wen, T., Kang, S., ... & Komarneni, S. (2021). Removal of heavy metals and dyes by clay-based adsorbents: From natural clays to 1D and 2D nano-composites. *Chemical Engineering Journal*, *420*, 127574.

[28]. Kula, P., Navrátilová, Z., Kulová, P., & Kotoucek, M. (1999). Sorption and determination of Hg (II) on clay modified carbon paste electrodes. *Analytica chimica acta*, *385*(1-3), 91-101.

# Prospective of biochar material production and process optimization using co-pyrolysis approach- A mini-review

**Mohamed Dafalla[1,2,*], Abrar Inayat[1,3,*], Farrukh Jamil[1], Chaouki Ghenai[1,3]**

1Biomass and Bioenergy Research Group, Center for Sustainable Energy and Power Systems Research, Research Institute of Sciences and Engineering, University of Sharjah, 27272 Sharjah, United Arab Emirates
2Department of Civil and Environmental Engineering, University of Sharjah, 27272 Sharjah, United Arab Emirates
3Department of Sustainable and Renewable Energy Engineering, University of Sharjah, 27272 Sharjah, United Arab Emirates
*Corresponding author email: mdafalla@sharjah.ac.ae; ainayat@sharjah.ac.ae

## Abstract
This mini-review explores the perspective of biochar material production using the co-pyrolysis approach, which involves the thermal decomposition of biomass and other carbonaceous materials in the absence of oxygen at low temperatures (300-500°C). The study investigates the co-pyrolysis of biomass with different materials such as plastics, tires, municipal solid waste, and other organic waste to produce a high biochar yield. The review focuses on the benefits of co-pyrolysis, including higher yield and better quality of biochar, as well as reduced environmental impact by using different waste materials as feedstock. The review also highlights co-pyrolysis challenges, such as process optimization, feedstock preparation, and product characterization. The study concludes that co-pyrolysis of biomass with different materials can be a promising approach for producing high-quality biochar with multiple applications. However, more research is needed to optimize the co-pyrolysis process and evaluate the economic feasibility of biochar production using a computation approach.

Keywords: co-pyrolysis; biochar material; biomass; slow pyrolysis

## 1. Introduction
Historically, fossil fuels have been central to energy provision, driving transportation, electricity generation, and climate control in buildings. However, their use has led to severe environmental issues, from habitat destruction to greenhouse gas emissions, as well as posing risks like oil spills [1]. The rising costs of fossil fuels have also impacted the global economy [2]. In contrast, renewable energy sources like solar, wind, and hydro offer cleaner alternatives, essential for reducing environmental damage, estimated at $2360 billion annually [3]. Thus, A planned transition to cleaner energy forms, such as natural gas and hydrogen, is crucial for both economic and environmental well-being. Collaboration is essential to accelerate the shift towards renewables, which have been shown to negatively correlate with fossil fuel consumption, emphasizing their role in resource conservation [4]. This aligns with broader goals of sustainable development and the Circular Economy, underlining the urgency of adopting renewable energy solutions to mitigate the harmful impacts of fossil fuels [5].

Renewable energy technologies like solar, bioenergy, and wind power are increasingly recognized for their environmental benefits, including reduced greenhouse gas emissions [6]. Despite their promise, challenges such as market limitations, information gaps, and raw material constraints need to be addressed through well-crafted policies [7]. Research has illuminated multiple benefits of renewables, ranging from energy security to climate change mitigation, prompting national governments to prioritize their adoption [8]. The EU, for instance, has ambitious targets to cut greenhouse gas emissions by up to 95% by 2050 [9]. In order to evaluate the 3E objectives of renewable energy progression in Taiwan, the fuzzy analytic hierarchy process (FAHP) is employed to address the complex, multi-objective challenge

[10]. Developments in sustainable energy were also discussed at the SEEP2014 conference in Dubai, including the Energy Quadr concept [9].

Historically, biomass has been a foundational energy source for various applications and currently accounts for 10-14% of global energy with the potential to increase substantially [11]. Its abundant availability, particularly in the form of lignocellulosic materials, makes it a promising candidate for sustainable energy production [12]. While traditional fuels still meet the energy demands of rural and underprivileged urban areas, alternative solutions like biogas and energy-efficient appliances could offer more sustainable uses of biomass [13]. Harnessing waste for electricity is also under exploration, but the growth of biomass energy must be balanced with environmental concerns like habitat conservation and food security [14]. Challenges specific to Indonesia, such as industry competition and environmental hazards, as well as the impact of the Covid-19 pandemic on renewable energy targets, further complicate its deployment [15]. In summary, while biomass presents a renewable and environmentally-friendly alternative to fossil fuels, its development must be approached with careful consideration of long-term sustainability and socio-economic impacts [16].

The conversion of biomass into biofuels can be achieved through three main types of processes: thermochemical, biochemical, and physical. Thermochemical conversion involves heating biomass through processes like combustion, gasification, and pyrolysis to release energy. The biochemical conversion uses microorganisms like bacteria and enzymes to break down the biomass into valuable products like ethanol and biogas. Physical transformation involves changing the physical form of the biomass to make it easier to transport and store, like through densification [17]. The three primary pathways for biofuel production are examined, and Table 1 presents a standard categorization of the procedures involved in the transformation of biomass into energy. Thermochemical conversion encompasses combustion, pyrolysis, gasification, and liquefaction processes. Meanwhile, biochemical conversion involves anaerobic digestion and fermentation techniques. Lastly, physicochemical conversion facilitates the creation of high-density biofuels such as biodiesel via esterification and transesterification processes [18]; [19].

Table 1: Primary conversion technologies and the related products ([17];[18];[20]).

| Process/Technology | | Feedstock | Usable end product |
|---|---|---|---|
| Thermo-chemical | Combustion | Flammable wastes | Heat, Electricity |
| | Pyrolysis | Organic waste | Bio-oil, char, syngas |
| | Gasification | Municipal solid waste | Syngas |
| | Liquefaction | Agricultural residues, Algal biomass | Fertilizer/biofuel, Syngas, Liquid fuels |
| | Torrefaction | Agriculture waste, forest waste | Torrified pellets or pieces |
| Biochemical | Anaerobic digestion | Sewage sludge, Municipal Solid waste, livestock manure, agriculture waste | Biogas |
| | Fermentation | Sugar-contented, agriculture waste | Liquid fuels (bioethanol) |
| | Composting | Biodegradable waste | Heat, Fertilizer |
| Physico-chemical | Esterification/Transesterification | Vegetable oils, Animal fats, Waste oils | Liquid fuels, Glycerol |

Pyrolysis is a thermal decomposition process that involves heating organic matter, such as biomass, to high temperatures without oxygen. This results in the breakdown of the material into simpler components, including biochar, bio-oil, and syngas. These products have various uses in fields such as energy production, agriculture, and waste management. Pyrolysis represents a significant advancement in the conversion of biomass to valuable resources, while simultaneously reducing waste [21]. It has the potential to serve as an eco-friendly and sustainable alternative to conventional fossil fuel-based energy production. The pyrolysis procedure comprises four primary stages, each occurring at distinct temperature levels. The initial phase entails dehydration of the incoming biomass at temperatures from 100-120°C, succeeded by the gas outlet distillation at 275°C. Exothermic reactions transpire within the 280°C to 350°C temperature range, extracting intricate combinations of chemical compounds, CO2, carbon monoxide, methane, C2H6, and H2 by severing the most fragile chemical bonds.
The final stage involves eliminating all volatile compounds through evaporation at 350°C, leading to increased concentrations of H2, CO, and carbon. Liquids condensed during the pyrolysis procedure can be isolated, and the estimated output from 100 kg of dry wood encompasses products such as charcoal, gas, wood oil, light tar, methyl alcohol, pitch, acetic acid, and esters.

Various methods of executing the pyrolytic process exist, which influence the generation of bio-oil, syngas, and carbon-rich byproducts. Carbonization, slow pyrolysis, fast pyrolysis, and flash pyrolysis all produce varying amounts of these products [22].The two primary pyrolysis methods are slow and fast, with slow pyrolysis characterized by a low heating rate and a temperature range of 300°C to 700°C, resulting in high-quality biochar as the main product. In contrast, fast pyrolysis utilizes rapid heating rates and extremely high temperatures, typically between 600°C and 1000°C, resulting in the production of bio-oil as the main output and biochar as a secondary output. Slow pyrolysis is energy-efficient and suitable for small-scale production, on the other hand, fast pyrolysis is a widely recognized and established technique for the production of bio-oil, which is also used for waste treatment.
However, it is crucial to maintain careful conditions during the process to obtain high-quality bio-oil [23]. The differences in the heating rate and maximum reaction temperature influence the quality of biochar produced, with slow pyrolysis reducing secondary pyrolysis and thermal cracking processes [24]. The two methods differ in terms of their output yields of biochar and bio-oil, with fast pyrolysis favoring bio-oil production. The vapor residence time in hot zones can be minimized to maximize bio-oil yield. Upgrading the liquid product and adjusting the applications to accommodate its unique characteristics and behavior poses a significant challenge in fast pyrolysis technology [25]. Table 2 provides an overview of the distinctions between slow and fast pyrolysis for biomass.

Co-pyrolysis of multiple feedstocks, including biomass, coal, plastics, tires, and sludge, has gained attention as an alternative approach for enhancing pyrolysis outputs [27]. Co-pyrolysis produces a unique product with combined properties, and the synergistic effect between different feedstocks significantly influences the process. This method has shown the potential to produce high-quality biochar and bio-oil, reduce production costs, and provide waste management alternatives [28];[29]. It can be used to convert various biomass, plastics, and waste materials into biofuels, chemicals, and other value-added products, Promoting the advancement of a circular and sustainable economy [26].

Studies on plastic pyrolysis have primarily concentrated on converting plastic to pyrolysis oil and gas, with PET being a common target [30];[31]. Both co-pyrolysis and traditional pyrolysis employ similar mechanisms, wherein the process takes place in a sealed reactor system under high temperatures and in the absence of oxygen. However, co-pyrolysis holds a distinct advantage in terms of waste volume reduction, cost savings for waste management, and environmental problem-solving.

Table 2. Slow and fast pyrolysis for biomass. [24] ; [26]; [27].

| Features | Slow Pyrolysis | Fast pyrolysis |
|---|---|---|
| Temperature (°C) | 300-700 | 600-1000 |
| Heating rates (°Cmin-1 ) | 0.1-10 | 10-10000 |
| Aeration | Oxygen-free or limited | Oxygen-free |
| Residence Time | Minutes-hours | Seconds |
| Target product | Biochar | Bio-oil |
| Bio-oil yield (%) | 20–50 | 60–75 |
| Biochar yield (%) | 25–35 | 10–25 |
| Gas yield (%) | 20–50 | 10–30 |
| Feedstock size | Briquette/ whole | Finely ground |
| Reactors | Fixed bed pyrolysis reactor, auger pyrolysis reactor | Bubbling fluidized bed, ablative reactor, rotary cone |
| Advantages | High biochar yield | high bio-oil yield |
| | Accepts a broad variety of feedstock | |
| Disadvantages | Due to elevated concentrations of CO, additional processing of gases is required | The yield of biochar is minimal, and the use of finely ground biomass feed with a particle size of 1-2mm is necessary. Furthermore, it is preferable to use biomass with low moisture content, ideally below 10%. |

Biomass, which is obtained from organic materials derived from plants and animals, can be classified into four categories, including agricultural residues, wood residues, municipal solid waste, and dedicated energy crops. Co-pyrolysis has successfully utilized these waste products, with plastic waste being a common source [30]. Although pyrolysis is a promising technology for generating renewable energy from lignocellulosic biomass, its profitability depends on the cost of the feedstock. Greenhouse gas emissions of pyrolysis products are also affected by the source and production of feedstock [32];[33]. Furthermore, the physicochemical properties of biochar produced through pyrolysis of different feedstocks under varying conditions can influence its agricultural and environmental performance [34]. Algae are a potential source of biomass for sustainable fuels, chemicals, and materials due to their diverse composition and low land footprint [35]. Co-pyrolysis, which combines different feedstocks, is a more effective method for bio-oil production than simple pyrolysis, and its outcome depends on the synergistic effects of different feedstocks [33]. Although slow pyrolysis, fast pyrolysis, and carbonization have been used to produce biochar, data gaps limit accurate predictions of its properties and performance [34].

This review evaluates the effectiveness of co-pyrolysis for biochar production by focusing on critical variables like feedstock preparation, end-product characterization, and process optimization. It investigates how varying pyrolysis temperatures and mixing ratios affect biochar yield and examines the synergistic effects of co-pyrolysis on different types of biomasses. The paper also contrasts slow and fast pyrolysis methods, assessing their respective merits and drawbacks. Finally, it anticipates future challenges and opportunities in waste pyrolysis. In doing so, this review offers a comprehensive overview of recent advancements in co-pyrolysis technologies and their applications.

## 2. Co-pyrolysis of Biomass

Biochar production utilizing an eco-friendly co-pyrolysis process has been investigated, to remove water pollutants from water such as the emoves phosphate and ammonium from water sources by mixing sewage sludge with walnut husk to produce biochar [36]. The optimal proportion of sewage sludge to the walnut shell was determined to be 3:1, yielding biochar possessing an elevated adsorption capacity for ammonium under neutral or slightly alkaline aquatic conditions. The addition of a walnut shell enhanced the pore structure of the biochars made from sewage waste, leading to enhanced adsorption performance due to the presence of multiple metal oxides, functional groups, and surface adsorption active sites [37]. Furthermore, the potential of sewage sludge-derived biochar in agriculture has been explored, despite challenges posed by heavy metal content. Tube furnace pyrolysis experiments were conducted to generate various types of biochars, including SB, ISB, KSB, and IKSB [38].

In another study, It was discovered that co-pyrolyzing lignin and sewage sludge improved the quality and output of biochar. The experiment explored the influence of pyrolysis temperature and mixing proportion, demonstrating that co-pyrolysis promoted mixture decomposition at lower reaction temperatures. At 800°C, alkali and alkaline earth metals (AAEMs) were shown to have a dominant catalytic effect, leading to significant pore development [39]. Additionally, the co-pyrolysis of sludge and biomass, including industrial sludge and rice straw, produced high-quality biochar with reduced heavy metal risks, offering a comprehensive utilization method for waste materials. The co-pyrolytic biochar exhibited improved pore structure and functional group abundance, with a 102.8% increase in specific surface area [40]. Moreover, sewage sludge and cotton stalks were co-pyrolyzed at different mixing ratios to produce biochars with both agronomic and environmental benefits as soil amendments. The mixing ratios were found to substantially impact biochar properties and their potential for improving soil quality [41].

Recent research has revealed that the co-pyrolysis of various waste materials holds promise for generating valuable outputs, including bio-oil, biochar, and syngas while addressing environmental concerns. One investigation examined co-pyrolyzing plastic waste and date seeds, revealing that increased date seed proportions and lower temperatures contributed to enhanced biochar production [42]. Another study explored co-pyrolyzing agricultural waste with minor plastic contamination, demonstrating the possibility of producing stable carbon forms while avoiding labor-intensive separation processes [43]. Furthermore, research on pyrolyzing cow manure and Amaranthus retroflexus L. stems for biorenewable hydrogen production showed promising results, with high fixed carbon content in the resulting biochar [44]. Co-pyrolyzing food waste and lignocellulosic biomass in a continuous-flow reactor has also been identified as an efficient method for the clean disposal and value-added utilization of biowaste, with temperature and blending ratio impacting product yields [45]. The characteristics of biochar are affected by both the nature of feedstocks and pyrolysis circumstances, and blending different feedstocks is crucial for developing versatile biochar suitable for various applications. A critical review on co-pyrolyzing lignocellulosic and macroalgae biomass for biochar production highlighted the potential of this technique and emphasized the need for exploring more biomass types and conducting comprehensive sensitivity analysis [46]. Additionally, co-pyrolyzing pig manure and Japanese knotweed has been investigated to enhance carbon properties, reduce heavy metal risks, and address bio-invasion issues. The resulting biochar demonstrated improved aromaticity, lower heavy metal bioavailability, and increased nutrient retention, indicating its potential as a soil amendment for restoration purposes [47]. Table 3 presents an overview of recent investigations into biomass pyrolysis, illustrating the diverse range of feedstocks employed in these studies.

Table 3: Pyrolysis of different biomass

| Type of pyrolysis (Temperature) | Feedstock | Pyrolizer | Ref |
|---|---|---|---|
| Moderate | Corncob and waste cooking oil | Fixed bed reactor | [55] |
| High | Sewage sludge and Walnut shell | Tubular furnace | [37] |
| Moderate | Sewage sludge and Camellia oleifera shell. | Fixed-bed quartz reactor | [38] |
| Moderate | Date seeds and Plastic waste | Fixed bed pyrolizer | [42] |
| High | Sewage sludge and cotton stalks | Electric heating furnace | [41] |
| Slow | Crop residues and plastic mulching sheets (LDPE), and plastic grow bags (LDPE) | Fixed bed pyrolysis reactor | [43] |
| High | Food waste with lignocellulosic biomass (wood bark) | Continuous-flow pyrolysis reactor | [45] |
| Moderate | Rubberwood sawdust (RWS) and sewage sludge (SS) | Moving bed reactor | [56] |
| Moderate | Chitin and oyster shell | Two-stage pyrolysis reactor (tubular reactors) | [57] |
| Moderate | Food waste and biowaste | Electric tube furnace | [58] |

Co-pyrolysis, the simultaneous thermal decomposition of multiple feedstocks such as biomass and other organic materials, generates biochar, bio-oil, and syngas and has attracted interest for its potential to enhance biochar properties and improve overall biomass pyrolysis efficiency [46]. The success of this process hinges on several factors, including temperature, heating rate, residence time, feedstock composition, and particle size, all of which influence the yield, composition, and characteristics of the final products. Higher pyrolysis temperatures generally promote gas and bio-oil formation while reducing biochar yield, yet they can also improve biochar properties like porosity, surface area, and carbon content [48]; [23]. The heating rate, the speed at which feedstocks are heated, affects product distribution, with faster rates favoring bio-oil and gas production and slower rates encouraging biochar yield [49];[29]. Residence time, or the length of pyrolysis exposure, is essential, as longer times increase biochar yield and enhance its physicochemical properties, while shorter times favor bio-oil and gas production [50]. The type and mixing ratio of feedstocks also plays a significant role in the resulting products, with synergistic properties improving biochars for specific applications like soil amendment or carbon sequestration [40]. Additionally, particle size impacts mass and heat transfer during co-pyrolysis, with smaller particles generally yielding higher bio-oil and gas production due to increased surface area and reduced mass transfer limitations [49]. Biomass pyrolysis typically occurs in an inert environment, although other gases such as steam may be used to alter the process. The heating rate is crucial in determining the pyrolysis type—fast, slow, or flash—with faster rates leading to greater gas production and reduced char due to rapid biomass fragmentation. Fast heating rates also boost bio-oil production by alleviating mass and heat transfer limitations and minimizing secondary reactions, which are further diminished by shorter residence times that enable prompt removal of organic vapors from reactors [31];[51]. Numerous studies have explored the impact of different parameters on the production and characteristics of biochar and bio-oil generated via the pyrolysis of various biomass types, highlighting the substantial effect of elements including temperature, heating rate, and particle size on the final products [52];[53];[54].

## 3. Optimization for Biochar Production

The pyrolysis process has gained attention as a favorable waste treatment method due to its product generation and energy recovery compared to other techniques. Multi-stage pyrolysis has been introduced, implementing multiple heating stages to maximize processing advantages and optimize operating parameters [59]. [60] Utilized response surface methodology (RSM) to refine experimental conditions, aiming to maximize hydrogen and biochar output derived from the microwave pyrolysis of oil palm fronds. In a separate study, the co-pyrolysis of oil shale and rubber seed shell was optimized using RSM and Design-Expert software, revealing that temperature and blending ratio significantly influenced the yield of pyrolytic products, leading to higher heating values and improved grades of pyrolytic oil.

These innovative approaches demonstrate the potential of pyrolysis optimization in waste treatment and product generation, highlighting its versatility and adaptability in various applications [61].

Table 4: Pyrolysis optimization for biochar production.

| Feedstock | Reactor | Parameters investigated | Range of parameters | Optimum condition | Optimums yield of Bio-char | Ref |
|---|---|---|---|---|---|---|
| Corncob & waste cooking oil | Fixed bed | Temperature CC/WCO mass ratios | 500C,550C,600C 1:0, 1:0.1, 1:0.5, 1:1 and 0:1 | Temperature of 550 C and CC/WCO ratio of 1:1 | 15.3 wt.% | [55] |

| | | | | | | |
|---|---|---|---|---|---|---|
| sewage sludge and Lignin | Muffle reactor | SS/lignin weight ratios Temperature | 10/0, 8/2, 6/4, 5/5, 4/6, 2/8, 0/10 400 °C, 600 °C and 800 °C | 2/8 800 °C | 54.22 wt.% | [39] |
| cow manure (CM) and stems of weed Amaranthus retroflexus L | Fixed bed | CM/AR ratios | 1:1, 2:1 and 4:1 | 1:1, | 38.06wt.% | [44] |
| Industrial waste sludge and Rice straw | Horizontal fixed-bed reactor | Temperature | 400 ◦C, 500 ◦C, 600 ◦C or 700 ◦C | 500 ◦C | - | [40] |
| Pig manure (PM) and Japanese knotweed (JK) | Slow pyrolysis reactor | Temperature PM/JK mass ratios | 400C,500C,600C, 700C 3:1, 1:1, and 1:3 | 1:1- 600 ◦C | apx 40 wt.% | [47] |
| Sewage sludge (SS) and Lignin | Fixed bed reactor | Temperature Retention time Weight ratio of lignin | 400 °C, 600 °C and 800 °C 30 min, 60 min, and 90 min 0, 50%, and 100% | 400 °C, 60 min, 100% | 75 wt.% | [66] |
| Mussel shells and Forestry residues | Tube furnace reactor | Temperature Nitrogen flow Woodto-mussel shell ratio | 400 - 525 °C 50- 300 ml/min 0 - 50 wt% | 400 °C, 156ml/min, 0 wt% | 31.6 wt.% | [65] |

Response Surface Methodology (RSM) is a statistical approach commonly used in engineering and science to optimize system and process responses by identifying the optimal conditions that maximize or minimize desired outputs [62]. In the specific context of co-pyrolysis, RSM helps to maximize the yield of the desired products while minimizing unwanted byproducts. By varying process parameters such as temperature, reaction duration, heating rate, and feedstock composition, RSM builds mathematical model that describes the relationship between input variables and system response [63]. The model is used to predict responses for any given set of input conditions and to identify optimal conditions using a desirability function that combines multiple variables into a single objective function. RSM offers several advantages, including the efficient identification of optimal process conditions, insight into key process parameters, and evaluation of process robustness. RSM serves as an effective instrument for enhancing co-pyrolysis procedures, thereby augmenting the quantity and caliber of the resulting pyrolysis products [64].

The growing interest in co-pyrolysis as an innovative approach for waste material treatment stems from its potential to mitigate environmental risks and generate valuable products. Several studies have employed response surface methodology (RSM) to optimize co-pyrolysis processes, analyze the resulting products, and explore synergistic effects between various waste materials. For instance, studies investigating the co-pyrolysis of fisheries by-products and forestry residues [65], sewage sludge and lignin [66], and chitin and oyster shell [57] have demonstrated the production of biochars with enhanced adsorption properties, improved syngas production, and increased Cu(II) adsorption capabilities.

These results highlight the potential of co-pyrolysis for generating value-added products from waste materials, particularly when RSM is employed to optimize the process. Moreover, other studies have explored the interaction between different waste materials during co-pyrolysis, yielding improved product quality through synergistic effects. Investigations into the co-pyrolysis of lentil husks and Chlorella vulgaris by [67], lignocellulosic biomass and low-quality coal [68], and waste biomass and refinery oily sludge [69], have revealed significant improvements in bio-oil quality, optimal gaseous product distribution, and enhanced energy recovery. Additionally, a study by [70] examining the co-pyrolysis of waste tires and waste plastics provided foundational knowledge for polymer waste pyrolysis and demonstrated the importance of RSM in guiding the development of co-pyrolysis technologies for waste treatment. Table 4 provides a summary of recent research endeavors focused on the optimization of pyrolysis processes for the generation of biochar. This table delineates the variables examined, the scope of these variables, the ideal conditions, the maximum biochar yield, and the feedstock employed in the studies. The use of RSM in these studies has not only optimized co-pyrolysis processes but has also facilitated a greater understanding of the synergistic effects and resulting product characteristics, emphasizing its significance in advancing waste treatment technologies and producing value-added products from various waste materials.

## 4. Challenges & Future Work

The co-pyrolysis process, which relies on a specific energy source for heating—potentially including fossil fuels—demonstrates significant potential in biochar generation. This energy aspect is crucial, particularly at high temperatures, and demands detailed scrutiny. Contemporary studies focus on biochar from sewage sludge, plastic waste, and organic materials, considering parameters like temperature, heating rates, and feedstock ratios. Further research, however, is required to address other vital variables such as moisture content, particle size, vapor residence time, carrier gas velocity, and reactor pressure. Evaluating the storage of biomass feedstock and interim products like biochar is essential. Co-pyrolysis, particularly when integrating plastic waste with lignocellulosic biomass, shows promise in waste reduction. Thus, rigorous studies are needed to elucidate kinetic parameters and response models for biochar optimization in different reactor setups.

## 5. Conclusion

Co-pyrolysis has emerged as a vital strategy for waste management, serving both renewable energy and environmental goals. While it can be an integral part of holistic waste management systems, its limitations as a standalone solution warrant attention. Technological hurdles, including the need for a deeper understanding of reaction kinetics and process parameter optimization, still persist. Nevertheless, co-pyrolysis holds promise for energy security and the sustainable production of biochar, a material gaining interest for its energy and environmental benefits. Utilizing diverse biomass resources in co-pyrolysis can be part of a broader energy-based biorefinery strategy. However, preconditioning biomass feedstock remains critical to biochar quality, as does the assessment of its physicochemical properties for specific applications.

## 6. References

[1] F. Martins, C. Felgueiras, M. Smitkova, and N. Caetano, “Analysis of fossil fuel energy consumption and environmental impacts in european countries,” *Energies*, vol. 12, no. 6, pp. 1–11, 2019, doi: 10.3390/en12060964.

[2] S. Shafiee and E. Topal, “A long-term view of worldwide fossil fuel prices,” *Appl. Energy*, vol. 87, no. 3, pp. 988–1000, 2010, doi: 10.1016/j.apenergy.2009.09.012.

[3] F. Barbir, T. N. Veziroğlu, and H. J. Plass, "Environmental damage due to fossil fuels use," *Int. J. Hydrogen Energy*, vol. 15, no. 10, pp. 739–749, 1990, doi: 10.1016/0360-3199(90)90005-J.

[4] S. A. Solarin, "An environmental impact assessment of fossil fuel subsidies in emerging and developing economies," *Environ. Impact Assess. Rev.*, vol. 85, no. January, p. 106443, 2020, doi: 10.1016/j.eiar.2020.106443.

[5] N. Abas, A. Kalair, and N. Khan, "Review of fossil fuels and future energy technologies," *Futures*, vol. 69, pp. 31–49, 2015, doi: 10.1016/j.futures.2015.03.003.

[6] N. L. Panwar, S. C. Kaushik, and S. Kothari, "Role of renewable energy sources in environmental protection: A review," *Renew. Sustain. Energy Rev.*, vol. 15, no. 3, pp. 1513–1524, 2011, doi: 10.1016/j.rser.2010.11.037.

[7] S. Jenniches, "Assessing the regional economic impacts of renewable energy sources - A literature review," *Renew. Sustain. Energy Rev.*, vol. 93, no. May, pp. 35–51, 2018, doi: 10.1016/j.rser.2018.05.008.

[8] P. A. Owusu and S. Asumadu-Sarkodie, "A review of renewable energy sources, sustainability issues and climate change mitigation," *Cogent Eng.*, vol. 3, no. 1, 2016, doi: 10.1080/23311916.2016.1167990.

[9] A. G. Olabi, "Energy quadrilemma and the future of renewable energy," *Appl. Energy*, vol. 108, pp. 1–6, 2016, doi: 10.1016/j.energy.2016.07.145.

[10] Y. C. Shen, G. T. R. Lin, K. P. Li, and B. J. C. Yuan, "An assessment of exploiting renewable energy sources with concerns of policy and technology," *Energy Policy*, vol. 38, no. 8, pp. 4604–4616, 2010, doi: 10.1016/j.enpol.2010.04.016.

[11] F. Rosillo-Calle, "A review of biomass energy - shortcomings and concerns," *J. Chem. Technol. Biotechnol.*, vol. 91, no. 7, pp. 1933–1945, 2016, doi: 10.1002/jctb.4918.

[12] B. K. Nahak *et al.*, "Advancements in net-zero pertinency of lignocellulosic biomass for climate neutral energy production," *Renew. Sustain. Energy Rev.*, vol. 161, no. March, 2022, doi: 10.1016/j.rser.2022.112393.

[13] U. K. Mirza, N. Ahmad, and T. Majeed, "An overview of biomass energy utilization in Pakistan," *Renew. Sustain. Energy Rev.*, vol. 12, no. 7, pp. 1988–1996, 2008, doi: 10.1016/j.rser.2007.04.001.

[14] C. B. Field, J. E. Campbell, and D. B. Lobell, "Biomass energy: the scale of the potential resource," *Trends Ecol. Evol.*, vol. 23, no. 2, pp. 65–72, 2008, doi: 10.1016/j.tree.2007.12.001.

[15] S. Yana, M. Nizar, Irhamni, and D. Mulyati, "Biomass waste as a renewable energy in developing bio-based economies in Indonesia: A review," *Renew. Sustain. Energy Rev.*, vol. 160, no. 5, p. 112268, 2022, doi: 10.1016/j.rser.2022.112268.

[16] L. R. Amjith and B. Bavanish, "A review on biomass and wind as renewable energy for sustainable environment," *Chemosphere*, vol. 293, no. January, p. 133579, 2022, doi: 10.1016/j.chemosphere.2022.133579.

[17] S. S. Siwal *et al.*, "Recovery processes of sustainable energy using different biomass and wastes," *Renew. Sustain. Energy Rev.*, vol. 150, no. June, p. 111483, 2021, doi: 10.1016/j.rser.2021.111483.

[18] A. Tursi, "A review on biomass: Importance, chemistry, classification, and conversion," *Biofuel Res. J.*, vol. 6, no. 2, pp. 962–979, 2019, doi: 10.18331/BRJ2019.6.2.3.

[19] W. M. Lewandowski, M. Ryms, and W. Kosakowski, "Thermal biomass conversion: A review," *Processes*, vol. 8, no. 5, 2020, doi: 10.3390/PR8050516.

[20] S. Zhang *et al.*, “Liquefaction of biomass and upgrading of bio-oil: A review,” *Molecules*, vol. 24, no. 12, pp. 1–30, 2019, doi: 10.3390/molecules24122250.

[21] G. Maschio, C. Koufopanos, and A. Lucchesi, “Pyrolysis, a promising route for biomass utilization,” *Bioresour. Technol.*, vol. 42, no. 3, pp. 219–231, 1992, doi: 10.1016/0960-8524(92)90025-S.

[22] A. Demirbaş and G. Arin, “An overview of biomass pyrolysis,” *Energy Sources*, vol. 24, no. 5, pp. 471–482, 2002, doi: 10.1080/00908310252889979.

[23] G. Duman, C. Okutucu, S. Ucar, R. Stahl, and J. Yanik, “The slow and fast pyrolysis of cherry seed,” *Bioresour. Technol.*, vol. 102, no. 2, pp. 1869–1878, 2011, doi: 10.1016/j.biortech.2010.07.051.

[24] H. Tan *et al.*, “A Review On The Comparison Between Slow Pyrolysis And Fast Pyrolysis On The Quality Of Lignocellulosic And Lignin-Based Biochar,” *IOP Conf. Ser. Mater. Sci. Eng.*, vol. 1051, no. 1, p. 012075, 2021, doi: 10.1088/1757-899x/1051/1/012075.

[25] A. V. Bridgwater, S. Czernik, and J. Piskorz, “An Overview of Fast Pyrolysis,” *Prog. Thermochem. Biomass Convers.*, vol. 30, pp. 977–997, 2008, doi: 10.1002/9780470694954.ch80.

[26] P. Roy and G. Dias, “Prospects for pyrolysis technologies in the bioenergy sector: A review,” *Renew. Sustain. Energy Rev.*, vol. 77, no. February, pp. 59–69, 2017, doi: 10.1016/j.rser.2017.03.136.

[27] H. Jouhara, D. Ahmad, I. van den Boogaert, E. Katsou, S. Simons, and N. Spencer, “Pyrolysis of domestic based feedstock at temperatures up to 300 °C,” *Therm. Sci. Eng. Prog.*, vol. 5, no. October 2017, pp. 117–143, 2018, doi: 10.1016/j.tsep.2017.11.007.

[28] M. S. Abbas-Abadi *et al.*, “Challenges and opportunities of light olefin production via thermal and catalytic pyrolysis of end-of-life polyolefins: Towards full recyclability,” *Prog. Energy Combust. Sci.*, vol. 96, no. May 2022, p. 101046, 2023, doi: 10.1016/j.pecs.2022.101046.

[29] G. Wang *et al.*, “A review of recent advances in biomass pyrolysis,” *Energy and Fuels*, vol. 34, no. 12, pp. 15557–15578, 2020, doi: 10.1021/acs.energyfuels.0c03107.

[30] C. C. Seah *et al.*, “Co-pyrolysis of biomass and plastic: Circularity of wastes and comprehensive review of synergistic mechanism,” *Results Eng.*, vol. 17, no. October 2022, p. 100989, 2023, doi: 10.1016/j.rineng.2023.100989.

[31] A. Al-Rumaihi, M. Shahbaz, G. Mckay, H. Mackey, and T. Al-Ansari, “A review of pyrolysis technologies and feedstock: A blending approach for plastic and biomass towards optimum biochar yield,” *Renew. Sustain. Energy Rev.*, vol. 167, no. June, p. 112715, 2022, doi: 10.1016/j.rser.2022.112715.

[32] W. Cai *et al.*, “Synergetic effects in the co-pyrolysis of lignocellulosic biomass and plastic waste for renewable fuels and chemicals,” *Fuel*, vol. 353, no. June, p. 129210, 2023, doi: 10.1016/j.fuel.2023.129210.

[33] N. Rathnayake *et al.*, “Co-pyrolysis of biosolids with lignocellulosic biomass: Effect of feedstock on product yield and composition,” *Process Saf. Environ. Prot.*, vol. 173, no. November 2022, pp. 75–87, 2023, doi: 10.1016/j.psep.2023.02.087.

[34] S. Li, S. Harris, A. Anandhi, and G. Chen, “Predicting biochar properties and functions based on feedstock and pyrolysis temperature: A review and data syntheses,” *J. Clean. Prod.*, vol. 215, pp. 890–902, 2019, doi: 10.1016/j.jclepro.2019.01.106.

[35] J. Sun, O. Norouzi, and O. Mašek, “A state-of-the-art review on algae pyrolysis for bioenergy and biochar production,” *Bioresour. Technol.*, vol. 346, no. August 2021,

2022, doi: 10.1016/j.biortech.2021.126258.
[36] B. A. Mohamed *et al.*, "Co-pyrolysis of sewage sludge and biomass for stabilizing heavy metals and reducing biochar toxicity: A review," *Environ. Chem. Lett.*, vol. 21, no. 2, pp. 1231–1250, 2023, doi: 10.1007/s10311-022-01542-6.
[37] Q. Yin, M. Liu, and H. Ren, "Biochar produced from the co-pyrolysis of sewage sludge and walnut shell for ammonium and phosphate adsorption from water," *J. Environ. Manage.*, vol. 249, no. August, p. 109410, 2019, doi: 10.1016/j.jenvman.2019.109410.
[38] X. Min *et al.*, "Combining impregnation and co-pyrolysis to reduce the environmental risk of biochar derived from sewage sludge," *Chemosphere*, vol. 290, no. November 2021, p. 133371, 2022, doi: 10.1016/j.chemosphere.2021.133371.
[39] Q. Dai, W. Xiang, Q. Liu, M. Wang, X. Zhang, and X. Zhang, "Co-pyrolysis biochar derived from sewage sludge and lignin: Synergetic effect and adsorption properties," *J. Environ. Chem. Eng.*, vol. 10, no. 3, p. 107898, 2022, doi: 10.1016/j.jece.2022.107898.
[40] B. Peng, Q. Liu, X. Li, Z. Zhou, C. Wu, and H. Zhang, "Co-pyrolysis of industrial sludge and rice straw: Synergistic effects of biomass on reaction characteristics, biochar properties and heavy metals solidification," *Fuel Process. Technol.*, vol. 230, no. October 2021, p. 107211, 2022, doi: 10.1016/j.fuproc.2022.107211.
[41] Z. Wang *et al.*, "Co-pyrolysis of sewage sludge and cotton stalks," *Waste Manag.*, vol. 89, pp. 430–438, 2019, doi: 10.1016/j.wasman.2019.04.033.
[42] A. Inayat *et al.*, "Co-pyrolysis for bio-oil production via fixed bed reactor using date seeds and plastic waste as biomass," *Case Stud. Therm. Eng.*, vol. 31, no. November 2021, p. 101841, 2022, doi: 10.1016/j.csite.2022.101841.
[43] D. Rathnayake *et al.*, "Investigation of biomass and agricultural plastic co-pyrolysis: Effect on biochar yield and properties," *J. Anal. Appl. Pyrolysis*, vol. 155, no. January, 2021, doi: 10.1016/j.jaap.2021.105029.
[44] J. V. Karaeva *et al.*, "CO-PYROLYSIS of agricultural waste and estimation of the applicability of pyrolysis in the integrated technology of biorenewable hydrogen production," *Int. J. Hydrogen Energy*, vol. 47, no. 23, pp. 11787–11798, 2022, doi: 10.1016/j.ijhydene.2022.02.057.
[45] C. Park, N. Lee, J. Kim, and J. Lee, "Co-pyrolysis of food waste and wood bark to produce hydrogen with minimizing pollutant emissions," *Environ. Pollut.*, vol. 270, p. 116045, 2021, doi: 10.1016/j.envpol.2020.116045.
[46] O. A. Fakayode, E. A. A. Aboagarib, C. Zhou, and H. Ma, "Co-pyrolysis of lignocellulosic and macroalgae biomasses for the production of biochar – A review," *Bioresour. Technol.*, vol. 297, no. November 2019, p. 122408, 2020, doi: 10.1016/j.biortech.2019.122408.
[47] J. Qiu *et al.*, "Improving biochar properties by co-pyrolysis of pig manure with bio-invasive weed for use as the soil amendment," *Chemosphere*, vol. 312, no. P1, p. 137229, 2023, doi: 10.1016/j.chemosphere.2022.137229.
[48] Y. Zhang *et al.*, "Effects of feedstock characteristics on microwave-assisted pyrolysis – A review," *Bioresour. Technol.*, vol. 230, pp. 143–151, 2017, doi: 10.1016/j.biortech.2017.01.046.
[49] V. Dhyani and T. Bhaskar, "A comprehensive review on the pyrolysis of lignocellulosic biomass," *Renew. Energy*, vol. 129, pp. 695–716, 2018, doi: 10.1016/j.renene.2017.04.035.
[50] H. Zhang, R. Xiao, H. Huang, and G. Xiao, "Comparison of non-catalytic and catalytic fast pyrolysis of corncob in a fluidized bed reactor," *Bioresour. Technol.*, vol. 100, no.

3, pp. 1428–1434, 2009, doi: 10.1016/j.biortech.2008.08.031.

[51] P. T. Williams, "Pyrolysis of waste tyres: A review," *Waste Manag.*, vol. 33, no. 8, pp. 1714–1728, 2013, doi: 10.1016/j.wasman.2013.05.003.

[52] V. O. Santos *et al.*, "Pyrolysis of acai seed biomass: Kinetics and thermodynamic parameters using thermogravimetric analysis," *Bioresour. Technol. Reports*, vol. 12, no. August, p. 100553, 2020, doi: 10.1016/j.biteb.2020.100553.

[53] D. Angin, "Effect of pyrolysis temperature and heating rate on biochar obtained from pyrolysis of safflower seed press cake," *Bioresour. Technol.*, vol. 128, pp. 593–597, 2013, doi: 10.1016/j.biortech.2012.10.150.

[54] T. Kan, V. Strezov, and T. J. Evans, "Lignocellulosic biomass pyrolysis: A review of product properties and effects of pyrolysis parameters," *Renew. Sustain. Energy Rev.*, vol. 57, pp. 1126–1140, 2016, doi: 10.1016/j.rser.2015.12.185.

[55] G. Chen *et al.*, "Co-pyrolysis of corn cob and waste cooking oil in a fixed bed," *Bioresour. Technol.*, vol. 166, pp. 500–507, 2014, doi: 10.1016/j.biortech.2014.05.090.

[56] L. Ali, A. Palamanit, K. Techato, A. Ullah, M. S. Chowdhury, and K. Phoungthong, "Characteristics of Biochars Derived from the Pyrolysis and Co-Pyrolysis of Rubberwood Sawdust and Sewage Sludge for Further Applications," *Sustain.*, vol. 14, no. 7, 2022, doi: 10.3390/su14073829.

[57] G. Kwon, D. W. Cho, H. Jang, S. Shiung Lam, and H. Song, "Synergistic effects of blending seafood wastes as Co-pyrolysis feedstock on syngas production and biochar properties," *Chem. Eng. J.*, vol. 429, no. September 2021, p. 132487, 2022, doi: 10.1016/j.cej.2021.132487.

[58] D. D. Chun, D. Ni, and A. Simson, "The effect of inherent inorganics and CO2 co-pyrolysis on biochar production from biowastes and their gasification reactivity," *Biomass and Bioenergy*, vol. 158, no. January, p. 106361, 2022, doi: 10.1016/j.biombioe.2022.106361.

[59] A. O. Oyedun, K. L. Lam, T. Gebreegziabher, and C. W. Hui, "Optimization of multi-stage pyrolysis," *Appl. Therm. Eng.*, vol. 61, no. 1, pp. 123–127, 2013, doi: 10.1016/j.applthermaleng.2013.03.043.

[60] M. Arafat Hossain, P. Ganesan, J. Jewaratnam, and K. Chinna, "Optimization of process parameters for microwave pyrolysis of oil palm fiber (OPF) for hydrogen and biochar production," *Energy Convers. Manag.*, vol. 133, pp. 349–362, 2017, doi: 10.1016/j.enconman.2016.10.046.

[61] B. Baruah and P. Tiwari, "Investigation of co-pyrolysis mechanism of oil shale and rubber seed shell, product yield and optimization of pyrolysis parameter using response surface methodology," *Fuel*, vol. 317, no. November 2021, p. 123441, 2022, doi: 10.1016/j.fuel.2022.123441.

[62] S. Mariyam, M. Alherbawi, S. Pradhan, T. Al-Ansari, and G. McKay, "Biochar yield prediction using response surface methodology: effect of fixed carbon and pyrolysis operating conditions," *Biomass Convers. Biorefinery*, no. 0123456789, 2023, doi: 10.1007/s13399-023-03825-6.

[63] N. Dutta, P. Mondal, and A. Gupta, "Optimization of process parameters using response surface methodology for maximum liquid yield during thermal pyrolysis of blend of virgin and waste high-density polyethylene," *J. Mater. Cycles Waste Manag.*, vol. 24, no. 3, pp. 1182–1193, 2022, doi: 10.1007/s10163-022-01392-y.

[64] M. Elkelawy, E. A. El Shenawy, H. Alm-Eldin Bastawissi, M. M. Shams, and H. Panchal, "A comprehensive review on the effects of diesel/biofuel blends with

nanofluid additives on compression ignition engine by response surface methodology," *Energy Convers. Manag. X*, vol. 14, no. July 2021, p. 100177, 2022, doi: 10.1016/j.ecmx.2021.100177.

[65] A. Krutof, H. Bamdad, K. A. Hawboldt, and S. MacQuarrie, "Co-pyrolysis of softwood with waste mussel shells: Biochar analysis," *Fuel*, vol. 282, no. April, p. 118792, 2020, doi: 10.1016/j.fuel.2020.118792.

[66] Q. Dai *et al.*, "Synergetic effect of co-pyrolysis of sewage sludge and lignin on biochar production and adsorption of methylene blue," *Fuel*, vol. 324, no. PA, p. 124587, 2022, doi: 10.1016/j.fuel.2022.124587.

[67] N. Kazemi Targhi, O. Tavakoli, and A. H. Nazemi, "Co-pyrolysis of lentil husk wastes and Chlorella vulgaris: Bio-oil and biochar yields optimization," *J. Anal. Appl. Pyrolysis*, vol. 165, no. April, p. 105548, 2022, doi: 10.1016/j.jaap.2022.105548.

[68] Z. Wu, Y. Li, D. Xu, and H. Meng, "Co-pyrolysis of lignocellulosic biomass with low-quality coal: Optimal design and synergistic effect from gaseous products distribution," *Fuel*, vol. 236, no. September 2018, pp. 43–54, 2019, doi: 10.1016/j.fuel.2018.08.116.

[69] G. Hu, J. Li, X. Zhang, and Y. Li, "Investigation of waste biomass co-pyrolysis with petroleum sludge using a response surface methodology," *J. Environ. Manage.*, vol. 192, pp. 234–242, 2017, doi: 10.1016/j.jenvman.2017.01.069.

[70] D. Li, S. Lei, G. Rajput, L. Zhong, W. Ma, and G. Chen, "Study on the co-pyrolysis of waste tires and plastics," *Energy*, vol. 226, p. 120381, 2021, doi: 10.1016/j.energy.2021.120381.

# Efficient Cooling for Agricultural Greenhouses in Hot Climates: A Novel Water Light Filter Design

**Abstract:** The use of agricultural greenhouses in hot climates requires transparent enclosures to enable cooling for optimal plant growth. However, the cooling process incurs high water and energy costs due to solar gain and heat trapping. To address this issue, a novel enclosure design featuring a layer of liquid or water solution between two commercial grade glass layers has been proposed. This water solution absorbs thermal radiation and can be circulated through an active heat dissipation system. Carefully selected dyes can also be added to enhance water filter performance, allowing sufficient photosynthetically active radiation (PAR) through to the plants while blocking more of the non-PAR radiation.

To evaluate the efficiency of the proposed filter, a shading filter was used as a benchmarking scheme. The tests were conducted using clear water, mineral oil, and water with four commercially available pigments. Results indicated a reduction in the thermal load ranging from 31-62%, while the reduction in the photosynthetically active portion of the solar spectrum ranged from 32-79%. The clear water filter performed the best, blocking 45% of the thermal load while allowing 69% of weighted PAR radiation. This allowed for an additional 14% of PAR radiation compared to an equivalent 45% transparent shading device.

This new liquid filter design is effective in reducing cooling costs for agricultural greenhouses in hot climates, while still allowing for optimal plant growth. The clear water filter was found to be the most effective in blocking non-PAR radiation while still allowing sufficient PAR for the plants. However, there is potential for future work to identify additives that could enhance the filter's overall performance. Engineering and technoeconomic development should be pursued to implement this technology for practical use in the field.

***Keywords – Greenhouses,* photosynthetically active radiation, Shading device, *liquid light filter***

**Jihad Hassan AlSadah**

Physics Department, King Fahd University of Petroleum and Minerals, Dhahran 31261, Saudi Arabia

jhalsadah@kfupm.edu.sa

**Acronyms**

| | |
|---|---|
| PAR | Photosynthetically Active Radiation (400-700nm) |
| PPM | Part per million |
| PPFD | Photosynthetic Photon Flux Density (PPFD) |
| DLI | Daily Light Integral |
| TGR | Thermal gain reduction |
| UV-VIS | Ultraviolet to visible region of the spectrum |
| NIR | Near infrared region |
| nm | Nano meter ($10^{-9}m$) |
| ASTM | the American Society for Testing and Material |
| TR | Thermal Reduction |

## 1. Introduction

Food production is a key objective for any human society which should be achieved more in localized level for both food security, nutritional value, and environmental considerations [1]. Horticulture, the science, and art of growing plants has been evolving since the dawn of human civilization. Through the process of photosynthesis, plants convert carbon dioxide and water into carbohydrates using solar energy. Plants use photosynthesis to transform sunlight energy into chemical energy in the form of carbohydrates, by combining carbon dioxide and water. Several factors affect the efficiency of the growth process, including water, fertilizers, lighting, temperature, humidity, and carbon dioxide concentration. While these factors are beyond the control of traditional open field agriculture, the advent of innovative technologies has paved the way for controlled greenhouse environments and plant factories that can ensure optimal growth conditions. These controlled environments not only offer greater productivity, but also enable cultivation in regions and seasons that were previously unviable for plant growth. As horticulture continues to evolve, it is anticipated that new technological advancements will emerge, further enhancing the productivity, efficiency and resilience of the practice in the face of global challenges such as climate change and population growth. The usage of greenhouses has experienced a significant surge in the last few decades. In China, for instance, 3% of their agricultural areas are now covered with greenhouses which account for almost 35% of the country's vegetable production [2].

Greenhouses are transparent enclosures originally intended to provide warmer temperatures in cold climates with passive or even active heating. Currently, all provide control over temperature. Advanced greenhouses, some called plant factories, almost all parameters are monitored and controlled including humidity, light level, and carbon dioxide concentration [3]. The temperature should be maintained within $10 - 35°C$ and optimally within $17 - 27°C$. In cold climates, the transparent enclosure increases the temperature of the greenhouse by trapping heat which might be sufficient on its own or in need of supplementary heating. However, in hot climates, the enclosure by providing some protection from hot temperature, it creates a doubly difficult challenge: insulate from hot environment and provide cooling for high thermal gain of the greenhouse [4]. This is typically resolved by evaporative cooling systems which consume modest energy but much water. In a recent study of evaporative cooling in Riyadh Saudi Arabia, cooling water consumption is around 5500 liters per square meter on annual basis which represents more than 75% of the total water consumption [5]. This level of water consumption is considered excessive when compared to Malaysia's maximum annual rainfall rate of 4000 mm per square meter (equivalent to 4000 liters/m$^2$).
Greenhouse covering servs both as volume boundary and as shading material. The material selected affects initial cost, lifetime as well as thermal gain and thermal losses in the greenhouse. Glass in 3-4mm thickness favors transmission (0.88-0.90) of PAR in 400-800nm. Polyethylene in 0.1-0.2 mm thickness is lower in cost and lifetime and favors transmission of infrared radiation but slightly lower PAR transmission (0.70-0.93) [6]. In Spain, shading has been used to reduce plant water and thermal

Journal of Physics: Conference Series **2751** (2024) 012025 doi:10.1088/1742-6596/2751/1/012025

losses and withering. Aluminized screens with shading in the range of 40-60% showed effectiveness in thermal and water stress but they slightly reduced the sugar content of the harvest (Brix metric). The shade screens were internal and installed bellow 200 μm polyethylene cover having 80% transmittance in the 400–800 [7].

An alternative approach to shading technology is the integration of photovoltaic energy generation into agricultural production. This can be achieved through the use of sparse photovoltaic panels installed on the roof [8]. Ravishankar et al have modeled the shading of organic solar cell material on both heat and PAR quantity and quality measured as red to blue ratio. The transmittance of PAR was around 29%-38% giving PPDF around 600 uMol/m2/s while the overall transmittance was around 50%. The estimated solar power conversion efficiency was between 8.1-10.2% [9].

This research project aims to develop a method for mitigating thermal gain in greenhouses situated in hot climates. The proposed concept involves enclosing a water solution between two layers of glass, which would absorb some of the solar heat while allowing most of the photosynthetically active radiation (PAR) through to the plants. The captured heat could either be utilized or dissipated using an active circulation method.

The objective of this project is to characterize the efficacy of this liquid filter in both thermal gain reduction (TGR) and PAR preservation. This will be achieved through a series of experimental measurements, utilizing a thermopile solar sensor and UV-Vis spectrometric techniques. The experiments involve testing different filters including glasses, plain water, water with commercial grade dies, and an oil filter.

Furthermore, the analysis will also be based on the published quantum yield of photons on PAR, which will provide accurate quantification of the filter's impact on photosynthesis. The goal is to ultimately develop a more efficient and sustainable method for greenhouse temperature control while promoting plant growth.

## 2. Literature Review

The intended filter operates on reducing the solar radiation with the aim of sparing PAR part of the spectrum and reducing the rest. The solar radiation spectrum can be represented in energy and quantum spectra. The general shape of the solar radiation is captured by the blackbody radiation model. The sun is modeled as a 5772K hot surface emitting radiation as per Plank's radiation law [10]. Equation (1) state the spectral radiance, $B_\lambda$, which is power per unit area per solid angle in units of watt per square meters per unit of wavelength:

$$B_\lambda\left(\frac{W}{m^2\ m\ sr}\right) = \frac{2\pi c^2 h}{\lambda^5}\frac{1}{e^{hc/\lambda kT} - 1} \tag{1}$$

Where the physical constants used, and their values are stated in **Error! Reference source not found.**.

**Table 1: physical constants used in computing solar spectral radiance [11]**

| Physical Constant | Value |
|---|---|
| Planck Constant | $h = 6.626\ 070\ 15\ \times 10^{-34} J\ s$ |
| Speed Of Light | $c = 299\ 792\ 458\ m/s$ |
| Boltzmann Constant | $K = 1.38065 \times 10^{-23}\ J/K$ |
| Avogadro Constant | $N_A = 6.02214 \times 10^{23} Mole^{-1}$ |
| Sun's Surface Temperature[1] | $T_{sun} = 5772\ K$ |
| Earth-Sun Mean distance | $R_{SE} = 1.496 \times 10^{11} m$ |
| Sun's radius | $R_{Sun} = 6.96340 \times 10^{8} m$ |
| Solar Constant | $P_{AM0} = 1366.1\ W/m^2$ |

[1] This the accepted temperature of the Sun's surface. However, the integral does not match the solar constant of $1366.1 W/m^2$which is only 0.18% higher. This could be remedied by changing the temperature to $T = 5774.6\ K$ or the assumed distances and sun's radius.

To obtain the solar spectrum just outside the Earth's atmosphere, the spectral irradiance measured just outside the sun's surface needs to be reduced by the inverse square distance between the Sun and the Earth and multiplied by the sun's radius squared as shown in Equation 2. The energy content of solar radiation affects the solar thermal gain of a greenhouse, but when characterizing solar radiation for photosynthesis, the number of photons becomes more important. To convert energy units to the number of photons, the energy in a photon is divided by the photon's energy for each pin in the spectra. Since the number of photons is large, it is converted to the number of moles using Equation 3.

$$B_{\lambda,AM0}\left(\frac{W}{m^2\, m}\right) = B_\lambda \left(\frac{R_{sun}}{R_{SE}}\right)^2 \tag{2}$$

$$B_{\lambda,AM0}\left(\frac{mol}{s\, m^2\, m}\right) = B_{\lambda,AM0}\left(\frac{J}{s\, m^2\, m}\right) * \frac{1}{E_{ph}} * \frac{mol}{N_A} \tag{3}$$

While the black body radiation model may provide a general approximation of the solar radiance, its accuracy is limited as the radiance that reaches Earth is influenced by the sun's atmosphere, leaving distinctive imprints on the spectrum. This specific spectrum is referred to as Air Mass zero (AM0) and has been recorded by various instruments including satellites measurements, with an accepted integral, called solar constant of 1366.1 $W/m^2$, [12]. The Earth's atmosphere plays a crucial role in reflecting some of the solar irradiance and absorbing certain spectral regions. It should be noted that the solar irradiance is not constant, as it varies depending on the location and time on Earth. To obtain a standard reference spectrum, Air Mass 1.5 is used with a fixed solar zenith angle of 48.2 degrees, as stated in [13]. The three data sets of solar irradiance spectra expressed in units of power intensity and in quantum flux intensity are plotted in Figure 1.

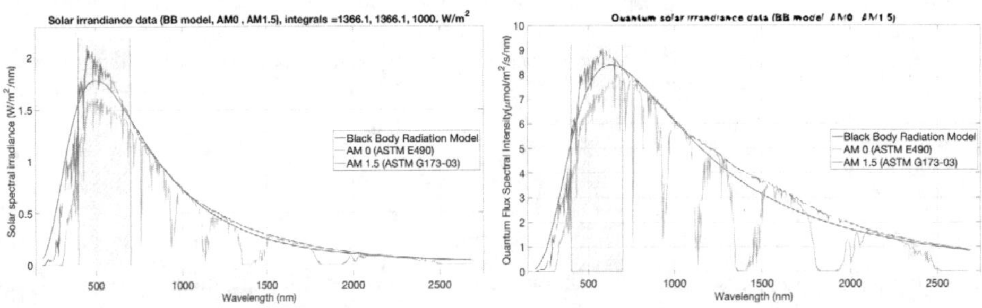


Figure 1 Solar irradiance modeled with PAR shown as a green rectangle: Left: power spectral intensity; Right:

Photosynthesis is a complex biological process that utilizes solar energy in a quantum way. The number of photons, rather than their energy content, is crucial in this process. The PAR photons are measured in moles per unit area per second $(mol/m^2/s)$. The rate of photosynthesis is measured by the amount of carbon dioxide molecules fixed in a given time expressed as moles per unit area per second $(mol_{CO_2}/m^2/s)$. The quantum nature of photosynthesis means that approximately nine moles of photons are required to fix one mole of $CO_2$, as depicted in Figure 2 (Right). However, it's important to note that the number of photons required can vary somewhat due to the production of CO2 through respiration. In general, the number of photons needed for each $CO_2$ molecule fixed ranges from 8 to 10 [14].

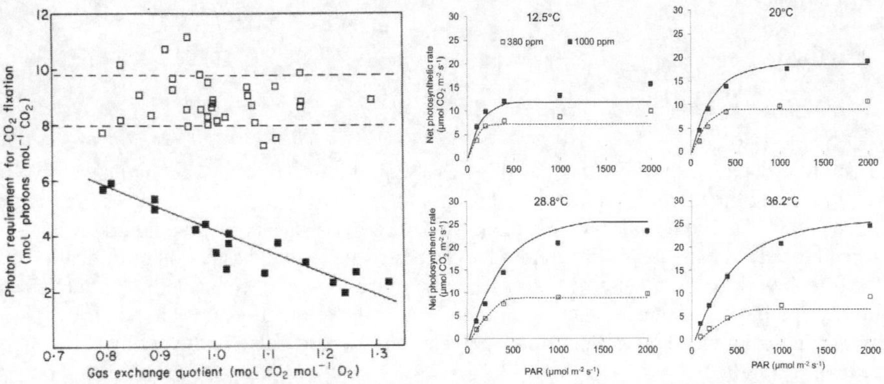


**Figure 2 Left: Photon requirement for fixing CO2 is about 9 moles of photons per mol of CO2 fixed affected by respiration rate [14]; Right: Significance of lighting level expressed in (μmole/m2/s) must be determined in context of temperature and CO2 concentration [15];**

Literature shows that the level of lighting required for proper plant growth depends on the plant type, growth stage, and environmental factors in which it is growing. The standard AM 1.5 spectrum typically contains approximately 2000 ($\mu mol/m^2/s$). Low or high temperatures can retard the photosynthesis process, with an optimal temperature of around $27°C$. Increasing $CO_2$ concentration in enclosed greenhouses can enhance photosynthesis by as much as two to three times, as shown in Figure 2 (Left). However, a key factor in plant growth is the minimum light requirement, which is crucial for determining the rate of photosynthesis. For optimal temperatures and normal $CO_2$ concentration, a minimum of approximately 400 ($\mu mol/m^2/s$) is needed, whereas at colder temperatures of 12.5°C, the minimum level reduces to 200 ($\mu mol/m^2/s$). At higher $CO_2$ concentrations of 1000 ppm, the minimum lighting levels significantly increase to 1360 ($\mu mol/m^2/s$) and 440 ($\mu mol/m^2/s$), respectively. These findings help explain why 50% meshes used in some greenhouses do not significantly impact plant growth.

The absorption of light during the photosynthesis process is carried out by the chlorophyll pigment, which has a distinct chemical absorption profile, as shown in Figure 3. However, the photosynthesis process in plant leaves has broader photochemical response compared the pigment. Studies conducted by McCree have shown that the photon utilization varies across different wavelengths, and he has reported the spectral quantum yield or the number of photons required for fixing a CO2 molecule, which has been averaged over a variety of plants and growth stages [16] [17]. To account for these variations, we can normalize the yield to the maximum and obtain the relative quantum yield, shown in Figure 3 on the right. This weighting function can be used to more accurately measure the filter effect on the Photosynthetically Active Radiation (PAR) radiation.

**quantum flux intensity**

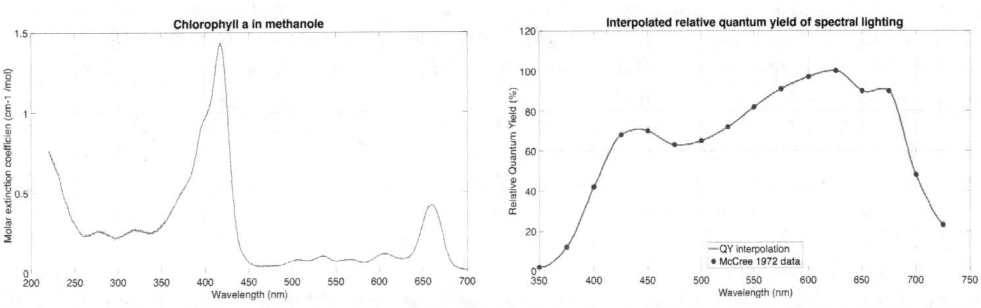


**Figure 3 Left: Chlorophyll a molar extinction coefficient [18]; Right: Typical quantum yield of photo synthesis [16]**

## 3. Methods

### *3.01 Experimental*

This study seeks to assess the efficiency of a novel liquid-based filter by accomplishing two main objectives: reducing the thermal load on greenhouses and mitigating light loss from the PAR region of the solar spectrum. The liquid solar filter is made up of two glass surfaces with a liquid layer sandwiched between them, where heated liquid is circulated for active dissipation or utilization. To apply the filter, it will be installed onto the surface of a greenhouse and connected to an active heat utilization or dissipation system, as shown in Figure 4. The experiment employs borosilicate glass with a 3.8mm thickness, mounted on a 10mm-thick polycarbonate frame filled with varying solutions, presented in Figure 5.

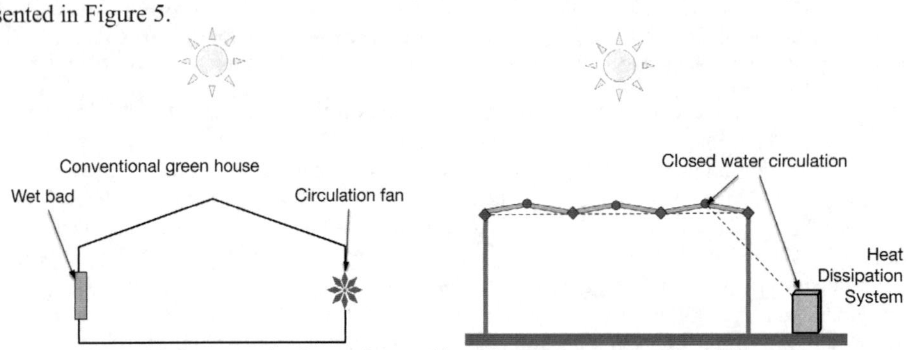


**Figure 4 Left: conventional green house with cooling; Right: proposed liquid filter ceiling.**

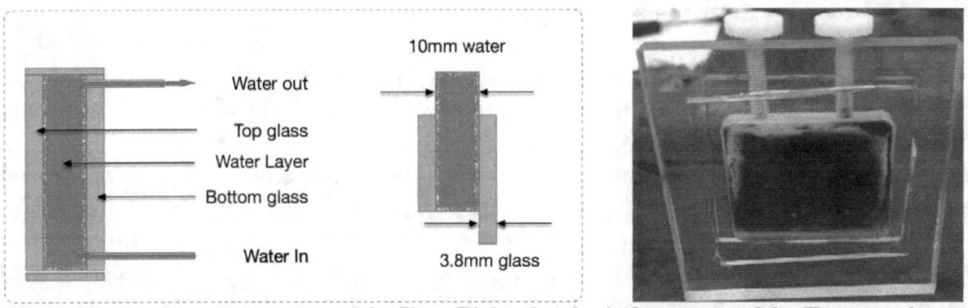


**Figure 5 Left: schematic structure of the filter; Right: the physical structure of the filter experiment**

When sunlight enters a greenhouse, it is absorbed by plants for photosynthesis or other surfaces, which raises the temperature in the space. But not all of the sun's energy is used equally by plants – certain wavelengths of light (PAR photons) are especially important for their growth. To measure both how much heat the greenhouse is absorbing and how much PAR light is available to plants, it's important to study these factors separately.

To determine the effectiveness of the filter, a solar simulator light source and a wideband thermopile detector are used. The filter is inserted between the light source and detector. Thermopile detector are essentially a series of thermocouples for increased accuracy and sensitivity compared to a single thermocouple The solar simulator we use is the 16S-Series 150W/300W Air Mass Solar Simulator, which can generate simulated solar light within the 280-4000nm range up to 8X suns in a small, horizontal beam [19].

The PMA2144 model by Solar Light Company, a Class II Pyranometer with a spectral band of 300-2800nm, was used to measure the intensity of the solar simulator. It is capable of reading between 200mW-2000W [20] [21]. The pyranometer is stationed to detect the horizontal beam of the simulator, with a distance of approximately 15cm from it. The simulator is calibrated to ensure that the pyranometer measures a consistent intensity of around 1000 $W/m^2$.

The measurement setup for determining the transmission percentage or thermal reduction percentage involves inserting the filter near the simulator exit port and allowing the detector to stabilize for a few seconds. The detected power is then measured with and without the filter, and the ratio is taken as the percentage of transmission or thermal reduction. Refer to Figure 6 for an illustration of the typical setup.

As a means of evaluating the impact of filters on PAR photons, we employed a Jasco V-670 UV-VIS-NIR spectrometer [22]. This instrument boasts a dual-beam, dual-lamb design, and boasts a sensitivity spectrum spanning from 190 to 2700 nm with an accuracy of $\pm$ 0.3 nm in the UV-Vis region and $\pm$ 1.5 nm in the NIR region. To conduct our analysis, we situated the filter in the reading beam pathway with the lid closed, as depicted in Figure 6. The spectrometer was set to scan the broadest possible range of 190-2700 nm with a resolution of 1 nm.

The spectrometer can be used to for estimating the transmission factor of a filter. Comparisons have shown that the estimates generated by the spectrometer agree with pyranometer readings within a range of 1-8%, making it an acceptable method for determining the general transmission value. For PAR transmission region measurements, the spectrometer is the only choice. We prefer using pyranometers to measure thermal reduction factors because they are simple and directly measure the thermal effect.

**Figure 6 Left: The filter is inserted in-between the solar simulator and pyranometer; Right: Filter is inserted in UV-Vis spectrometer in the beam pathway**

*3.02 Computational*

The sampling rate and extent of the ASTM data for the AM0 and AM1.5 spectra differ from the regular sampling rate of a UV-Vis spectrometer. To overcome this, the data has been resampled to 1 nm steps using the modified cubic linear interpolation algorithm due to Akima implemented in MATLAB software [23]. Additionally, the range of the data has been extrapolated in the 190-280nm region for the AM1.5 dataset to enable computing across the entire spectral range of the spectrometer. The pyranometer's reported sensitivity range of 300-2800nm is lower than the full solar spectrum, which may introduce errors in thermal reduction estimates. Analysis was conducted on the energy content of the solar spectrum across the original dataset and two restricted ranges (190-2700nm and 300-2800nm) as shown in Range restriction causes a 3-4% impact on black body radiation and AM0 spectra. However, the effect on AM1.5 spectrum is minor, resulting in only a 0.77% reduction difference.

The solar spectrum is divided into three bands: the pre-PAR band, PAR (400-700nm), and the post-PAR band. The PAR region holds 43% of the total energy content, as shown in Figure 7. An ideal filter should block all non-PAR photons and allow all PAR photons to pass through. This breakdown

sets a maximum limit of 57% thermal reduction for the perfect filter. When evaluating a filter's effectiveness, it must lower the energy content of the beam while still retaining as much of the PAR band as achievable. Both criteria play a significant role in assessing the filter's performance.
The effectiveness of the PAR spectral region in powering the photosynthesis process is not uniform. Similar to the photovoltaic effect, photosynthesis utilizes the quantum nature of solar energy, where the number of photons, rather than their energy content, is crucial. To assess the filter transmission of PAR photons, the relative quantum yield, a weighting factor to the quantum solar spectrum in the PAR region, must be used. Since the relative quantum yield function extends past the PAR region of 400-700nm, integration must also be extended to 375-725nm. For any filter, the transmission profile as a function of wavelength is multiplied by the relative quantum yield and quantum solar spectrum in $\mu\ mole/m^2/s$, then integrated using Equation 4. While the total photon flux density remains significant, the total is divided by the respective flux with a perfect flux that transmits all PAR photons with no attenuation. This ratio is a dimensionless number and is referred to as the weighted PAR or wPAR metric, which allows for a comparison of different filter performances.

$$N_{ch} = \int_{\lambda_f=375nm}^{\lambda_f=725nm} N(\lambda)\cdot Tr(\lambda)\cdot QY(\lambda)d\lambda = \sum_{i=375nm\to725nm} I_i\left(\frac{\mu\ mole}{m^2 s\ nm}\right) * T_i * QY_i * nm \quad (4)$$

$$wPar = \frac{\sum_{i=375nm\to725nm} I_i\left(\frac{\mu\ mole}{m^2 s\ nm}\right) * T_i * QY_i * nm}{\sum_{i=375nm\to725nm} I_i\left(\frac{\mu\ mole}{m^2 s\ nm}\right) * 1 * QY_i * nm} \quad (5)$$

**Table 2: Solar Spectra Band Analysis**

| Parameter ↓, Spectum Type → | BB | AM0 | AM1.5 |
|---|---|---|---|
| Maximum Range of Data (nm) | unlimited | 119.5-1000,000 | 280-4,000 |
| Computing range (nm) | 1-1000,000 | 119.5-1000,000 | 280-4,000 |
| Total Integral ($W/m^2$) | 1363.63 | 1366.09 | 1000.37 |
| Restricted Integral to 190-2700nm | 1324.09 | 1330.79 | 992.62 |
| Percentage | 97.10% | 97.42% | 99.23% |
| Restricted Integral to 300-2800nm | 1285.52 | 1319.20 | 992.62 |
| Percentage | 94.27% | 96.57% | 99.23% |
| Pre-PAR (W/m2) | 165 | 106 | 46 |
| PAR (W/m2) | 502 | 532 | 431 |
| Post-PAR (W/m2) | 660 | 692 | 516 |
| Pre-PAR (%) | 12.1% | 7.7% | 4.6% |
| PAR (%) | 36.8% | 38.9% | 43.1% |
| Post-PAR (%) | 48.4% | 50.7% | 51.6% |

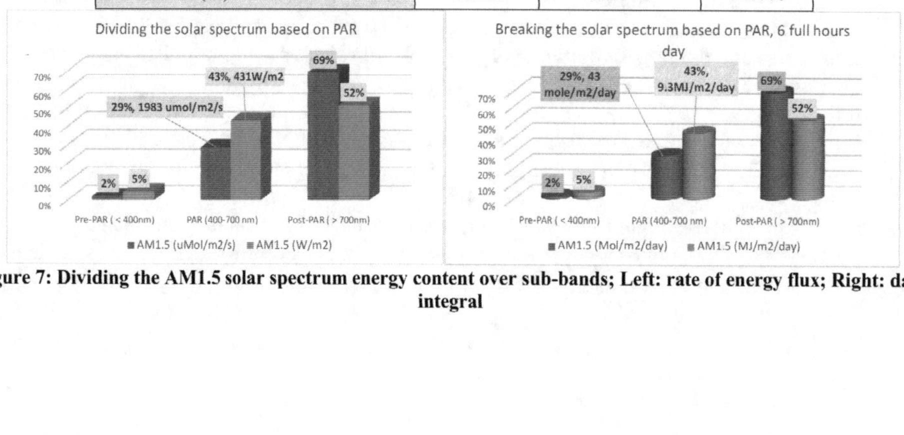


**Figure 7: Dividing the AM1.5 solar spectrum energy content over sub-bands; Left: rate of energy flux; Right: daily integral**

**Table 3: Content description of the commercially available consumer grade dyes.**

| | Yellow | Blue | Green | Red |
|---|---|---|---|---|
| Name | Shank's Yellow Food Color | Shank's Blue Food Color | Shank's Green Food Color | Shank's Red Food Color |
| Contents | Water, Glycerin, Yellow Color, Citric Acid, Preservative (Sodium Benzoate) | Water, Glycerin, Blue Color, Citric Acid, Preservative (Sodium Benzoate) | Water, Glycerin, Blue and Yellow Color, Citric Acid, Preservative (Sodium Benzoate) | Water, Glycerin, Red Color, Citric Acid, Preservative (Sodium Benzoate) |
| Color | Yellow Color (E 102)<br>Yellow Color (E 110) | Blue Color (E 133) | Blue Color (E 133)<br>Yellow Color (E 102) | Red Color (E 129) |

## 4. Results and Discussions

The liquid filters underwent testing using both pyranometer and spectrometer devices. The first set of filters consists of glass filters (single and double layers), as well as filters containing water and oil within glass. The second set of filters were created by mixing deionized water with dyes at a concentration of 1000ppm. As this study is exploratory, common consumer-grade dyes were used and their names are listed in Table 3. The transmission profiles for both sets of filters can be seen in Figure 8.

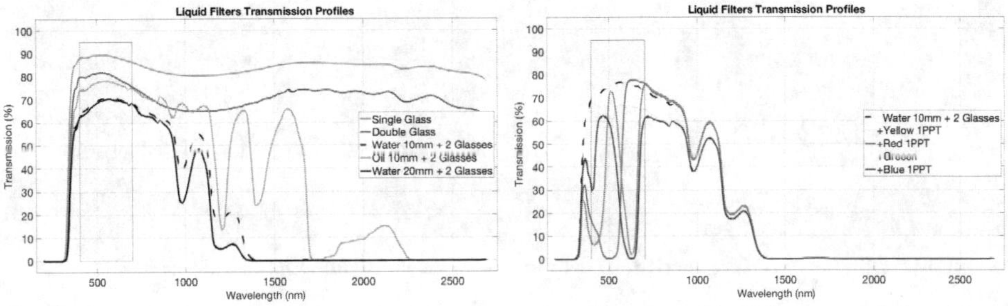


**Figure 8 Transmission profile of the liquid filters and materials; Right: water liquid filter transmission with added dies**

Double glass is more attenuating compared to a single glass. Clear mineral oil is more transmitting in PAR region compared to water but less attenuating in NIR region. The colored solution appears to be adequately concentrated (1000 ppm) as the high values are similar to those of water, and the low values are close to zero. Unfortunately, there are some significant gaps in the PAR region, which reduces the amount of available light for photosynthesis in plants. To calculate the wPAR (weighted PAR) metric for each filter, one must multiply the filter transmission profile with the relative quantum yield factor and the AM1.5 quantum solar spectrum before integrating the result. Finally, the quotient is obtained by dividing the integrated result by 100%T filter, as illustrated in Figure 10 and Equation 5.

Table 4 lists the computed wPAR values and measured thermal reduction values, while in Figure 11 visually illustrates these values. In order to evaluate filters based on two metrics, we utilize a shading filter as a model. This shading device consists of an opaque fraction and open fraction, which is demonstrated in Figure 11 using stripe shading filters with transmission fractions of 70% and 30%. The shading device does not possess a spectral signature, meaning that if the filter blocks 40% of the light, then the thermal reduction factor must be 40% and the wPAR factor must be 60%. It is important to note that the total of wPAR and thermal reduction factors equals unity. Naturally, a general filter is not required to adhere to this principle. A well-designed filter may block less in the PAR region and more in the non-PAR region, whereas a poorly designed filter will block more PAR fraction and permit more non-PAR fraction.

**Figure 9: The shading filter reduces the amount of solar light passing through by a specific percentage, regardless of the wavelength. Left: 70% T; Right: 20% T**

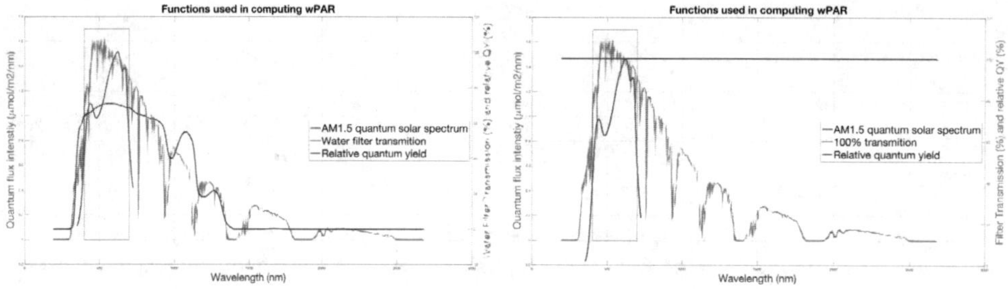


**Figure 10: To compute wPAR metric for a filter, the AM1.5 quantum solar spectrum the relative quantum yield and the filter transmission are multiplied and then divided by a no filter (100% T) value.**

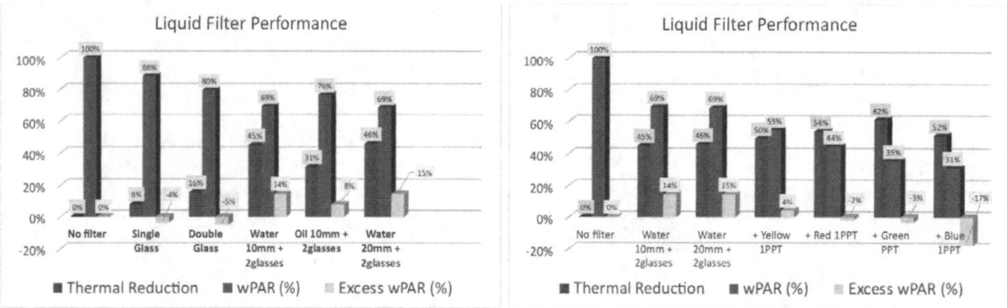


**Figure 11 Left: liquid filter performance metrics for: Left: first set of filters; Right: second set of filters.**

The shading device serves as a benchmark for evaluating the effectiveness of any spectral filter. Figure 12 illustrates a two-dimensional phase space featuring two key metrics - wPAR and Thermal Reduction. The shading filter's efficacy is represented by the diagonal line, wherein there is a trade-off between wPAR and thermal reduction. Filters that perform below this line are ineffective and should be replaced with a shading device. A high-quality filter should attenuate non-PAR regions more than the PAR region, surpassing the shading device's performance.

Take a filter's thermal reduction factor to be 65% with wPAR factor to be 50%. A shading device with thermal reduction factor of 65% will transmit only 35% in PAR. So, this filter outperform the shading device by 15% excess wPAR. Another way to state the difference, is to compare the filter again to a shading device of 50% transmission. This second shading device will have 50% thermal reduction factor only. The filter outperforms this shading device by blocking addition 15% on top of the 50%. This is defined here as excess thermal reduction.

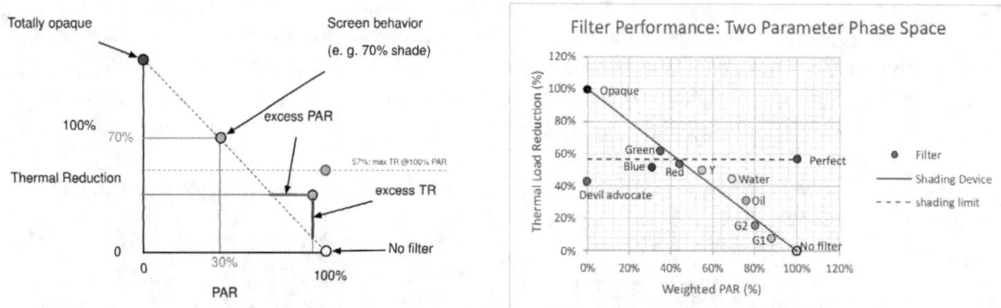


**Figure 12 Left: Thermal reduction vs PAR transmission model; Right: model applied to the liquid filters**

Let the thermal reduction factor for a filter be 65% and the wPAR factor be 50%. A shading device with a thermal reduction factor of 65% will allow only 35% transmission in PAR. Therefore, this filter provides better performance than the shading device, as it allows for 15% excess wPAR.

Journal of Physics: Conference Series **2751** (2024) 012025 doi:10.1088/1742-6596/2751/1/012025

Alternatively, we can compare the filter to a shading device with a 50% transmission rate, which would only have a thermal reduction factor of 50%. In this case, the filter blocks an additional 15% on top of the 50%, which is referred to as excess thermal reduction. The numerical values for both excess wPAR and excess thermal reduction are equal.

The performance metrics for the liquid filters have been charted on the 2D performance map, as shown in Figure 12 on the right. The glass and three colored filters had negative excess wPAR, meaning they fell below the shading device line. On the other hand, the water, oil, and yellow colored solution had positive excess wPAR. Clear water performed better than both oil and the yellow-colored aqueous solution, with a 14% excess wPAR compared to 8% and 4%, respectively. Increasing the water layer thickness had a minimal effect on the absorption of non-PAR, but it did not decrease wPAR.

The excess wPAR should not be the sole evaluation factor of a filter, as there exists a critical minimum level of wPAR necessary for the plants' wellbeing. Previously highlighted in the introduction, the standard AM 1.5 spectrum typically encompasses 1983 $(\mu mol/m^2/s)$. At optimum temperatures of approximately 27°C, the minimal recommended flux is $400(\mu mol/m^2/s)$, approximately 20% below the current average $CO_2$ concentration of 421 ppm. The increase in PAR does not necessarily translate to a higher rate of photosynthesis, as other limitations, predominantly $CO_2$, become limiting factors.

**Table 4: Measurements of the filters. All liquids are contained within 2 glasses.**

| # | Filter | Thermal Reduction (%) | wPAR (µMol/m2/s) | wPAR (%) | Excess wPAR (%) | Excess wPAR (µMol/m2/s) | excess TR (%) |
|---|---|---|---|---|---|---|---|
| 1 | No filter | 100% | 1662 | 100% | 0% | 0 | 0% |
| 2 | Glass Single | 8% | 1469 | 88% | -4% | -61 | -4% |
| 3 | Glass Double | 16% | 1227 | 74% | -11% | -178 | -11% |
| 4 | Water 10mm | 45% | 1153 | 69% | 14% | 239 | 14% |
| 5 | Oil 10mm | 31% | 1271 | 76% | 8% | 131 | 8% |
| 6 | Water 20mm | 46% | 1140 | 69% | 15% | 241 | 15% |
| 7 | Yellow 1PPT 10mm | 50% | 909 | 55% | 4% | 69 | 4% |
| 8 | Red 1PPT 10mm | 54% | 739 | 44% | -2% | -26 | -2% |
| 9 | Greeen 1PPT 10mm | 62% | 589 | 35% | -3% | -51 | -3% |
| 10 | Blue 1PPT 10mm | 52% | 516 | 31% | -17% | -290 | -17% |

Plants experience variations in temperature and solar spectrum throughout the year in realistic greenhouse conditions. In the Arabian region, such as in the city of Dhahran, Saudi Arabia, seasonal changes occur, as demonstrated in

**Table 5: Seasonal Changes to solar radiation to Dhahran (Lat: 26.29, Long: 50.11) [24] [25] [26]**

| # | Day | Day in 365 | Noon LST | Note | Total noon (W/m2) | Relative (%) |
|---|---|---|---|---|---|---|
| 1 | March 20, 2023 | 79 | 11:47 | March equinox | 949.1 | 95% |
| 2 | June 21, 2023 | 172 | 11:41 | June solstice | 1042.5 | 104% |
| 3 | Sep 23, 2023 | 266 | 11:32 | September equinox | 948.4 | 95% |
| 4 | Dec 22, 2023 | 356 | 11:37 | December solstice | 663.6 | 66% |

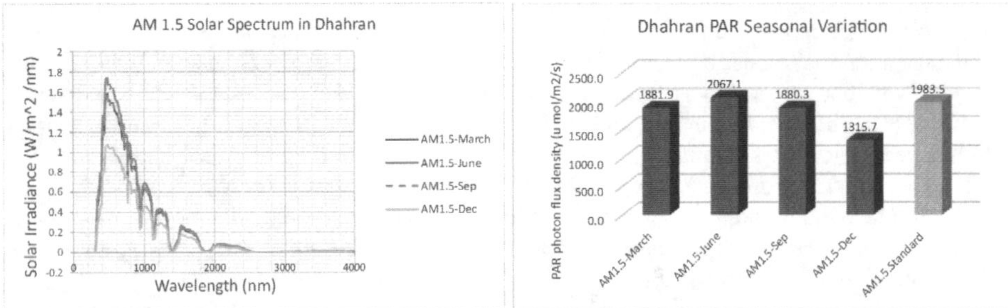


**Figure 13 Left: Noon AM1.5 Solar spectra at 4 key days in the year in Dhahran, Right: Respective PAR quantum content**

## 5. Conclusion

In hot climates, agricultural greenhouses require transparent enclosures to enable cooling the space for optimal plant growth. However, the cooling process is of high water and energy costs due to the high level of solar gain and heat trapping. To address this issue, a novel enclosure design has been proposed featuring a layer of liquid or water solution between two commercial grade glass layers (4-10-4mm). This water solution absorbs thermal radiation and can be circulated through an active heat dissipation system or utilized. Additionally, carefully selected dyes can be added to enhance the water filter performance.

The filter must efficiently allow sufficient photosynthetically active radiation (PAR) through to the plants while also blocking some non-PAR radiation. To evaluate filters based on these two metrics, we utilize a shading filter as a benchmarking scheme. The shading filter blocks a portion of the area and allows the remaining fraction through regardless of spectral characteristics. A well-designed filter should block less in the PAR region and more in the non-PAR region, surpassing the shading device's performance while maintaining the same PAR allowance.

We conducted tests using clear water, oil, and water with four commercially available pigments. The liquid filter showed a reduction in thermal load ranging from 31-62%, while reduction in the photosynthetically active portion of the solar spectrum ranged from 32-79%. The clear water filter performed the best, blocking 45% of the thermal load while allowing 69% of weighted PAR radiation. This allowed for an additional 14% of PAR radiation compared to an equivalent 45% transparent shading device. These preliminary results indicate that the proposed method is effective in achieving the intended task.

A potential avenue for future work could involve searching for additional chemical additives that further improve the performance of the clear water. However, it is advisable to prioritize the development of the engineering aspects of the system and conduct a technoeconomic analysis to increase the technology readiness level and enable practical implementation.

## Acknowledgments

The author would like to express his gratitude to KFUPM, specifically the Physics Department, the Renewable Energy and Power Systems Center, and the Center for Refining and Advanced Chemicals, for their valuable support.

## 6. References

[1] B. Krishna , G. M. Dias and others, "When too much isn't enough: Does current food production meet global nutritional needs?," *PLoS ONE,* vol. 13(10), p. e0205683, 2018.

[2] "China/ Bright future for photovoltaic greenhouses," Beijing Kingpeng International Agriculture Corporation, 5 Jan 2018. [Online]. Available: https://www.hortidaily.com/article/6040183/china-bright-future-for-photovoltaic-greenhouses/. [Accessed 05 May 2023].

[3] T. Kozai, G. Niu and J. Masabni, Plant Factory Basics, Applications and Advances, Academic Press, 2022.

[4] C. Kittas, N. Katsoulas and T. Bartzanas, "Greenhouse climate control in mediterranean greenhouses," *Cuadernos de Estudios Agroalimentarios (CEA03),* vol. Julio, pp. 89-114, 2012.

[5] I. Tsafaras, J. B. Campen, C. Stanghellinia , H. F. Zwart, W. Voogt, K. Scheffers, A. AlHarbi and K. AlAssaf, "Intelligent greenhouse design decreases water use for evaporative cooling in arid regions,," *Agricultural Water Management,* vol. 250, p. 106807, 2021.

[6] S. Ghani , F. Bakochristou, E. ElBialy, S. Gamaledin, M. Rashwan, A. Abdelhalim and S. Ismail, "Design challenges of agricultural greenhouses in hot and arid environments – A review," *Engineering in Agriculture, Environment and Food,* vol. 12, no. 1, pp. 48-70, 2019.

[7] A. J. Callejón-Ferre, F. Manzano-Agugliaro, M. Díaz-Pérez, A. Carreño-Ortega and J. Pérez-Alonso, "Effect of shading with aluminised screens on fruit production and quality in tomato (Solanum lycopersicumL.) under greenhouse conditions," *Spanish Journal of Agricultural Research,* vol. 7, pp. 41-49, 2009.

[8] L. Lua, M. E. Ya'acobb, M. Shamsul Anuar and M. N. Mohtar, "Comprehensive review on the application of inorganic and organic photovoltaics as greenhouse shading materials," *Sustainable Energy Technologies and Assessments,* vol. 52, p. 102077, 2022.

[9] E. Ravishankar, C. Melodi and et al, "Balancing crop production and energy harvesting in organic solar-powered greenhouses," *Cell Reports Physical Science,* vol. 2, no. 3, p. 100381, 2021.

[10] M. Planck, "On the Theory of the Energy Distribution Law of the Normal Spectrum," *Verhandl, Dtsch, phys Ges,* vol. 6, p. 168, 1900.

[11] Wolfram Research, Inc. (2023). Mathematica Version 12., "Wolfram Mathematica Physical Constant database,," 2023.

[12] American Society for Testing and Materials, "2000 ASTM Standard Extraterrestrial Spectrum Reference E-490-00 | Grid Modernization | NREL," The National Renewable Energy Laboratory, U.S. Department of Energy, 2000. [Online]. Available: https://www.nrel.gov/grid/solar-resource/spectra-astm-e490.html. [Accessed 22 04 2023].

[13] American Society for Testing and Materials, "Reference Air Mass 1.5 Spectra | Grid Modernization | NREL," National Renewable Energy Laboratory, U.S. Department of Energy, 2003. [Online]. Available: https://www.nrel.gov/grid/solar-resource/spectra-am1.5.html. [Accessed 22 4 2023].

[14] B. A. OSBORNE and R. J. GEIDER, "THE MINIMUM PHOTON REQUIREMENT FOR PHOTOSYNTHESIS," *New Phytologist,* vol. 106, pp. 631-644, 1987.

[15] W. Weerakkody and L. Suriyagoda,, "Estimation of leaf and canopy photosynthesis of pot chrysanthemum and its implication on intensive canopy management," *Scientia Horticulturae,,* vol. 192, pp. 237-243,, 2015.

[16] K. J. McCREE, "The action spectrum, absorptance and quantum yield of photosynthesis in crop plants,," *Agricultural Meteorology,,* vol. 9, pp. 191-216, 1971.

[17] J. R. Evans, "The Dependence of Quantum Yield on Wavelength and Growth Irradiance," *Australian Journal of Plant Physiology,* vol. 14, no. 1, pp. 69 - 79.

[18] S. Jacques and S. Prahl, "OMLC," https://omlc.org/index.html, 2017. [Online]. Available: https://omlc.org/spectra/PhotochemCAD/html/122.html. [Accessed 20 April 2023].

[19] Solar Light Company LLC, "16S-Series Air Mass Solar Simulators | Class A Emission Spectra," Solar Light Company LLC, 2022. [Online]. Available: https://solarlight.com/product_category/solar-simulators/16s-series-air-mass-simulators/. [Accessed 12 5 2023].

[20] Solar Light Company LLC, "Pyranometer | PMA 2144 Class II Pyranometer," Solar Light Company LLC, 2022. [Online]. Available: https://solarlight.com/product/pma-2144-class-ii-pyranometer/. [Accessed 12 05 2023].

[21] Solar Light Company LLC, "Model PMA2100 Dual-Input Data Logging Radiometer," Solar Light Company LLC, 2022. [Online]. Available: https://solarlight.com/product/datalogging-radiometer-pma2100/. [Accessed 12 05 2022].

[22] JASCO CORPORATION, JASCO CORPORATION, [Online]. Available: https://www.jasco.hu/wp-content/uploads/2011/12/V670_Leaflet.pdf. [Accessed 12 5 2023].

[23] H. Akima, "A method of bivariate interpolation and smooth surface fitting based on local procedures.," *Journal of the ACM (JACM),* vol. 17, no. 4, pp. 589-602, 1970.

[24] latitude.to, "GPS coordinates of Dhahran, Saudi Arabia. Latitude/ 26.2886 Longitude/ 50.1140," [Online]. Available: https://latitude.to/map/sa/saudi-arabia/cities/dhahran. [Accessed 14 04 2023].

[25] Wikipedia, "Winter solstice - Wikipedia," Wikipedia, [Online]. Available: https://en.wikipedia.org/wiki/Winter_solstice. [Accessed 14 4 2023].

[26] NREL: SMARTS, "SMARTS/ Simple Model of the Atmospheric Radiative Transfer of Sunshine | Grid Modernization | NREL," NREL: SMARTS, [Online]. Available: https://www.nrel.gov/grid/solar-resource/smarts.html. [Accessed 1 8 2022].

# Removal of rhodamine b and alizarin red from aqueous solutions by oxidized carbon nanotubes: kinetics and isotherm Study

**Abdelrahman K. A. Khalil**[1], **Ismail W. Almanassra**[1], **Anjaneyulu Chatla**[1], **Muataz Ali Atieh**[1,2], **Tahar Laoui**[1,3*]

[1] Research Institute of Sciences & Engineering (RISE), University of Sharjah, Sharjah P.O. Box 27272, United Arab Emirates.

[2] Chemical and Water Desalination Engineering (CWDE) Program, College of Engineering, University of Sharjah, Sharjah P.O. Box 27272, United Arab Emirates.

[3] Department of Mechanical and Nuclear Engineering, College of Engineering, University of Sharjah, Sharjah P.O. Box 27272, United Arab Emirates

[*]Corresponding author: tlaoui@sharjah.ac.ae

**Abstract.** Alizarin Red and Rhodamine B are widely used dyes in the textile, paper, and plastic industries. However, the disposal or release of these dyes into the environment can negatively impact on both the environment and human health. In the present study, oxidized carbon nanotubes (OCNT) were used to eliminate Alizarin Red and Rhodamine B from contaminated water in batch adsorption experiments. The adsorption outcomes suggested that OCNT has the potential to be an effective material for the removal of these dyes from contaminated water. The kinetics modeling revealed that the adsorption process of both dyes onto OCNT follows the pseudo-second-order model, indicating a chemisorption process. Moreover, the OCNT has indicated fast kinetics in which the equilibrium was achieved in 4 h. The isotherm study demonstrated that the Freundlich isotherm model best fits the experimental data, suggesting that dyes' adsorption onto OCNT is a monolayer adsorption process. The maximum adsorption capacities were 124.7 and 614.7 mg/g for Rhodamine B and Alizarin Red, respectively. The adsorption of these dyes was found to be more efficient at pH values of 6-8. This study suggests that OCNT can be an effective adsorbent for removing Alizarin Red and Rhodamine B from contaminated water.

## 1. Introduction

Industrial operations are responsible for a huge proportion of water contaminants that damage the ecosystem, and their ever-increasing nature has contributed to the current situation of the worldwide environment [1]. Industrial processes such as pulp, petrochemicals, refining, petroleum production, and paper production, battery manufacturing, pigment and paint industries, pesticide, tanneries, herbicide, fertilizer industries, as well as metallurgical and mining plants, require a large amount of chemical input and inadequate resource use of the input components results in a large number of wastes that enter the environment.

Various pollutants were found in the industrial wastewaters, including dyes, phosphates heavy metal ions, sulfates, hydrocarbon and organic compounds, and other organic and inorganic contaminants [2-4]. Water pollution caused by dyes is a significant environmental concern worldwide. Dyes are broadly utilized in industries like leather, textile, printing, and paper, leading to the discharge of huge quantities of dye-containing wastewater into water bodies. The presence of dyes in water can adversely affect aquatic ecosystems, human health, and water quality [5,6]. Rhodamine B (RHB) and Alizarin Red (AZR) are two specific dyes that can contribute to water pollution. They can adversely affect fish, algae, and other aquatic organisms. The dyes can disrupt their physiological processes, impair reproduction, and ultimately lead to ecological imbalances.

Different technologies have been used to remove RHB and AZR from drinking water, including ion exchange, precipitation/coprecipitation, adsorption, and membrane filtration [7,8]. Adsorption is one of the most used separation methods due to its low cost, convenience of use, and can remove low concentrations of pollutants. Carbon-based materials have emerged as promising candidates because of their superior textural characteristics, tunable structure, and excellent adsorption capabilities. Carbon-based materials, such as carbon nanotubes graphene oxide, and, activated carbon, have been extensively investigated for removing organic dyes from water sources [9,10]. However, despite the extensive research conducted on carbon-based materials, a research gap still exists regarding using oxidized carbon nanotubes to remove rhodamine B and alizarin red. Oxidized carbon nanotubes possess unique surface chemistry and functional groups that can enhance the adsorption capacity and selectivity towards target pollutants [11]. Previous studies have mainly focused on applying pristine carbon nanotubes or other carbon-based materials, neglecting the potential of oxidized carbon nanotubes for dye removal.

Therefore, the aim of our study is to explore the effectiveness of oxidized carbon nanotubes as an adsorbent for the elimination of rhodamine B and alizarin red from water using batch mode experiments. By exploring the adsorption capacity, thermodynamics, and kinetics of these oxidized carbon nanotubes,

## 2. Experimental

### *2.1. Materials*

Oxidized carbon nanotubes (OCNT) type (Multi-Walled Carbon Nanotubes COOH) purity 95%, 1.5% ash, and 3.86% functional content were purchased from Cheap Tubes Company (Grafton, USA). Rhodamine B and Alizarine red were bought from Sinopharm Chemical Reagent Co. (Shanghai, China). Deionized (DI) water was utilized to prepare all solutions.

### *2.2. Characterization*

The textural characteristics were studied from nitrogen adsorption isotherm at -196 °C utilizing NOVATECH LX2 analyzer, Anton Paar. Prior the $N_2$ testing, the sample was degassed at a temperature 300 °C for 6 hours. FE-SEM (Field emission scanning electron microscopy, Thermo scientific Apreo C) was utilized to explore the morphology and microstructure of the developed adsorbent. The FE-SEM was operated with an EDS (energy dispersion spectrometer). The accelerating voltage was 15 kV. The crystallinity of the hybrids was determined by XRD D8 (X-ray diffractometer, Advance, Bruker,) with a wavelength $\lambda$ = 1.54056 Å at generator current of 40 mA and generator voltage of 40 kV. Infrared spectra were measured between 4000 and 400 $cm^{-1}$ using a Jasco FT-IR (Model: 6300) instrument. Thermogravimetric analysis (TGA) of the OCNT was achieved by Netzsch analyzer (Jupiter, STA 449 F5).

### *2.3. Adsorption experiments*

The adsorption tests were conducted in batch mode where 50 mL of dye solution was mixed with certain amount of OCNT particles in 100 mL conical flasks using a Kuhner Lab-Shaker LS-X (Model MAZ10661LAB). To analyze the samples, they were collected and passed through a 0.45 μm PTFE membrane filter. The concentration of dyes was determined using a UV-VIS Spectrophotometer (UV-2550, Shimadzu, Kyoto, Japan) at specific wavelengths (554 nm for RHB and 421 nm for AZR). The adsorption capacity (Q), measured in mg of adsorbate per g of OCNT, and the removal efficiency (RE) expressed as a (%), were calculated using the following formulas:

$$Q = \frac{(C_i - C_e) \times V}{W} \tag{1}$$

$$RE(\%) = \frac{(C_i - C_e)}{C_i} \times 100 \tag{2}$$

Where $C_i$ and $C_e$ are the initial and exit dye loadings (mg/L), V is the dye solution volume (L), and W is the OCNT mass (g). The adsorption trials were repeated two times, and mean numbers were presented. The effect of OCNT amount was examined using OCNT concentrations between 0.2 and 2 g/L with 200 rpm agitation. Unless mentioned otherwise, the adsorption experiments were shaken for 24 h at pH 5.6.

### *2.4. Adsorption modeling*

To determine the isotherms of adsorption, 125 mg of OCNT were mixed with 50 mL of dye solution with loadings ranging from 50 to 200 mg/L and 50-1000 mg/L for RHB and AZR, respectively at 25 °C, stirring speed of 200 RPM and pH 5.6 for a day. The equilibrium data were scrutinized utilizing Langmuir, Freundlich, Temkin, Jovanovic, and Redlich–Peterson (RP) isotherm models. The expressions and parameters of the applied models were presented in table 1.

**Table 1.** Isotherm models used to explain the surfactant equilibrium data.

| Isotherm model | Mathematical equation | Related parameters |
|---|---|---|
| Langmuir | $q_e = \frac{X_m h C_e}{(1 + hC_e)}$ | *$X_m$,* Langmuir maximum dye uptake (mg/g). *h* , Langmuir constant (L/mg) |
| Freundlich | $q_e = K_F C_e^{\frac{1}{n_F}}$ | *$K_F$* Freundlich adsorption uptake coefficient (mg/g/(mg/L)$^{1/n}$ and $n_F$coefficient of adsorption intensity |
| Temkin | $q_e = \frac{RTln(A_t C_e)}{b}$ | *T* temperature (K), *R* ideal gas constant (J/mole/K), *$A_t$* equilibrium binding or adsorption uptake coefficient (L/mg), *b* Temkin constant |
| Jovanovic | $q_e = q_m(1 - exp^{-(K_J C_e)})$ | $q_e$, the adsorbed amount at equilibrium, $q_m$ the maximum adsorption capacity, K Jovanovic constant, $C_e$ equilibrium concentration. |
| RP | $q_e = \frac{K_R C_e}{1 + b_R C_e^{n_R}}$ | $K_R$ and $b_R$ RP model constants (L/g and mg$^{-1}$, respectively), $n_R$ RP exponent. |

## 3. Results and discussion

### *3.1. OCNT characterization*

Figure 1a illustrated the BET nitrogen adsorption-desorption isotherm analysis providing valuable information about the textural characteristics of OCNT. In this study, the BET analysis revealed a surface area (SA) of 302.779 m$^2$/g, indicating a significant amount of available surface for adsorption. The high SA of OCNT was attributed to their unique tubular structure, which provides a large external SA. The pore volume of 1.74 cc/g indicated the presence of substantial pore space within the OCNT structure. The presence of pores allows for the accommodation of nitrogen molecules during the adsorption process, enhancing the overall adsorption uptake of OCNT. The calculated pore radius of 14.5053 nm suggested the presence of mesopores within the OCNT structure. Mesopores are pores with diameters typically ranging from 2 nm to 50 nm. The presence of mesopores in OCNT can offer a suitable environment for the uptake of larger molecules or compounds, allowing for the efficient removal of dyes with larger molecular sizes. Figure 1b illustrated the XRD analysis of the OCNT. The (002) plane corresponds to the basal plane of graphite, which was the plane where carbon atoms were arranged in a hexagonal lattice. The (100) plane corresponds to a plane perpendicular to the OCNT axis and passes through the center of the tube. In OCNT, the (002) plane can provide information about the degree of graphitization and the quality of the graphene-like layers in the OCNT. The (002) peak in the XRD pattern of OCNT can shift to higher or lower angles depending on the functionalization process, affecting the interlayer spacing between the graphene-like layers.

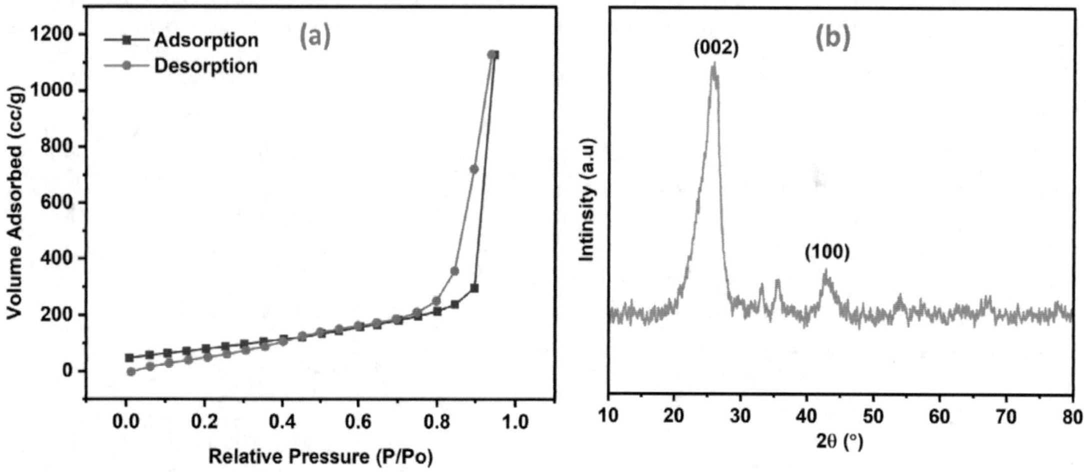


**Figure 1.** (a) BET $N_2$ isotherm, b) XRD of OCNT

On the other hand, the (100) plane can provide information about the orientation and alignment of the OCNT in a sample. In well-aligned OCNT samples, the (100) peak will be intense, indicating a high degree of alignment. However, in samples where the CNTs were randomly oriented, the (100) peak will be weaker and broader.

According to figure 2a, the FT-IR analysis of OCNT revealed several functional groups. The absorption peaks at 3300-3500 $cm^{-1}$ were attributed to the O-H stretching vibrations, demonstrating the existence of hydroxyl groups (-OH) on the surface of the OCNT [12]. The peaks at 2800-3000 $cm^{-1}$ corresponded to C-H stretching vibrations, indicating the presence of sp3 hybridized carbon atoms bonded to hydrogen atoms. The peaks observed at 1500-1700 $cm^{-1}$ were given to C=O stretching vibrations, which suggest the presence of carbonyl groups (-C=O) on the surface of OCNT. The peaks detected at 1300-1500 $cm^{-1}$ corresponded to C-C stretching vibrations, indicating the existence of sp2 hybridized carbon-carbon bonds. Finally, the peaks observed at 500-800 $cm^{-1}$ corresponded to C-H bending vibrations [13].

The TGA utilized to explore the thermal stability and decomposition behavior of materials by measuring their weight changes as a function of temperature. The results of the TGA analysis of OCNT shown in figure 2b can provide important information about their thermal stability and composition. According to the results, when the temperature reached 485°C, the weight loss of the OCNT declined to 90%, suggesting that there was a significant mass loss below this temperature, likely due to the removal of any adsorbed or chemically bound water, as well as the degradation of any organic impurities on the OCNT's surface. When the temperature reached 600°C, the OCNT lost 60% of their weight, indicating that there was further decomposition of the OCNT, likely due to the breaking of carbon-carbon bonds in the OCNT, which leads to the release of gaseous products such as carbon monoxide and carbon dioxide. From 600°C until 800°C, the OCNT lost an additional 73% of their weight, leaving a residual weight of 27% of the total sample weight, indicating that the remaining mass was primarily composed of carbon, which could indicate the presence of residual amorphous carbon or the formation of graphite-like structures. Overall, the TGA results suggest that the OCNT has relatively good thermal stability up to 485°C but starts to decompose significantly above that temperature.

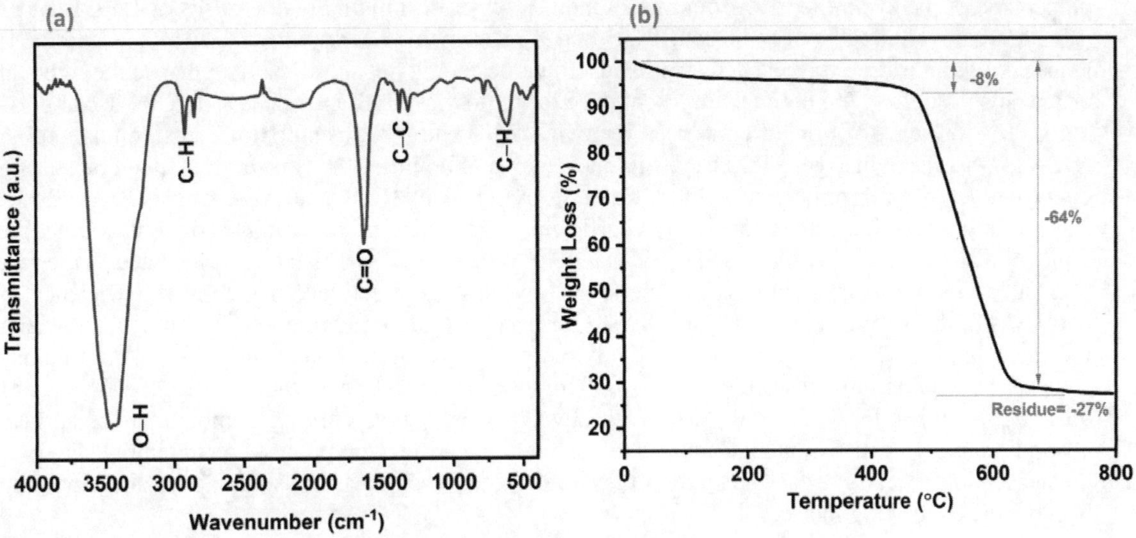


**Figure 2.** a) FT-IR spectrum of OCNT; b) TGA analysis of OCNT.

Figure 3 displayed the FE-SEM images of the OCNT that provide information about their morphology, structure, surface features, and alignment. OCNT were cylindrical structures made of carbon atoms arranged in a hexagonal lattice. In an FE-SEM image, OCNT appear as thin, elongated structures with a tubular shape. The diameter of OCNT can range from a few nanometers to tens of nanometers, depending on the specific synthesis method and conditions. The surface of OCNT can also exhibit various morphologies depending on the synthesis method and any post-treatment processes.

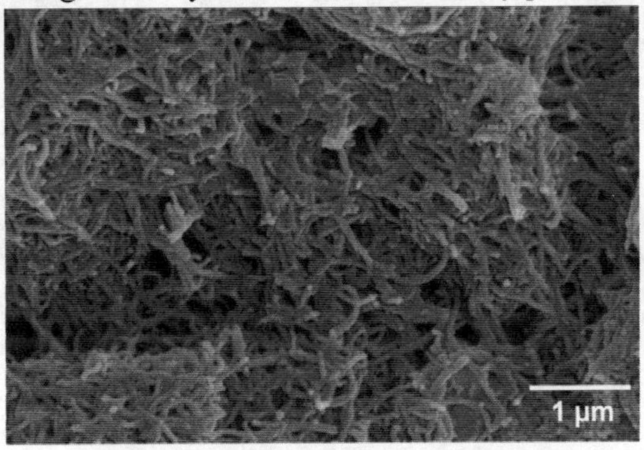


**Figure 3.** FE-SEM image of OCNT.

For example, OCNT can be functionalized with different chemical groups to modify their surface properties, resulting in different surface morphologies. In the FE-SEM image, one may observe the presence of different surface features such as bumps, wrinkles, and defects indicative of the morphology of the OCNT.

*3.2. Effect of adsorption parameters*

Figure 4a illustrated the influence of OCNT dose on the elimination of RHB and AZR. The outcomes implied that the removal effectiveness of both RHB and AZR raised as the OCNT dose increased. This could be attributed to the fact that increasing the adsorbent dose improved the SA available for adsorption, which in turn lead to an improvement in the quantity of available adsorption sites. At lower doses (0.2-0.5 g/L), the adsorption locations on the OCNT surface may not be sufficient to eliminate all dye molecules in the solution, resulting in lower removal efficiencies. However, as the dose increases to 1 g/L and beyond, the adsorption sites become more plentiful and can effectively remove a significant percentage of the dye molecules from the solution. It was worth noting that at doses of 1.5 g/L and 2 g/L, the removal effectiveness of both RHB and AZR remains nearly constant at 99.8% and 99%, respectively, indicating further increases in adsorbent dose may not significantly enhance the removal efficiency. Therefore, the optimal adsorbent dose for removing RHB and AZR using OCNT can be 1-2 g/L. The influence of time of contact on removing RHB and AZR using oxidized carbon nanotubes (OCNT) was shown in Figure 4b. The results showed that removing both dyes increases with increasing contact time. After 1 minute of contact time, the removal of RHB and AZR was 12% and 33%, respectively, which indicates a low removal efficiency. However, after 20 minutes of contact time, the removal of RHB and AZR increased to 37% and 49%, respectively. This can be attributed to the fact that OCNT requires sufficient contact time to absorb the dyes effectively. The removal efficiency of both dyes further increased to 80.8% and 89.8%, respectively, after 240 minutes of contact time, which indicates that OCNT was capable of effectively removing dyes from water. Furthermore, the findings showed that equilibrium was achieved after 600 minutes of contact time, with a removal efficiency of 96.7% and 99.8% for RHB and AZR, respectively. This suggests that a combination of surface adsorption and pore diffusion mechanisms may govern the adsorption process. At the initial stages of the adsorption process, surface adsorption may be dominant, while at later stages, pore diffusion may become the controlling factor.

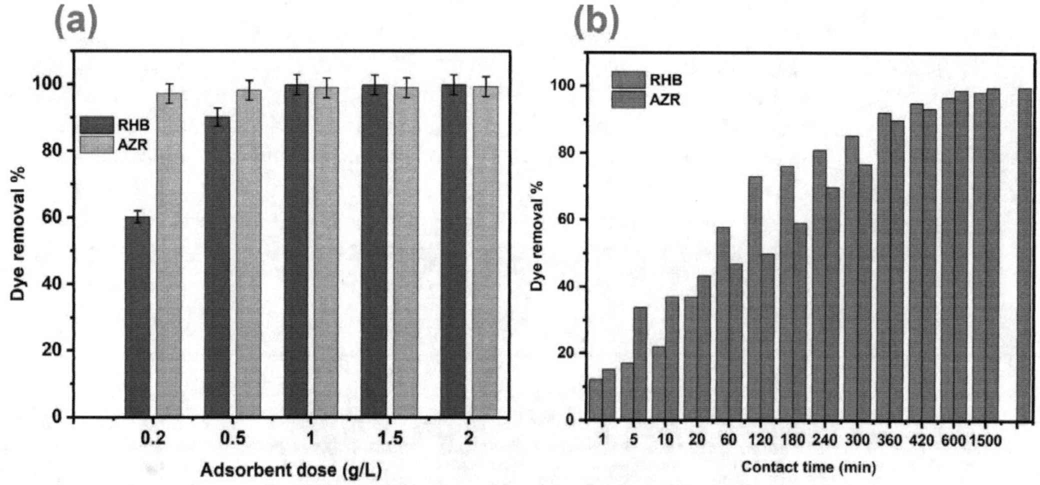


**Figure 4.** Effect of adsorption parameters for RHB and AZR removal by OCNT; a) The impact of OCNT dose; b) Effect of contact time.

### *3.3. Adsorption isotherm*

Figure 5 presents the dye adsorption isotherm on OCNT, and we utilized the Langmuir, Freundlich, Temkin, Jovanovic, and RP adsorption models to match the data and gain a deeper understanding of the adsorption mechanism. Upon analyzing the results, we observed that the investigational data were most accurately described by the Freundlich models, with R-squared ($R^2$) values of 95% and 97% for RHB and AZR, respectively. These high $R^2$ values indicate a good fit between the Freundlich models and the obtained data for both dyes. On the other hand, the Langmuir and RP models also exhibited reasonable agreement with the experimental data, suggesting different adsorption mechanisms for removing dyes by OCNT. These outcomes offer worthy insights into the adsorption process of dyes on OCNT. The results suggest the creation of a monolayer of RHB and AZR on the surface of OCNT, potentially indicating some level of heterogeneity in the adsorption sites of OCNT.

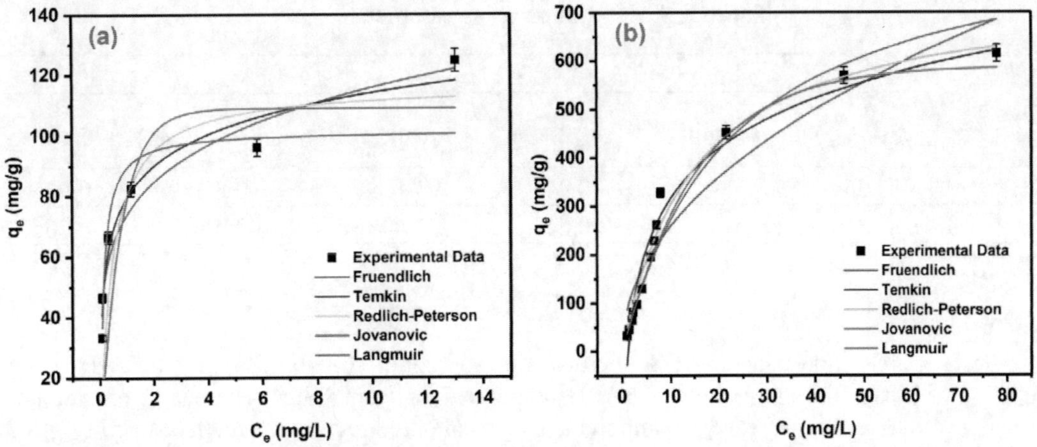


**Figure 5.** (a) Adsorption isotherm models fit of RHB removal by OCNT; (b) Adsorption isotherm models for AZR removal by OCNT.

### *3.4. Adsorption kinetics*

Figure 6 presents the dye adsorption kinetics on OCNT, and we applied the pseudo-first order (PFO) and pseudo-second order (PSO) adsorption models to examine the experimental data and gain a more information about of the mechanism of adsorption. Upon analyzing the results, we found that the obtained data were most accurately described by the PSO model. The $R^2$ values obtained were 94% for RHB indicating reasonably good fits between the PSO model and the obtained data for RHB dye. The PSO model implies that the rate of adsorption was dependent on the loading of the dye and the available adsorption sites on OCNT. Moreover, these findings suggest that a chemical adsorption process might be involved in the removal of RHB by OCNT. However, for AZR, the $R^2$ for both kinetic models were between 0.647 and 0.766, implying that that neither of these models adequately describes the adsorption behavior of the AZR dye on the OCNT material. Further investigation and analysis may be required to understand the kinetics of the AZR dye adsorption and identify a suitable kinetic model that accurately describes the experimental data. Moreover, it indicates that the removal process of the AZR dye involves different adsorption mechanisms. This suggests that the adsorption behavior of the AZR dye on the OCNT material was more complex and cannot be solely described by PFO and PSO kinetic models. Table 2 indicates the adsorption kinetics factors of the RHB and AZR removal by OCNT. The findings indicate that the rate of adsorption was influenced by the concentration of the dyes and the accessibility of adsorption locations on OCNT surfaces.

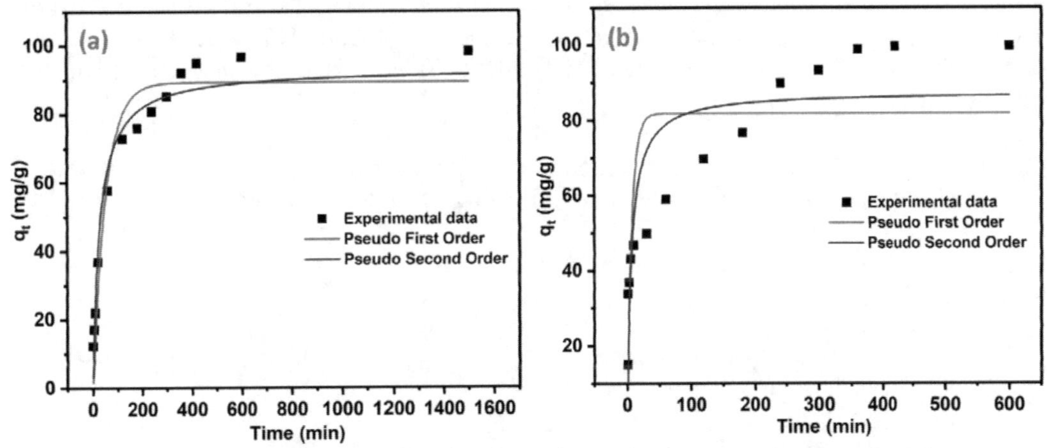


**Figure 6.** Adsorption kinetic models fit of (a) RHB and (b) AZR, elimination by OCNT.

**Table 2.** Adsorption kinetics parameters.

| Dye | PFO | | | PSO | | |
|---|---|---|---|---|---|---|
| | $q_e$ (mg/g) | $k_1$ ($min^{-1}$) | $R^2$ | $q_e$ (mg/g) | $k_2$ (g $mg^{-1}$ $min^{-1}$) | R2 |
| RHB | 89.353 | 0.019 | 0.941 | 93.189 | 0.000389 | 0.943 |
| AZR | 81.815 | 0.135 | 0.647 | 87.498 | 0.00175 | 0.766 |

## 4. Conclusion

The present research investigated the properties and adsorption capabilities of oxidized carbon nanotubes (OCNT) for the elimination of RHB and AZR dyes from water. The characterization of OCNT using various techniques revealed important insights into its structure and thermal stability. XRD analysis showed the graphitization and alignment of OCNT, while FT-IR analysis identified the presence of functional groups on OCNT surface. TGA analysis indicated that OCNT exhibited good thermal stability up to 485°C. FE-SEM images provided information about the morphology and surface features of OCNT, highlighting their tubular structure and various surface morphologies. The adsorption parameters were investigated, and it was found that the elimination efficiency of RHB and AZR enhanced with increasing OCNT dose. The optimal adsorbent dose for both dyes was determined to be 1-2 g/L. The removal of dyes also boosted with longer contact time, indicating the importance of sufficient contact time for effective adsorption. The equilibrium was attained after 600 minutes. Furthermore, the adsorption capacity of OCNT increased with improving original dye concentration, suggesting a higher driving force for the remediation process. However, a plateau in the adsorption uptake was attained as the available adsorption locations became saturated. Adsorption isotherm analysis demonstrated that the Freundlich model was the optimum fit to the obtained data, indicating a heterogeneous distribution of adsorption sites on OCNT surfaces. Regarding adsorption kinetics, the PSO model described the experimental data more accurately than the PFO model, suggesting that the adsorption rate was relying on the concentration of the dye and the availability of adsorption sites on OCNT surfaces. Overall, this research demonstrated the potential of OCNT as an effective material for the elimination of RHB and AZR dyes from water. The findings contribute to the understanding of OCNT properties and offer valuable insights for the design of efficient adsorption processes in water treatment applications.

## References

[1] Khalil AKA, Dweiri F, Almanassra IW, et al. Mg-Al Layered Double Hydroxide Doped Activated Carbon Composites for Phosphate Removal from Synthetic Water: Adsorption and Thermodynamics Studies. Sustainability 2022;14:6991.

[2] Park D, Yun Y, Jo JH, Park JM. Biosorption process for treatment of electroplating wastewater containing Cr (VI): Laboratory-scale feasibility test. Ind Eng Chem Res 2006;45(14):5059-5065.

[3] Ho Y, Ofomaja AE. Biosorption thermodynamics of cadmium on coconut copra meal as biosorbent. Biochem Eng J 2006;30(2):117-123.

[4] Bailey SE, Olin TJ, Bricka RM, Adrian DD. A review of potentially low-cost sorbents for heavy metals. Water Res 1999;33(11):2469-2479.

[5] Maheshwari K, Agrawal M, Gupta AB. Dye pollution in water and wastewater. Novel Materials for Dye-containing Wastewater Treatment 2021:1-25.

[6] Pollution and treatment of dye waste-water. IOP Conference Series: Earth and Environmental Science: IOP Publishing; 2020.

[7] Waheed A, Mansha M, Kazi IW, Ullah N. Synthesis of a novel 3, 5-diacrylamidobenzoic acid based hyper-cross-linked resin for the efficient adsorption of Congo Red and Rhodamine B. J Hazard Mater 2019;369:528-538.

[8] Özdemir M, Durmuş Ö, Şahin Ö, Saka C. Removal of methylene blue, methyl violet, rhodamine B, alizarin red, and bromocresol green dyes from aqueous solutions on activated cotton stalks. Desalination and Water Treatment 2016;57(38):18038-18048.

[9] Bal G, Thakur A. Distinct approaches of removal of dyes from wastewater: A review. Materials Today: Proceedings 2022;50:1575-1579.

[10] Kyzas GZ, Deliyanni EA, Matis KA. Graphene oxide and its application as an adsorbent for wastewater treatment. Journal of Chemical Technology & Biotechnology 2014;89(2):196-205.

[11] Wang S, Ng CW, Wang W, Li Q, Hao Z. Synergistic and competitive adsorption of organic dyes on multiwalled carbon nanotubes. Chem Eng J 2012;197:34-40.

[12] Abushawish A, Almanassra IW, Namboorimadathil S, et al; High-efficiency removal of hexavalent chromium from contaminated water using nitrogen-doped activated carbon : kinetics and isotherm study. Mater Chem Phys 2022;291:126758.

[13] Jaber L, Ihsanullah I, Almanassra IW, et al. Adsorptive Removal of Lead and Chromate Ions from Water by Using Iron-Doped Granular Activated Carbon Obtained from Coconut Shells. Sustainability 2022;14:10877.

Journal of Physics: Conference Series **2751** (2024) 012027 doi:10.1088/1742-6596/2751/1/012027

# The Effect of Antioxidants on the Stability of Perovskite Solar Cells

**Fatima Alzubaidi[1,2], Di Zhang[2] Nouar Tabet[1,3] ***

*[1]Center for Advanced Materials Research (CAMR), Research Institute of Sciences and Engineering (RISE), University of Sharjah, P.O. Box 27272, Sharjah, United Arab Emirates*
*[2]Department of Sustainable and Renewable Energy Engineering, University of Sharjah, P.O. Box 27272, Sharjah, United Arab Emirates*
*[3]Department of Applied Physics & Astronomy, University of Sharjah, P.O. Box 27272, Sharjah, United Arab Emirates*

*falzubaidi@sharjah.ac.ae

**Abstract**. The passivation of defects present in the bulk and surface of the perovskite absorbing layers of solar cells is currently the subject of intense research. To address this issue, dopants can be used to minimize the density of defects that occur at the surface and grain boundaries of perovskite films. Herein, we introduced two new antioxidants (Morin and Quercetin) in the precursor solutions to passivate the defects present in the $Cs_{0.1}FA0_{0.9}PbI_3$ perovskite absorbing layer. The positive impact of antioxidants on the efficiency of perovskite cells has been recently reported. However, more experimental data are needed to identify the most effective passivating agent. The time resolved photoluminescence measurements revealed that the added molecules increased the photocarrier lifetime by 120-194%. Our results showed that the optimum concentrations are 2.223 mM for Morin and 0.0156 mM for Quercetin. These results demonstrate that the new molecules are effective passivating agents and have the potential to improve the device performance. The variation of the cell power conversion efficiency as function of the concentration of the antioxidants will be discussed.

## 1. Introduction

In recent years, the uptake of renewable sources has increased in an attempt to reduce dependence on fossil fuels for power generation and to cover the increased energy demand caused by the global population growth [1]. Solar energy is considered as a significant renewable energy source because its cleanliness, abundance and versatility [2]. The silicon-based solar cells currently hold the lead in terms of high-power conversion efficiency (PCE), reaching 26%. However, perovskite solar cells (PSCs), a novel class of advanced solar technology, are becoming a possible substitute for silicon-based solar cells that can approach 25.7 % PCE [3]. In general, the efficient photovoltaic performance of these solar cells is a result of the PSCs' intrinsic properties, which include their large absorption spectra, adjustable bandgap, defect tolerance, fast charge carrier dynamics, compatibility with multiple charge extraction layers and extended carrier diffusion length compared with the film thickness [4]. The structure of perovskite cell consists of an absorber layer, such as $MAPbI_3$ and $CsFAPbI_3$ that is inserted between the electron-transport layer (HTL), a material with a good electron extraction capability, and the hole-transport layer (ETL), a material with efficient hole extraction [5].

The efficient production of high-performance PSCs requires the control of defect assisted carrier recombination that can occur at grain boundaries and an effective carrier extraction at the interfaces [6]. The chemical passivation of defects in the bulk of the absorbing layer of PSCs typically refers to the reduction of the electric activity of trap states by using appropriate passivating agents to the perovskite precursor solution [7],[8].
Antioxidants are molecules that are found naturally in plants. These molecules have been used to minimize the harmful free radicals' effect by neutralizing them. Free radicals are molecules that exhibit high reactivity and instability due to the presence of at least one unpaired valence electron, and they are seen as dangling bonds [9]. In many recent studies, antioxidant additives were employed to passivate defects and increase the performance of PSCs [10]. As an example, Xu et al. [11] used ascorbic acid (vitamin C) in the $MA_{0.5}FA_{0.5}Pb_{0.5}Sn_{0.5}I_3$ perovskite solution to enhance the quality and elevate the efficiency of mixed Sn–Pb based PSC. The devices with the ascorbic acid additive showed an efficiency of 14 % compared with 12% for the standard with a noticed improved stability. In a study conducted by Zhuang et al. [12], lycopene, derived from a tomato, was introduced to $Cs_{0.05}(FA_{0.83}MA_{0.17})_{0.95}PbI_{0.83}Br_{0.17}$ perovskite precursor. A markable improvement in the PCE was observed where it increased from 20.57% to 23.62% and the rate of degradation in terms of the absorption ability dropped from 0.12 to 0.02 after adding lycopene. Ban et al. [13] showed that introducing tea polyphenol into the mixture of $CsSn_{0.5}Pb_{0.5}I_2Br$ resulted in an improved electrical potential value and extended the lifetime of charge carriers.
3, 5, 7, 2′, 4′-pentahydroxyflavone (Morin) and 3,3′,4′,5,7-Pentahydroxyflavone (Quercetin) are antioxidant molecules that belong to a plant pigment group called flavonoids [14]. These molecules are extracted from fruits, vegetables and herbs and commonly used in the pharmaceutical and cosmetic industry for their ability of protect skin under UV exposure and protect human bodies from unstable cells [15],[16].
Herein, we introduced the use of two new antioxidants, namely Morin and Quercetin, by adding them to $Cs_{0.1}FA_{0.9}PbI_3$ perovskite precursor solution. The structural and optical characteristics of the layers were examined using several techniques. Moreover, the stability of the films was tested after 500h under 35% relative humidity exposure. The films produced either lacking or containing the antioxidant additives were referred to as the standard and passivated films, respectively.

## 2. Methodology

### *2.1. Materials and Chemicals*

Formamidinium iodide (FAI, 99.9%), cesium iodide (CsI, 99.9%) and morin (98%) were obtained from Great-cell solar, ACROS Organics and SRL, respectively, and lead(II) iodide ($PbI_2$, 99%) and Quercetin (95%) were acquired from Sigma-Aldrich. All solvents utilized in this study including dimethyl sulfoxide (DMSO), N,N-dimethylformamide (DMF) and antisolvents such as chlorobenzene (CB) were anhydrous and bought from Sigma-Aldrich.

### *2.2. Perovskite Precursor and Film Formation*

$Cs_{0.1}FA_{0.9}PbI_3$ perovskite films were fabricated via a one-step solution-processing method within an $N_2$-filled glovebox. The preparation of the solution involved dissolving stoichiometric amounts of (FAI, 185.9 mg), ($PbI_2$, 553.4 mg), and (CsI, 31.2 mg) in a mixture solvent containing in a mixed solvent (0.800 mL DMF and 0.200 mL DMSO, volume ratio = 1:4). The vial containing the precursor was tightly sealed and stirred for 4 h at 25 °C to ensure complete precursor dissolution.
A volume of 50 μL of the precursor solution was deposited onto pre-cleaned 1.5 cm x 1.5 cm glass substrates using spin-coating. This specific volume selection is crucial for optimal film formation on this substrate size. Using a larger volume could lead to solution spilling over the edges of the substrate and onto the backside during the high-speed spinning process. This unwanted coating on the backside can hinder film characterization and potentially waste materials. Therefore, 50 μL offers a balance

between ensuring complete coverage of the substrate while preventing solution overflow and minimizing waste.

After that the glass substrate was spin-coated in a two-step process. The first spin step was performed at 1000 rpm for 10 seconds to spread the solution evenly. Subsequently, the spin speed was increased to 5000 rpm for 30 seconds to form a uniform thin film. During the second spin step (after 20 seconds), 0.30 mL of antisolvent CB was rapidly dispensed onto the centre of the substrate to promote rapid crystallization and improve film morphology. Finally, the spin-coated films were annealed at 150°C for 20 minutes to complete perovskite film formation.

For the fabrication of modified perovskite films, a specific volume of an antioxidant solution (10 mg Morin in 1 mL DMSO and 10 mg Quercetin in 2.6 mL DMSO) was pre-mixed with the perovskite precursor solution and stirred for an additional 10 minutes before deposition. The subsequent deposition process remained identical to that described for standard samples.

### *2.3. Sample Characterization Techniques*

The various properties of perovskite films were investigated using several analytical techniques. Time-resolved photoluminescence (TRPL) spectroscopy was employed to measure the photoluminescence lifetime of the films. This measurement was performed with a Mini-tau Fluorescence Spectrometer (Edinburgh Instruments) using a pulsed laser source with a wavelength of 405 nm. The X-ray diffraction (XRD) patterns were conducted utilizing Bruker D8 ADVANCE diffractometer equipped with Cu-Kα X-ray source emitting at 0.15406 nm wavelength. Shimadzu UV-2600i spectrometer was employed to acquire the UV-vis absorption spectra. High-resolution images of the perovskite surface films were acquired from a field emission scanning electron microscope (FESEM) manufactured by Thermo Scientific.

## 3. Results and Discussion

The XRD patterns of the perovskite layer after and before adding the passivating agents are displayed in Figure 1. We observed that the positions of the diffraction peaks of the passivated perovskite were the same as for the unmodified films. This suggests that the antioxidant dopants did not alter the perovskite crystalline structure and lattice parameter. The characteristic peaks of the samples at (100), (110), (111) and (200) diffraction peaks of the samples labeled in the figure indicate a cubic crystal structure of the perovskite material with an average distance between atomic planes of 6.299 Å. XRD analysis did not identify any diffraction peak at $2\theta = 11.6°$ which indicates that the film is free of the unwanted phase of $Cs_{0.1}FA_{0.9}PbI_3$ [17],[18].

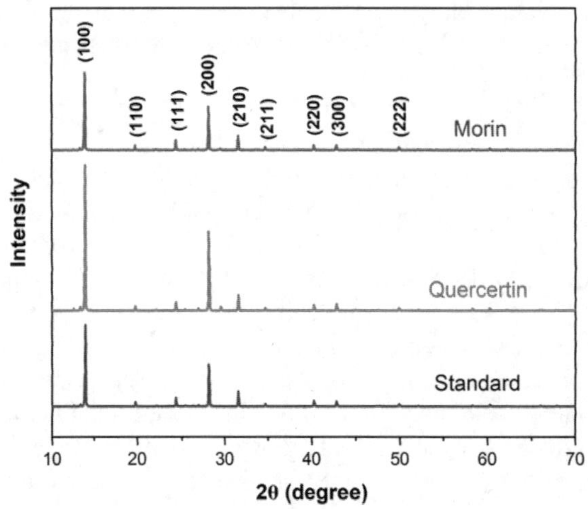


**Figure 1:** XRD analysis of pristine and antioxidant-treated perovskite films

The UV–vis absorption spectra of the fabricated perovskite films, before and after the antioxidant treatments and the corresponding Tauc plots are presented in Figure 2a and 2b, respectively. Tauc plots are used to determine the optical bandgap (Eg) of the perovskite films. Analysis of these plots revealed a uniform Eg value of approximately 1.57 eV for all tested samples, regardless of passivation treatment.

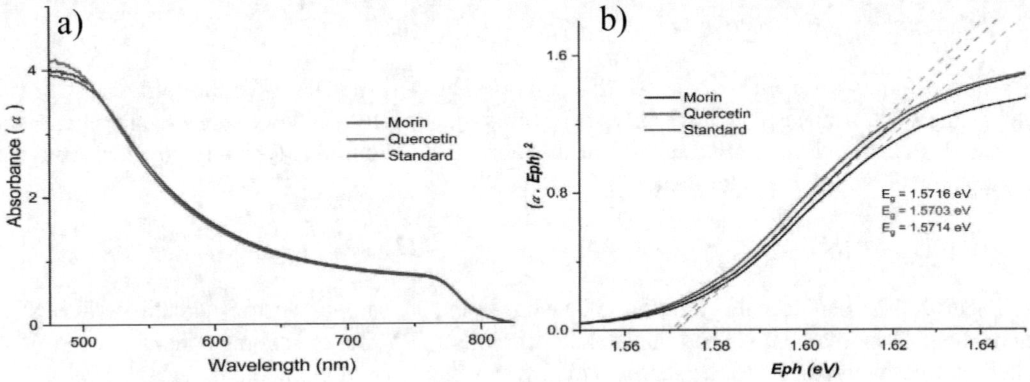


**Figure 2:** a) UV-vis absorption spectra of both treated and untreated perovskite films b) the corresponding Tauc plots derived from the UV-vis data.

TRPL spectra of the fabricated perovskite films for pristine (untreated) perovskite films and films incorporating antioxidant additives during processing are shown in Figure 3a, b.

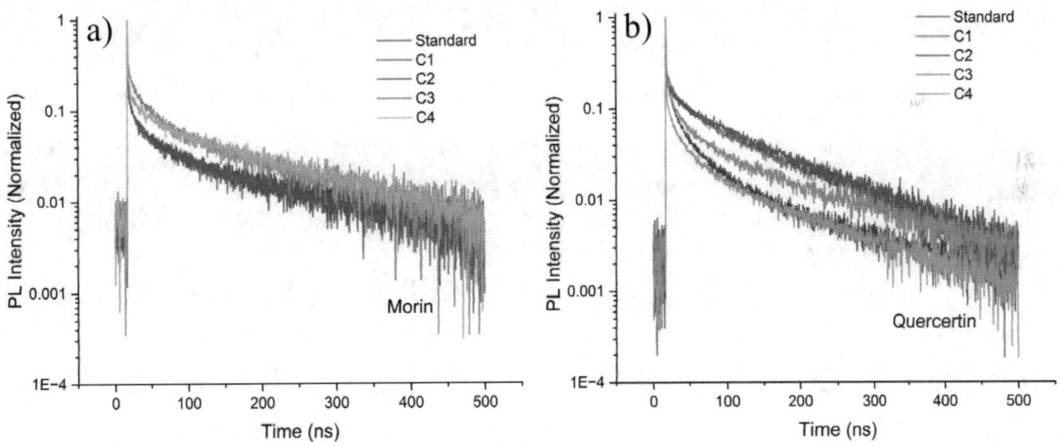


**Figure 3:** a) TRPL curves of the perovskite material fabricated with various concentrations of Morin b) Quercetin concentration

The photoluminescence lifetime of the perovskite films was determined by fitting the TRPL signal using a triexponential decay model (equation provided below) [19].

$$I_{TRPL} = \sum_i^n q_i e^{-t/\tau_i} \qquad (1)$$

Where $q_i$ and $\tau_i$ are respectively the amplitude coefficient and lifetime of term $i$ and n is the number of terms that is used. In most of the published works, the TRPL results are fitted using a single or double exponential model with no direct and clear correlation of the physical parameters that govern the photoluminescence signal decay. In our case, we have used a 3-term exponential model to fit our experimental curves. Equation 2 is the 3-term exponential model used here.

$$I_{TRPL}(t) = q_1 e^{-t/\tau_1} + q_2 e^{-t/\tau_2} + q_3 e^{-t/\tau_3} \quad (2)$$

Analysis of the fitting parameters indicates that $\tau_1$ is primarily responsible for the initial, fast decay observed in the TRPL transients. Conversely, $\tau_3$ governs the long-lived, slower decay component that extends towards the tail of the TRPL signal. The average carrier lifetime is typically obtained from the TRPL data using the following equation [20]:

$$\tau_A = \sum_{i=1}^{n} q_i \tau_i / \sum_{i=1}^{n} q_i \quad (3)$$

Tables 1 and 2 provide the concentration of each passivated sample with Quercetin and Morin, respectively, the fitting parameters and the average lifetime ($\tau_A$). Also, the average change in $\tau_A$ of the passivated samples with respect to the standard one was calculated and included in the tables below.

**Table 1:** The molarity of different concentrations, fitting parameters, and the change of for standard and passivated samples (Quercetin)

| Concentration | Molarity mM | $\tau_1$ (ns) | $A_1$ | $\tau_2$ (ns) | $A_2$ | $\tau_3$ (ns) | $A_3$ | $\tau_A$ (ns) | $\Delta\tau_A$ (ns) |
|---|---|---|---|---|---|---|---|---|---|
| Standard | 0 | 5.59 | 1347.98 | 26.95 | 930.04 | 129.54 | 277.49 | 26.82 | - |
| C1 | 0.00785 | 5.67 | 478.92 | 34.1 | 484.67 | 153.94 | 397.3 | 59.08 | 120% |
| C2 | 0.01557 | 6.25 | 248.56 | 35.88 | 274.26 | 141 | 479.7 | 78.83 | 194% |
| C3 | 0.05053 | 4.15 | 652.82 | 25.69 | 598.41 | 148.1 | 315.1 | 41.34 | 54% |
| C4 | 0.10047 | 4.75 | 916 | 26.05 | 845.55 | 142.11 | 374.79 | 37.28 | 39% |

**Table 2:** The molarity, fitting parameters, and the change of for standard and passivated samples (Morin)

| Concentration | Molarity mM | $\tau_1$ (ns) | $A_1$ | $\tau_2$ (ns) | $A_2$ | $\tau_3$ (ns) | $A_3$ | $\tau_A$ (ns) | $\Delta\tau_A$ (ns) |
|---|---|---|---|---|---|---|---|---|---|
| Standard | 0 | 2.25 | 485.65 | 19.13 | 328.30 | 169.04 | 202.87 | 40.98 | - |
| C1 | 0.414 | 2.78 | 733.11 | 22.20 | 305.95 | 179.21 | 195.95 | 35.58 | -13% |
| C2 | 1.241 | 2.56 | 630.29 | 21.55 | 274.57 | 168.89 | 201.73 | 37.59 | -8% |
| C3 | 1.654 | 3.53 | 358.33 | 26.46 | 374.49 | 172.71 | 203.08 | 49.42 | 21% |
| C4 | 2.233 | 3.40 | 274.58 | 37.58 | 260.19 | 269.11 | 209.65 | 90.18 | 120% |

It is worth noticing that $\tau_1$, $\tau_2$, and $\tau_3$ are mainly fitting parameters with no straightforward correlation with specific recombination processes such as the band to band and SHR defect assisted recombination. However, Baloch et al.[21] developed a model that clarifies the effect of the key physical parameters that control the recombination mechanisms in the bulk and at the interface of the perovskite layers. The model shows that the initial decay of the signal is essentially related to the band-to-band recombination and the recombination at the interfaces while the tail of the TRPL signal is mainly affected by defect assisted recombination in the bulk.

In our case, the band-to-band recombination is not expected to change significantly as the antioxidant molecules did not affect the perovskite crystalline structure as determined XRD. However, there was an observed fluctuation on $\tau_1$ in our results, and it is probably related to a combination of the thickness and the surface recombination process. Our results show 120% and 194% increase of the average carrier lifetime in the treated films with Morin and Quercetin, respectively.

To test the films stability, the samples were kept in a petri dish covered with an opaque lid that blocks the light for 3 weeks (~500 hours) and placed in the dry box where the relative humidity (RH%) and the temperature were monitored and maintained to be 35% ± 5% and 23˚C ± 0.5˚C, respectively. Figure 4 illustrates the change in the normalized absorbance over time for the standard and the antioxidant modified perovskite layers deposited on the top of glass substrates. As shown, the absorbance decreased significantly for Morin passivated films between 550 and 450 nm in particular whereas Quercetin exhibits a more stable absorption intensity compared with the other samples.

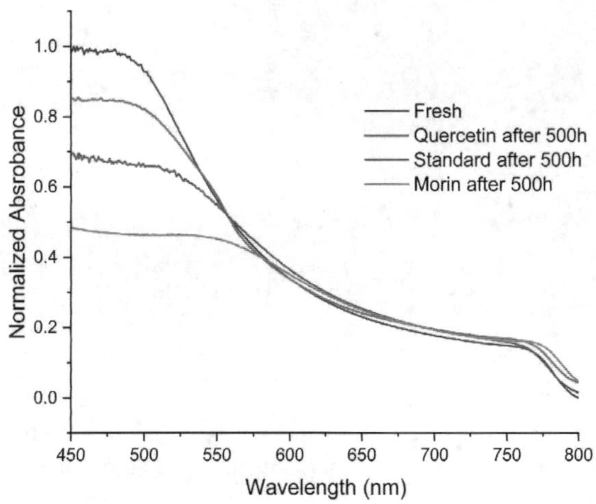


**Figure 4:** The absorption spectra for standard and perovskite films passivated with Quercetin or Morin before and after leaving the samples under dark at 23˚C ± 0.5˚C and in 35% ± 5% relative humidity atmosphere for 3 weeks.

Tracking the normalized absorbance at 450 nm simply enabled us to determine the rates of degradation under these conditions. The reduction of the absorbance was about 30%, 50% and 15% for standard, morin and quercetin passivated perovskite samples respectively, indicating that Quercetin molecule can effectively slow down the degradation rates compared to standard films. However, it was found that the Morin additives affects the perovskite films negatively, as the rate of degradation was nearly 20% less than the standard one. The observed increase in the degradation rate of Morin passivated samples might be a result of using the hydrate Morin powder that contains water molecules, which induce the degradation of the perovskite layers [22].

To characterize the perovskite's morphology, the top-view FESEM images of standard and optimized passivated perovskite layers are given in Figure 5. It was noticed that the average grain size is almost the same in the standard and Quercetin samples (~0.5 mm) and slightly larger in Morin sample (~ 0.7 mm). The observed increase of the average lifetime in both Morin and Quercetin samples is probably due to the drop in the number of the active traps at the grain boundaries [23],[24].

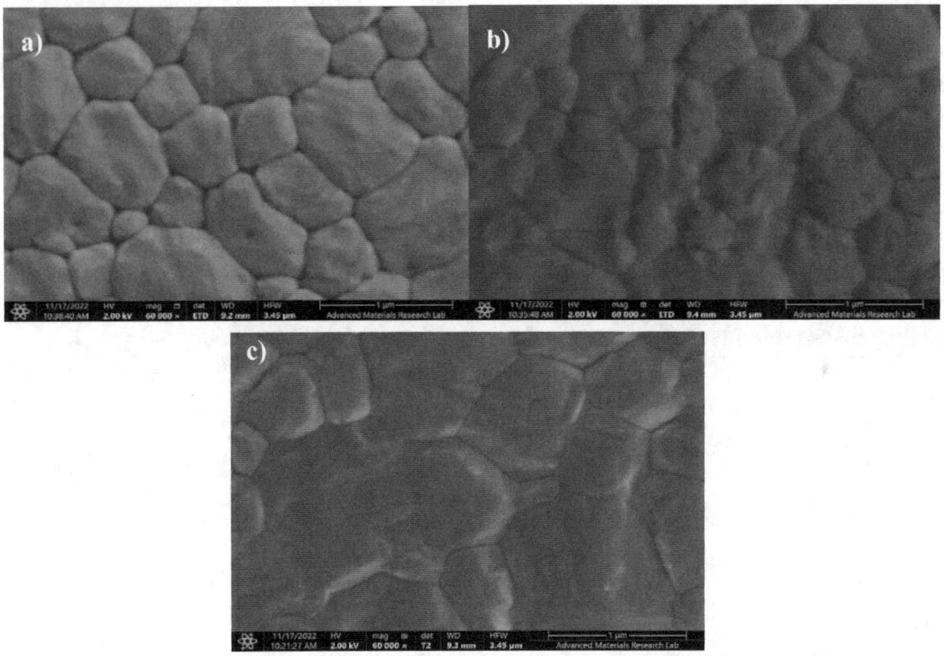


**Figure 5:** FESEM images for a) standard b) Quercetin and c) Morin samples

## 4. Conclusion

In this work, Morin and Quercetin, two naturally occurring antioxidants, were added to $Cs_{0.1}FA_{0.9}PbI_3$-based perovskite absorbing layer in an attempt to study their impact on the longevity of the films under environmental conditions. As a result, a significant change in the time resolved photoluminescence was achieved where 120% and 194% were the increment percentage for morin and quercetin, respectively. The degradation rates of the standard and non-passivated perovskite layer after 500h storage in 35% relative humidity with encapsulation were recorded based on the UV-vis absorption spectra. We found that morin accelerates the degradation of the perovskite while quercetin molecule enhances its stability. This study confirms that antioxidants offer the possibility to enhance perovskite solar cell performance. More experimental data need to be collected and reported to develop a deeper understanding of the detailed mechanism behind the observed enhancement.

**Acknowledgments**

We would like to acknowledge Dr. Ali El-Keblawy from the applied biology department and Dr. Sameh Soliman from the pharmacy college at the University of Sharjah for providing us with the chemicals.

## References

[1] Sims R E H 2004 Renewable energy: a response to climate change *Solar Energy* **76** 9–17

[2] Brown G F and Wu J 2009 Third generation photovoltaics *Laser & Photon. Rev.* **3** 394–405

[3] Momblona C, Gil-Escrig L, Bandiello E, Hutter E M, Sessolo M, Lederer K, Blochwitz-Nimoth J and Bolink H J 2016 Efficient vacuum deposited p-i-n and n-i-p perovskite solar cells employing doped charge transport layers *Energy Environ. Sci.* **9** 3456–63

[4] Park N-G 2015 Perovskite solar cells: an emerging photovoltaic technology *Materials Today* **18** 65–72

[5] Castro-Hermosa S, Dagar J, Marsella A and Brown T M 2017 Perovskite solar cells on paper and the role of substrates and electrodes on performance *IEEE Electron Device Lett.* **38** 1278–81

[6] Green M A, Ho-Baillie A and Snaith H J 2014 The emergence of perovskite solar cells *Nature Photon* **8** 506–14

[7] Song T-B, Chen Q, Zhou H, Jiang C, Wang H-H, (Michael) Yang Y, Liu Y, You J and Yang Y 2015 Perovskite solar cells: film formation and properties *J. Mater. Chem. A* **3** 9032–50

[8] Wang Q and Abate A 2018 Strategies toward Stable Perovskite Solar Cells *Adv. Mater. Interfaces* **5** 1800264

[9] Shinde A, Ganu J and Naik P 2012 Effect of Free Radicals & Antioxidants on Oxidative Stress: A Review *J Dent Allied Sci* **1** 63

[10] Cao J and Yan F 2021 Recent progress in tin-based perovskite solar cells *Energy Environ. Sci.* **14** 1286–325

[11] Xu X, Chueh C-C, Yang Z, Rajagopal A, Xu J, Jo S B and Jen A K-Y 2017 Ascorbic acid as an effective antioxidant additive to enhance the efficiency and stability of Pb/Sn-based binary perovskite solar cells *Nano Energy* **34** 392–8

[12] Zhuang X, Zhou D, Liu S, Sun R, Shi Z, Liu L, Wang T, Liu B, Liu D and Song H 2022 Learning From Plants: Lycopene Additive Passivation toward Efficient and "Fresh" Perovskite Solar Cells with Oxygen and Ultraviolet Resistance *Advanced Energy Materials* **12** 2200614

[13] Ban H, Sun Q, Zhang T, Li H, Shen Y and Wang M 2020 Stabilization of Inorganic $CsPb_{0.5}Sn_{0.5}I_2Br$ Perovskite Compounds by Antioxidant Tea Polyphenol *Sol. RRL* **4** 1900457

[14] Olayinka E T, Ore A, Adeyemo O A and Ola O S 2019 The role of flavonoid antioxidant, morin in improving procarbazine-induced oxidative stress on testicular function in rat *Porto Biomedical Journal* **4** e28

[15] Yong H J and Ahn J J 2018 Antioxidant and skin protection effect of morin upon UVA exposure *biomed dermatol* **2** 12

[16] Biswas P, Dey D, Biswas P K, Rahaman T I, Saha S, Parvez A, Khan D A, Lily N J, Saha K, Sohel M, Hasan M M, Al Azad S, Bibi S, Hasan Md N, Rahmatullah M, Chun J, Rahman Md A and Kim B 2022 A Comprehensive Analysis and Anti-Cancer Activities of Quercetin in ROS-Mediated Cancer and Cancer Stem Cells *IJMS* **23** 11746

[17] Liu Y, Akin S, Hinderhofer A, Eickemeyer F T, Zhu H, Seo J, Zhang J, Schreiber F, Zhang H, Zakeeruddin S M, Hagfeldt A, Dar M I and Grätzel M 2020 Stabilization of Highly Efficient and Stable Phase-Pure $FAPbI_3$ Perovskite Solar Cells by Molecularly Tailored 2D-Overlayers *Angew. Chem. Int. Ed.* **59** 15688–94

[18] Zhou N, Shen Y, Zhang Y, Xu Z, Zheng G, Li L, Chen Q and Zhou H 2017 CsI Pre-Intercalation in the Inorganic Framework for Efficient and Stable $FA_{1-x}Cs_xPbI_3$ (Cl) Perovskite Solar Cells *Small* **13** 1700484

[19] Péan E V, Dimitrov S, De Castro C S and Davies M L 2020 Interpreting time-resolved photoluminescence of perovskite materials *Phys. Chem. Chem. Phys.* **22** 28345–58

[20] Li Y, Natakorn S, Chen Y, Safar M, Cunningham M, Tian J and Li D D-U 2020 Investigations on Average Fluorescence Lifetimes for Visualizing Multi-Exponential Decays *Front. Phys.* **8** 576862

[21] Baloch A A B, Alharbi F H, Grancini G, Hossain M I, Nazeeruddin Md K and Tabet N 2018 Analysis of Photocarrier Dynamics at Interfaces in Perovskite Solar Cells by Time-Resolved Photoluminescence *J. Phys. Chem. C* **122** 26805–15

[22] Zhang W, Xiong J, Li J and Daoud W A 2020 Impact of Temperature-Dependent Hydration Water on Perovskite Solar Cells *Sol. RRL* **4** 1900370

[23] Agarwal S and Nair P R 2017 Pinhole induced efficiency variation in perovskite solar cells *Journal of Applied Physics* **122** 163104

[24] Sherkar T S, Momblona C, Gil-Escrig L, Ávila J, Sessolo M, Bolink H J and Koster L J A 2017 Recombination in Perovskite Solar Cells: Significance of Grain Boundaries, Interface Traps, and Defect Ions *ACS Energy Lett.* **2** 1214–22

# Flexural response of sandwich structures comprising topologically perturbed cores

**Omar Al Osman[1*], Maen Alkhader[1], Wael Abuzaid[1]**

[1] Department of Mechanical Engineering, American University of Sharjah, Sharjah United Arab Emirates.

*Corresponding author: b00059046@aus.edu

***Abstract*****—**Enhancing the flexural stiffness of sandwich structures comprising honeycomb cores through perturbing their cellular cores is explored computationally in this work. In-plane sinusoidal perturbations are imposed on the ligaments of hexagonal cores to increase their second moment of inertia and flexural load-carrying capacity. The sensitivity of sandwich structures' flexural strength, peak load and specific energy absorption to the imposed perturbation is investigated. Introduced perturbations are controlled through their amplitude and frequency. Results show that increasing the perturbation frequency and amplitude generally increases the flexural stiffness and peak loads of sandwich structures with perturbed cores without compromising their lightweight properties. However, low frequencies and amplitudes had a negligible effect on the sandwich structures' mechanical properties.

Keywords—aluminum sandwich panel, perturbations, finite element analysis, three-point bending test, dynamic behavior

## 1.Introduction

Composite sandwich structures are used in various engineering applications due to their lightweight properties. They can deliver high stiffness, high energy absorption capacity, and high flexural stiffness at low weights. Some of the engineering applications that significantly utilize sandwich structures include aerospace [1], civil infrastructure [2,3], automobile [4], and marine applications [5–7].

A composite sandwich structure is comprised of two parallel thin stiff sheets and a lightweight core. The two thin sheets are bonded to the top and bottom sides of the core, often by viscoelastic adhesive layers such as epoxy resin[8], silicon[9] or polyurethane [10]. In a correctly designed sandwich structure, the two stiff sheets provide the primary load-carrying capacity of the sandwich structure, while the core is mainly responsible for increasing the structure's second moment of inertia and ensures that the sheets stay parallel to each other [11]. The core has secondary roles, such as providing damping properties and transverse shear strength capacity. Although cores provide secondary load-carrying capacity, improving their properties can enhance the response of sandwich structures [12]. The latter is especially true in situations involving impact. When a sandwich structure is subjected to localized impact loads, e.g., birds impacting wind turbines, the core beneath the stiff sheet can experience compressive stresses. These can damage the core if they exceed its compressive stress load carrying capacity, which is relatively small (order of a few MPa) [13]. Therefore, enhancing the compressive properties of cores has attracted many research efforts [14].

The mechanical response of cellular cores subjected to compressive loads can be categorized into three phases [15]: elastic region, plateau and densification. In the plateau phase, the core undergoes progressive plastic deformations due to the collapse of its cell walls, leading to exhibiting high energy absorption capacity. Increasing the plateau stress of a core enhances its structural properties. Typically, tailoring the properties of cellular cores, including their plateau stress, can be realized by controlling their relative density, topological features and constituent material [15–18]. The relative density is defined as the ratio of the density of the core to the density of its constituent material. Increasing the relative density of cellular cores increases the cross-sectional area of their cell walls, increasing the cell walls' resistance to bending, buckling, and stretching. Accordingly, increasing relative density ultimately leads to higher energy absorption and enhanced mechanical properties. However, this route is not favored as it is associated with a substantial increase in weight. Utilizing constituent materials with higher structural properties-to-weight ratios constitutes another option to improve the structural properties of cellular solids. However, this option is not practical due to the limited material options. The latter renders tailoring the topology of cellular solids as the most promising route to improve their properties. Particularly as topology can be tailored without comprising the weight. Guided by this conclusion, multiple efforts tried to improve the properties of cellular solids by modifying their topology. Examples include adding hierarchical features, increasing nodal connectivity, and using complex geometries (e.g., triply periodic minimal surfaces). Most of these efforts utilized finite element simulations to assess topological features' effect on cellular solids' properties. Finite element simulations were chosen due to their accuracy, low cost and ability to explore very complex geometries that would be hard to assess experimentally. However, many previous efforts resulted in topological features that imposed manufacturing restrictions and required expensive additive manufacturing techniques [19].

This study uses a manufacturing-friendly approach to tailor the topology of cellular honeycomb cores and enhance their structural properties. In this approach, sinusoidal perturbations are imposed on the ligaments of the cellular cores to increase their second moment of inertia, flexural stiffness and buckling strength. Finite element analysis is used to assess the effect of the imposed perturbations on the properties of the cellular core and the sandwich structure it serves in. Perturbed cellular honeycomb cores could be manufactured in a similar fashion to that of traditional honeycombs through the corrugation process.

## 2.Methodology

Three-point bend tests are used to assess the flexural behavior of the sandwich structures comprising the perturbed honeycomb cores. The three-point bend tests are conducted numerically using finite element methods, namely using the finite element software ABAQUS. From the simulations, the flexural strength, flexural modulus and energy absorption capacity are determined. The three-point bend simulations followed ASTM standards. Thus, 20 mm diameter rollers are used to load the sandwich structure. The support rollers are 120 mm apart, and the loading roller is positioned at an equal distance from the support rollers.

### *2.1. Geometry and materials*

The sandwich structures analyzed are comprised of aluminum alloy sheets and an aluminum honeycomb core. Aluminum alloy 5052 is used to model all aluminum components. Sandwich structures usually comprise composite sheets. However, as this work is interested in the effect of the core and for computational practicality, aluminum sheets are used. The honeycomb core of the sandwich structure is modified by applying sinusoidal perturbations. Fig.1 presents the analyzed sandwich structure and its perturbed honeycomb core. This figure superposes the perturbed honeycomb structure over the regular honeycomb structure to demonstrate the difference between them.

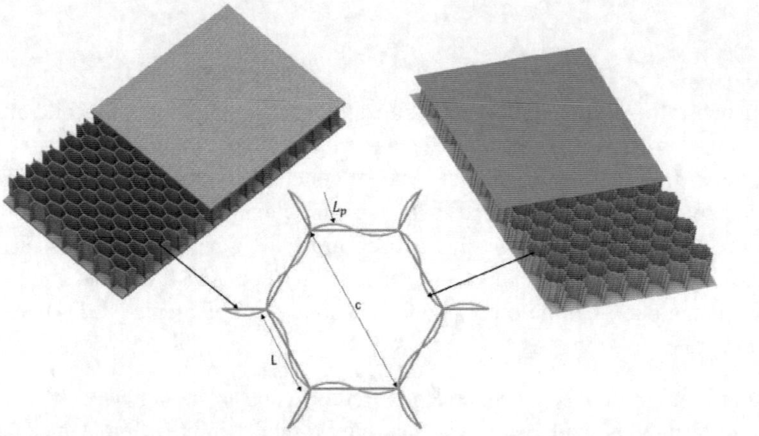


Figure 1: Sandwich structures with traditional (left) and perturbed (right) honeycomb cores. A comparison between the perturbed and standard honeycomb cores is shown in the middle.

Upon applying the periodic sinusoidal perturbations to the traditional honeycomb, the flexural properties of the sandwich structure are obtained using simulations. The frequency and amplitude of the applied sinusoidal perturbations are varied to investigate the effect of perturbation parameters and find the optimum configuration. Two perturbation amplitudes are used, 0.2 mm and 0.15mm. In addition, two perturbation frequencies are used, 1 and 0.5 cycles per cell wall length. Therefore, including the case without perturbation, a total of 5 honeycomb core configurations are considered. The honeycomb core chosen, before perturbation, has a cell wall thickness (t), cell size (c) and strut length (L) of 0.00762 mm, 6.35 mm and 3.175mm, respectively. The height of the honeycomb core is 12.5 mm. The sandwich structure, i.e., sheets and core, has a length of 200 mm and a width of 50 mm. The face sheets have a thickness of 1 mm. Thus, the total height of the sandwich structure is 14.5 mm.

The relative density ($\rho^*$) of the honeycomb core, before perturbation, is calculated as 2.71%. The relative density of a cellular solid structure is the ratio of the density of the cellular solid to the density of its constituent material. The sandwich structure, sheets and core, are assumed to be made from aluminum 5052. The periodic sinusoidal perturbations introduced to the core, see Fig.1, are based on the displacement function

$$u = A \sin\left(\frac{n\pi x}{L}\right) \qquad (1)$$

where A is the amplitude of the wave, n is the number of waves per strut, and L is the strut length of the traditional honeycomb. The displacement function (u) translates the points of each cell wall perpendicular to the cell wall axis. Introducing perturbations increases the overall length of the struts and, consequently, increases the relative density of the core. To evaluate the increase in struts' length, their length after the perturbation is computed using

$$L_p = \int_0^L \sqrt{1 + \left(\frac{An\pi}{L}\right)^2 \cos^2\left(\frac{n\pi x}{L}\right)}\, dx \qquad (2)$$

Based on the perturbation amplitude and magnitudes used, the maximum strut length after the perturbation is 5.32mm. Comparing this value to the strut length before perturbation, which is 5.28mm, demonstrates a small increase in length. Due to this increase in length, the relative density of the perturbed honeycomb is 0.51% higher than that of the unperturbed honeycomb.

*2.2 Finite Element Model*

Numerical simulations utilized ABAQUS's explicit solver as it can automatically and effectively model the contacts between the rollers and the sandwich structure. To ensure convergence and reduce the computational cost, a finer mesh is applied at areas of contact and large deformation gradients. Based on mesh sensitivity analysis, it was determined that a 0.5 mm element size is required for areas subjected to impact or exhibiting large deformation gradients. In contrast, a 1mm element size is sufficient to model the rest of the beam. Four node shell elements (S4R) are used to model both the thin aluminum sheets and the walls of the honeycomb core. On the other hand, four-node rigid elements (R3D4) are used to model the rollers.

Although the sandwich structures' sheets and honeycomb core are assumed to be made from the same aluminum alloy, experimental evidence shows that they might exhibit different mechanical properties [20]. The latter relates to the different manufacturing technologies used to fabricate each of them. An experimental assessment of a sandwich structure whose sheets and traditional honeycomb core are made from AL5052 showed a difference between the properties exhibited by the sheets and those of the core [20]. Table 1 shows the material properties in which the sheets exhibited an elastic-perfectly plastic response. On the other hand, the core material exhibited a hardening plastic response that can be described by the Johnson-Cook model. Material properties shown in Table 1 are used to model the sheets and the walls of the perturbed cores in this work. To simplify the modeling process, the sheets are assumed to be perfectly bonded to the core. Thus, the finite element model does not consider the adhesive layer that typically bonds the sheets to the core. This assumption might not be accurate at very large deformations or in situations involving large cyclic loading. However, under 3-point bend loading, the load under the loading roller is compressive, and debonding between the sheets and core is not usually observed in experiments [20]. Thus, modeling the adhesive sheet is not required, and the sheets can be considered fully bonded to the core. Adding on, the aluminum core and sheets are assumed to have isotropic material properties and the perturbed waves are assumed to not have any imperfections, which could be a result of the manufacturing process.

Table 1: Material Properties

| Material | $\rho(kg/m^3)$ | $E(MPa)$ | $\mu$ | A(MPa) | B(MPa) | n | m |
|---|---|---|---|---|---|---|---|
| Al5052-Core | 2680 | 73,000 | 0.33 | 148 | 345 | 0.18 | 0.859 |
| | | | | Yield Strength (MPa) | | | |
| Al5052-Sheets | 2680 | 53,500 | 0.33 | 114 | | | |

The interactions between the rollers and the sheets of the sandwich structure are modeled using Surface-to-Surface interactions, which assumed the friction coefficient between the rollers and sheets to be 0.2. The boundary conditions used to simulate the 3-point bend test are shown in Fig.2. The two lower rollers are constrained in all directions, while the top roller is allowed to move along the vertical direction only.

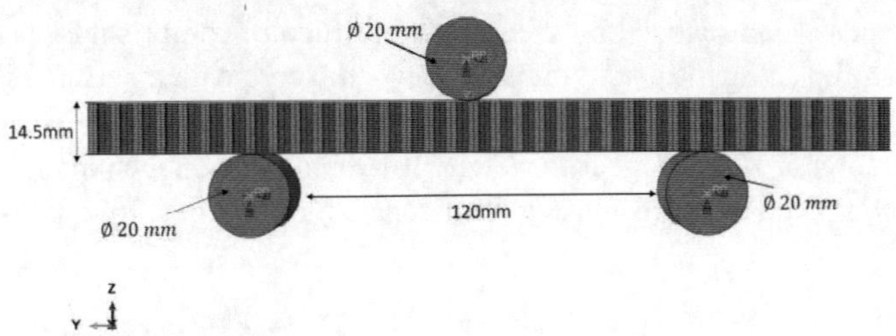


Figure 2: Boundary conditions

A displacement rate of 50mm/s is applied on the top roller to load the sandwich structure. This rate is used in this work as it was used to load sandwich structures similar to the ones used in this work but with traditional cores [20]. Having a benchmark allows for evaluating the effect of applying perturbations to the cores of sandwich structures.

## 3.Results

The deformation modes, peak loads, and specific energy absorption are reported for the sandwich structures corresponding to the 5 different cores considered, i.e., 4 perturbed and 1 traditional. The average macroscopic deformation of the 5 sandwich structures is similar. Fig.3 represents the typical observed deformed shape of the modeled sandwich structures. As the load is applied at the middle of the panel, a narrow indentation forms under the indenting roller. The figure also shows the local deformations at the point of contact. The face sheet thickness plays a crucial role in distributing the local forces and deformations.

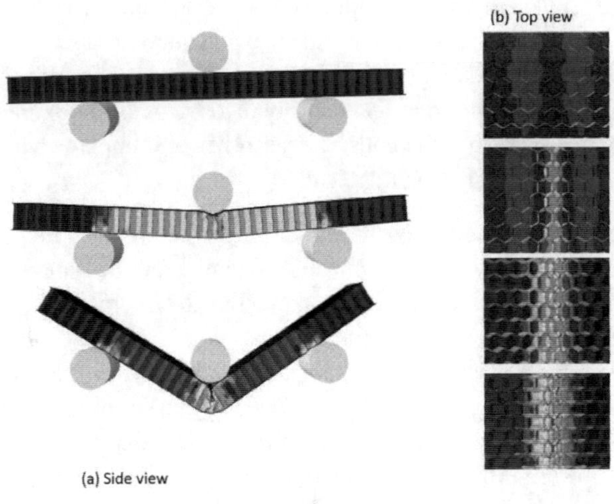


Figure 3: Deformation of a sandwich structure under 3-poing bend loading, showing localized deformation zone along the sandwich panel (side view) and under the loading roller (top view)

The load-displacement curves of the 5 sandwich structures considered are shown in Fig.4. To validate the finite element model's accuracy, the force-displacement result corresponding to the sandwich structure with

Journal of Physics: Conference Series **2751** (2024) 012028 doi:10.1088/1742-6596/2751/1/012028

the traditional honeycomb core is compared to that of a similar panel that was experimentally tested [20]. The result of this work agrees well with the experimental results; a maximum difference of 3% is observed. According to Fig.4, the peak load realized by the sandwich structure comprising a traditional honeycomb core is 2.17kN. However, replacing the conventional honeycomb core with a perturbed one can increase the peak load to 2.95kN. This increase is related to the increased second moment of inertia and improved stiffness. The perturbed cores are 0.3% higher in weight than their traditional counterparts, and this increase in weight is negligible compared to the increase in peak load which reached 36%.

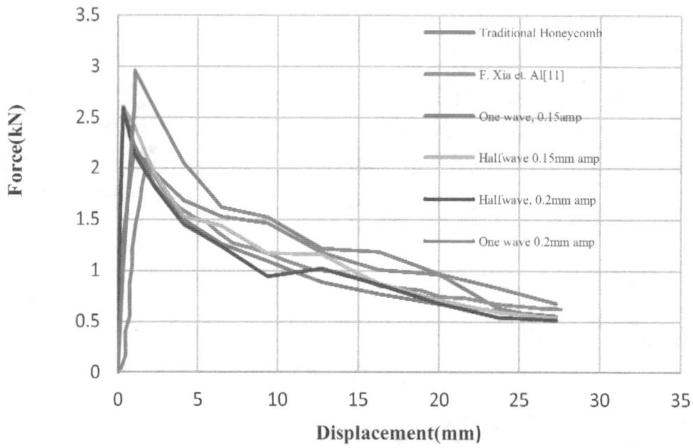


Figure 4: Load-displacement curves, showing the response of all modeled sandwich structures.

The best response is observed at higher perturbation amplitudes and frequencies. Fig.4 shows that the perturbation effect is more relevant in the loading phase. Once the peak load is reached, the force-displacement curves of all sandwich structures showed an exponential decay, and they relatively overlapped once the load decreased below the 1 kN mark. This behavior suggests that once the honeycomb cores start to crumble, the introduced perturbations no longer contribute to the overall flexural strength of the sandwich structure. After the applied force crushes the cells, the perturbed and unperturbed honeycomb cells behave similarly; thus, the final load for all panels is identical.

To further understand the effect of modifying hexagonal honeycombs using sinusoidal perturbations, the flexural stiffness and specific energy absorption are calculated for the sandwich structures comprising perturbed cores. Fig.5 shows the flexural stiffness of the sandwich structures. This figure also compares the results from the sandwich structures with perturbed honeycomb to those of sandwich structures with traditional honeycomb. Fig.5 shows that, in general, an increase in amplitude or frequency increases flexural stiffness. The sandwich structure with the highest frequency and amplitude, namely one wave and an amplitude of 0.2 mm, showed a 35.7% higher flexural stiffness than that of the sandwich structure with the traditional honeycomb. A lower perturbation frequency, namely 0.5 waves per cell wall, did not result in similar substantial improvements. Results suggest that the relationship between frequency, amplitude and flexural stiffness is non-linear.

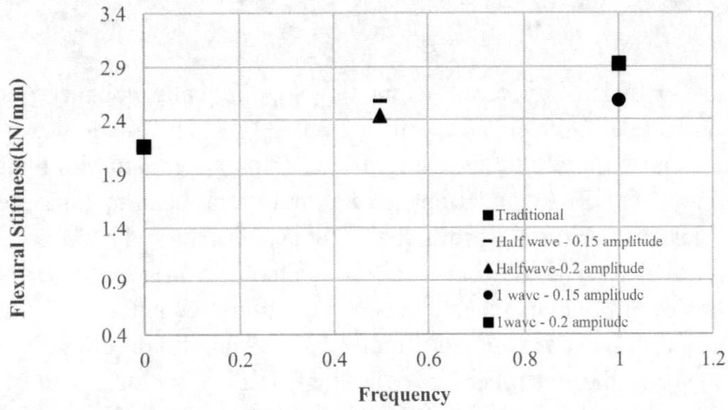


Figure 5: Flexural stiffness of sandwich structures comprising perturbed honeycomb cores.

Fig.6 shows the specific energy absorbed by the sandwich structures comprising perturbed and unperturbed honeycomb cores. Specific energy absorption is calculated by integrating the area under the force-displacement curve and normalizing the outcome by the mass of the sandwich structure.

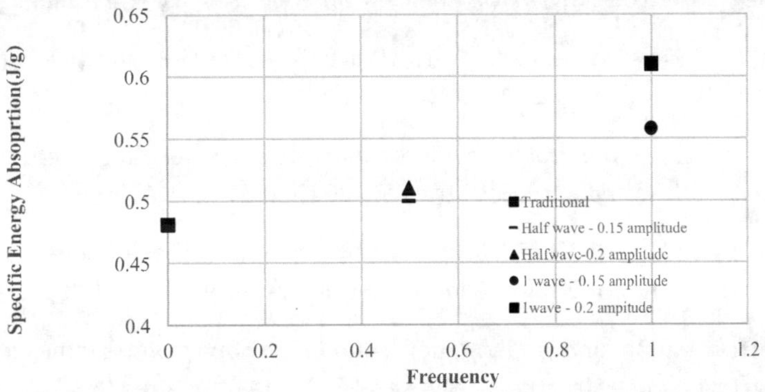


Figure 6: Specific energy absorption of sandwich structures comprising perturbed and unperturbed honeycomb cores.

This figure shows a general trend; the specific energy absorption increases as the perturbation amplitude and frequency used are increased. However, since the imposed sinusoidal perturbations do not contribute to the structural stiffness after the honeycomb starts to collapse, i.e., after peak load, they have a lower enhancing effect on the specific energy absorption as compared to their effect on peak load. For instance, the specific energy absorption of the sample with the highest perturbation frequency and amplitude is 26% higher than that of the sample with a traditional honeycomb. This increase, though substantial, is lesser than that exhibited by peak load. Accordingly, unlike the flexural stiffness and peak load, the specific energy absorption considers the phase where the perturbated and traditional honeycomb cores behave similarly.

## 4.Conclusion

The flexural response of sandwich structures comprising topologically enhanced cores is investigated numerically using the commercial finite element software ABAQUS. The topologically enhanced cores are created by imposing sinusoidal perturbations with various frequencies and amplitudes to an otherwise traditional hexagonal honeycomb. The simulations used a three-point bend loading scenario to investigate the sandwich structures' flexural response. Results show that using perturbed honeycomb cores in sandwich structures instead of traditional honeycomb cores can enhance their flexural response without affecting their weight. Moreover, results show that increasing the perturbation frequency and amplitude, in general, increases the flexural properties. At the highest frequency and amplitude considered in this work, the sandwich structure comprising the perturbed core, as compared to the one with an unperturbed core, exhibited 36%, 35.7% and 26% higher peak load, flexural stiffness and specific absorption energy, respectively. The level of improvement is lower at lower perturbation amplitudes or frequencies.

## References

[1] S.E. Sadiq, M.J. Jweeg, and S.H. Bakhy, in *IOP Conf Ser Mater Sci Eng* (IOP Publishing, 2020), p. 022126.

[2] G. Morcous, Y. Cho, A. El-Safty, and G. Chen, "Structural behavior of FRP sandwich panels for bridge decks," KSCE Journal of Civil Engineering **14**(6), 879–888 (2010).

[3] B. Kong, C.S. Cai, and X. Kong, "Thermal behaviors of concrete and steel bridges after slab replacements with GFRP honeycomb sandwich panels," Eng Struct **56**, 2041–2051 (2013).

[4] S. Hou, L. Ren, D. Dong, and X. Han, "Crashworthiness optimization design of honeycomb sandwich panel based on factor screening," Journal of Sandwich Structures & Materials **14**(6), 655–678 (2012).

[5] G. Palomba, G. Epasto, L. Sutherland, and V. Crupi, "Aluminium honeycomb sandwich as a design alternative for lightweight marine structures," Ships and Offshore Structures, 1–12 (2021).

[6] M. Alkhader, X. Zhai, and F.-P. Chiang, "Experimental investigation of the synergistic effects of moisture and freeze-thaw cycles on carbon fiber vinyl-ester composites," J Compos Mater **52**(7), 919–930 (2018).

[7] A. Afshar, M. Alkhader, C.S. Korach, and F.-P. Chiang, "Effect of long-term exposure to marine environments on the flexural properties of carbon fiber vinylester composites," Compos Struct **126**, 72–77 (2015).

[8] U. Farooq, M.S. Ahmad, S.A. Rakha, N. Ali, A.A. Khurram, and T. Subhani, "Interfacial mechanical performance of composite honeycomb sandwich panels for aerospace applications," Arab J Sci Eng **42**(5), 1775–1782 (2017).

[9] A. Kyner, K. Dharmasena, K. Williams, V. Deshpande, and H. Wadley, "Response of square honeycomb core sandwich panels to granular matter impact," Int J Impact Eng **117**, 13–31 (2018).

[10] M.E. Çetin, "Investigation of carbon nanotube reinforcement to polyurethane adhesive for improving impact performance of carbon fiber composite sandwich panels," Int J Adhes Adhes **112**, 103002 (2022).

[11] A.K. Kaw, *Mechanics of Composite Materials* (CRC press, 2005).

[12] A. Ahmed, M. Alkhader, and B. Abu-Nabah, "In-plane elastic wave propagation in aluminum honeycomb cores fabricated by bonding corrugated sheets," Journal of Sandwich Structures & Materials **21**(8), 2949–2974 (2019).

[13] M. Alkhader, and M. Vural, "A plasticity model for pressure-dependent anisotropic cellular solids," Int J Plast **26**(11), 1591–1605 (2010).

[14] M. Alkhader, and M. Vural, "Mechanical response of cellular solids: Role of cellular topology and microstructural irregularity," Int J Eng Sci **46**(10), 1035–1051 (2008).

[15] L. Gibson J., and M. Ashby F., *Cellular Solids: Structure and Properties* (1997).

[16] M. Alkhader, W. Abuzaid, M. Elyoussef, and S. Al-Adaileh, "Localized strain fields in honeycomb materials with convex and concaved cells," European Journal of Mechanics - A/Solids **80**, 103890 (2019).

[17] M. Alkhader, M. Nazzal, and K. Louca, "Design of bending dominated lattice architectures with improved stiffness using hierarchy," Proceedings of the Institution of Mechanical Engineers Part C-Journal of Mechanical Engineering Science **233**(11), 3976–3993 (2019).

[18] A. Ahmad, M. Alkhader, B.A. Abu-Nabah, and T.A. Venkatesh, "Effect of deformation mechanisms on in-plane wave propagation in periodic cellular materials," Waves in Random and Complex Media, 1–23 (2021).

[19] D. Kong, Y. Zhang, and S. Liu, "Convective heat transfer enhancement by novel honeycomb-core in sandwich panel exchanger fabricated by additive manufacturing," Appl Therm Eng **163**, 114408 (2019).

[20] F. Xia, Y. Durandet, P.J. Tan, and D. Ruan, "Three-point bending performance of sandwich panels with various types of cores," Thin-Walled Structures **179**, 109723 (2022).

Journal of Physics: Conference Series **2751** (2024) 012029 doi:10.1088/1742-6596/2751/1/012029

# LIPSS formation on Ni surface using mixed near-infrared laser beam harmonics

**Baha Khalil**[1], **Ali S Alnaser**[1,2]

[1] Materials Science and Engineering Program, College of Arts and Sciences, American University of Sharjah, PO Box 26666, Sharjah, United Arab Emirates

[2] Department of Physics, American University of Sharjah, PO Box 26666, Sharjah, United Arab Emirates

E-mail: b00025382@aus.edu, aalnaser@aus.edu

**Abstract.** Generation of laser-induced periodic surface structures (LIPSS) is investigated with femtosecond laser pulses with $1030nm$ wavelength and its second and its harmonics. The second ($\lambda_{2\omega} = 515nm$) and third harmonics ($\lambda_{3\omega} = 343nm$) beams in addition to their combination were utilized to control the period of the LIPSS on a Ni surface. The resulting periodicity of the formed structures was $\sim 0.42\mu m$, $0.22\mu m$, and $0.33\mu m$ respectively for the second harmonic beam, third harmonic beam, and their combination. This method demonstrates a direct way to control the periodicities on a metal target by employing an individual laser beam wavelength to induce a variable range of LIPSS periods.

## 1. Introduction

Ultra-fast pulsed lasers facilitate the LIPSS formations with minimal thermal effects due to their small-scale pulse duration. Femtosecond lasers are widely used in studying the ripples formation, a process that is highly dependent on the operating condition such as the environment atmosphere, wavelength, repetition rate, scanning speed, and average pulse power.

The induced LIPSS formation has been functionalized under various studies to achieve surface color control [1], low surface friction with hydrophobicity enhancement [2][3], magnetic properties [4], and many other promising applications. The response to nano-texturing functionalization depends on the period and depth of the formed ripples, both of which can be controlled through the laser's mode of operation.

Generally the LIPSS formation on the surface are observed in two different groups of ripples, low spatial-frequency LIPSS (LSFL) in which the surface period is relatively equal to the irradiated laser wavelength with perpendicular orientation, and a high-spatial frequency LIPSS (HSFL) that are formed with less than half of the irradiated laser wavelength with parallel orientation. The nature of the formed ripples was formulated under the efficacy factor theory ($\eta$ theory) which contributed the inhomogeneous energy deposition on the surface to roughness of the sample, with the condition that the height of the roughness is considerably less than the incident beam wavelength [5][6]. However, the $\eta$ theory is a direct electromagnetic approach which fails to consider the changes in the surface roughness function, the contribution of the laser fluence, and the transient dynamics of the materials properties such as the refractive index $\tilde{n}$ [7]. By introducing the effect of surface plasmon (SP) and the dielectric constant ($\epsilon_d$) transient dynamics under the influence of the laser pulse energy, [8] introduced a relationship between the LIPSS period ($\Lambda$), laser beam wavelength ($\lambda$), incident angle ($\theta$), and the material SP:

$$\Lambda = \frac{\lambda}{\left(\frac{\lambda}{\lambda_{SP}} + sin(\theta)\right)} \tag{1}$$

At a direct incident angle, $\Lambda$ becomes a function of the SP wave period. The SP wavelengths for metals and dielectrics are given by::

$$\lambda_{SP} = \lambda\left(\frac{\epsilon' + \epsilon_d}{\epsilon\epsilon_d}\right)^{\frac{1}{2}} \tag{2}$$

Hence, the model links the LIPSS period to the transient dynamics of $\epsilon$ under the impact of the laser energy deposition. The excited surface $\epsilon$ is modeled using Drude model which depends on the excited electron density $n_e$. Therefore, the beam fluence is considered as the driver of the model, as $n_e$ would increase at higher pulse energy or with a more focused spot size. At a larger fluence, where the ablation threshold is crossed, the energy deposition would increase the crater size, which would decrease the overall LIPSS surface period. Therefore, the formation of LSFL is contributed to the SP excitation with a period estimated with the changes in the excited state of the dielectric constant. In multi-pulse surface structuring, a recurrence relation is established between the surface ripples generated by the first pulse, the enhanced surface plasmon (SP) adapted to the new geometry, and the interaction of the local SP field with subsequent laser pulses. This interaction sharpens the surface morphology over time and can reduce the SP wavelength with an increase in pulse number [9]. On the other hand, the origin of HSFL is still under debate with hypotheses relating the formation of HSFL ripples to the change in non-thermal melting [10][11], interference of the laser beam with the surface scattered field [12], and the role of pressure applied in a controlled environment to generate HSFL [13].
Usually, researchers rely on the fundamental wavelength of the irradiated laser to control the periodicity of the formed ripples. Consequently, the overall controllability of surface morphology is limited to the available laser wavelength. Therefore, in this work we explore the possibility of isolating the irradiated laser beam harmonics through the use of nonlinear crystals to form LIPSS with lower ripples period and further combine the harmonics generated to control the ripples periodicity. Combining the laser beam with its harmonics components provided a higher spatial frequency ripples as the wavelength of each harmonic is a factor of the original beam wavelength. In our previous work [14], we explored the LIPSS formation when combining the laser beam with its second and their harmonic components, and in this work we continue to explore the effect of laser harmonics by investigating the LIPSS formed when combining the second and third harmonic components without the fundamental beam.

## 2. Experimental Setup

We tuned the optical setup as shown in Figure 1. The setup utilized a fiber-based femtosecond laser source with a 50 kHz repetition rate to produce Gaussian beam pulses with a duration of 260 femtoseconds and a central wavelength of $\lambda_w = 1030nm$. The laser pulse's average power was regulated with a HWP and the TFP. The incident beam's second and third harmonic components were filtered using the NC's and focused using a $+25mm$ focal length lens after combining the two beams. The speed of the XY stage was controlled to optimize the accumulation of pulses along the irradiation path.
The experimental work consists of three scenarios: applying the second harmonic beam and the third harmonic beam separately, and then combining both beams together on the Ni target. The peak fluence of the beams was calculated by dividing the power density supplied by the laser beam over the circular area size of the pulse shape, which was found to be approximately 0.60 $\mu m$.

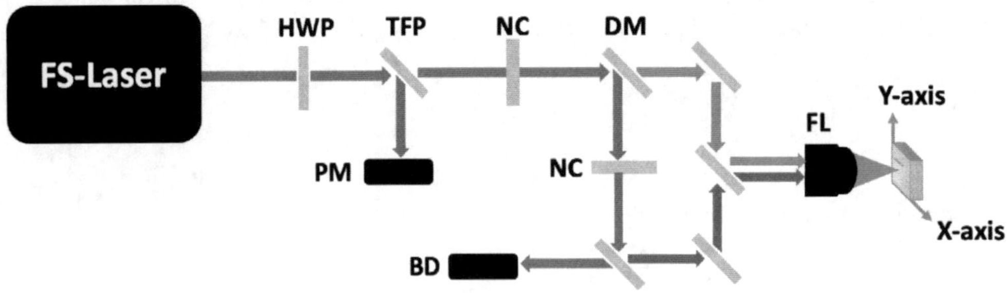


**Figure 1.** Experimental setup: **FS-laser**: femtosecond laser generating $1030nm$ ablating pulses at $50khz$ repetition rate with $260fs$ pulse duration. **HWP**: half-wave plate, **TFP**: thin film polarizer, **PM**: power meter, **NC**: nonlinear crystal, **TDC**: Time delay compensator, **DM**: dichroic mirror. **BD**: beam dump. **FL**: Focusing lens.

## 3. Results

Using laser beam harmonics generated by non-linear crystals, LIPSS were formed on the surface with an average periodicity ($\Lambda$) of $425nm$ for the second component and $225nm$ for the third component. Summary of the results are presented in Table 1. These LSFLs were within the $62-82\%$ range of the irradiated beam wavelength, which is typical for metallic surfaces [15] [16]. When a laser beam is directed at a metallic target, the resulting electron density on the metallic surface is greater than that on non-metallic surfaces [17]. This increased electron density is thought to generate stronger SP fields, which then cause the formation of ripples that have a wavelength shorter than that of the laser beam ($\Lambda \leq \lambda$).

**Table 1.** Fluence and periodicity summary

| Setup | Beam wavelength | Pulse Fluence ($Jcm^{-2}$) | Scanning Speed ($mms^{-1}$) | Surface Periodicity $\Lambda(nm)$ |
|---|---|---|---|---|
| $\lambda_{2\omega}$ | $515nm$ | 0.25 | 5 | 425 |
| $\lambda_{3\omega}$ | $343nm$ | 0.09 | 5 | 225 |
| $\lambda_{2\omega} + \lambda_{3\omega}$ | $515nm + 343nm$ | $\lambda_{2\omega}0.07/\lambda_{3\omega}0.05$ | 5 | $300 - 350$ |

Irradiating the Ni surface with the combination of the second and third harmonic components results in a surface periodicity range of $300 - 350nm$. These higher-frequency ripples, which are smaller than the LIPSS period when using the second harmonic component individually and larger than the LIPSS period when using the third harmonic component individually, suggest that the second harmonic component beam pulses were dominant during the ablation process. This dominance is indicated by the orientation following the polarization of the second harmonic component, as discussed in [17]. Figure 2 illustrates the SEM images taken for the LIPSS formed using the three different scenarios. These results suggest that the second harmonic component pulses lead the third harmonic component pulses with a time delay in the range of picoseconds. The periodicity of $300 - 350nm$" ripples, compared to the $425nm$ periodicity, suggests that the later pulses (third harmonic component) suppressed the ablation mechanism due to the interaction of the later pulse with the plasma generated by the initial pulse. This suppression mechanism is dependent on the time delay between the two pulses, as discussed in [18].

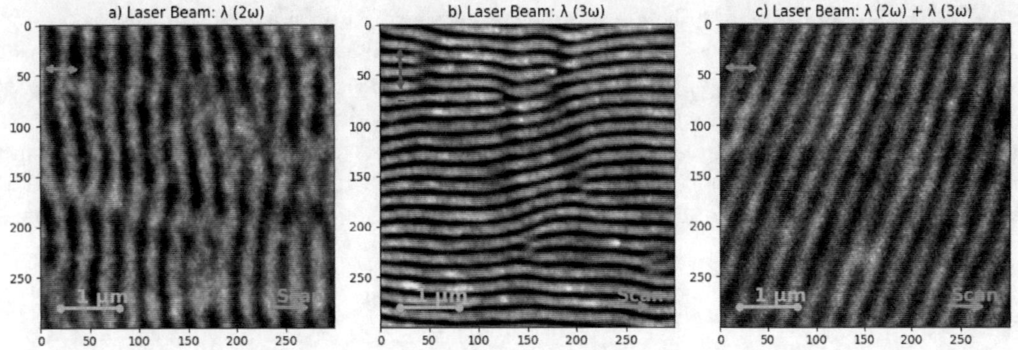


**Figure 2.** SEM images of LIPSS formed using the laser harmonics components. SEM for a) $\lambda_{2\omega}$, b) $\lambda_{3\omega}$, c) $\lambda_{2\omega} + \lambda_{3\omega}$

Figure 3 presents the spatial 2D-Fast Fourier Transform (2D-FFT) spectrum of the SEM images. The comparison between Figure 3a and Figure 3b clearly shows a higher magnitude of spatial frequency in the case of LIPSS generated by the third harmonic beam, compared to those generated by the second harmonic beam. This difference in spatial frequency reflects the distinct characteristics of the two beams' interaction with the surface. Additionally, the 90° rotation in the spectrum indicates the perpendicular nature of the second harmonic beam's polarization relative to that of the third harmonic beam, providing further insights into their interaction with the target material.

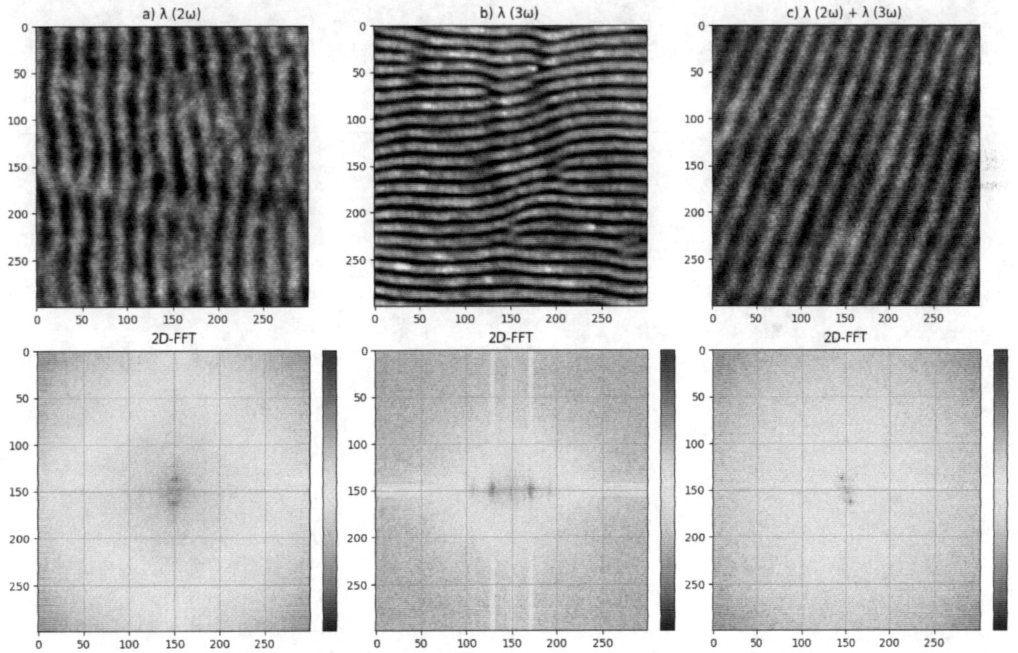


**Figure 3.** 2D-FFT of the corresponding SEM images. The color scale ranges from blue to red, indicating the minimum to maximum absolute values of the spatial frequency between the ripples. 2D-FFT for a) $\lambda_{2\omega}$, b) $\lambda_{3\omega}$, c) $\lambda_{2\omega} + \lambda_{3\omega}$

In addition to the above results, we can notice the overlap region between the two components and the higher frequency ripples generated at the edges of the formed structure (Figure 4). According to the SEM images obtained, a clear mirror direction of the ripples is seen between the second harmonic component (lower part of the image) and the third harmonic component ripples. Figure 4b illustrates the disappearance of the finer ripples due to the third harmonic component beam once the fluence of the second harmonic is increased. This indicates that an equal intensity of both beams does not produce LIPSS on the surface with the same response.

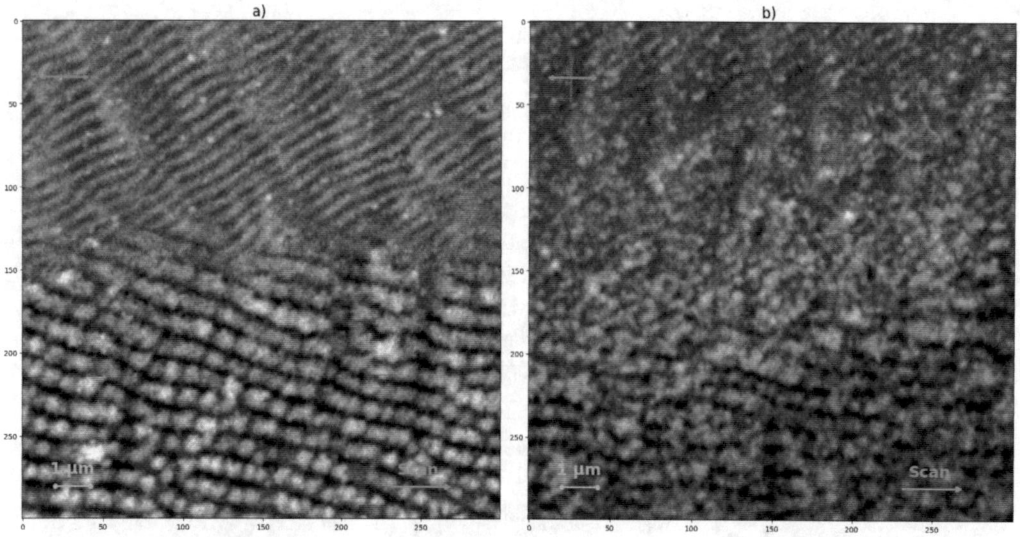


**Figure 4.** SEM images of LIPSS formed with different mix of fluences of each harmonic component.
a) Laser beam intensity at $I_{\lambda_{2\omega}} = 0.04\ Jcm^{-2}$, $I_{\lambda_{3\omega}} = 0.08\ Jcm^{-2}$
b) Laser beam intensity at $I_{\lambda_{2\omega}} = 0.08\ Jcm^{-2}$, $I_{\lambda_{3\omega}} = 0.08\ Jcm^{-2}$ ,

## 4. Conclusion

In this work, we build upon our previous findings [14] involving the utilization of the third harmonic component of a femtosecond laser beam generated in air to produce LIPSS on a Ni surface. Both second and third harmonic components were generated using BBO crystals and employed individually and in combination to generate LIPSS. The resulting LIPSS formation on the Ni surface led to surface periodicities of approximately $\sim 0.42\mu m$ for the second harmonic beam, $\sim 0.22\mu m$ for the third harmonic beam, and $\sim 0.30 - 0.35\mu m$ for the combined beam comprising both the second and third components.

This work is compared to previous results involving the ablation of a Ni surface using the third harmonic component beam generated in air. It highlights the capability to create fine structures on metal surfaces using a single laser beam, achieving a factor of 2-5 reduction in wavelength compared to the fundamental beam wavelength by harnessing the low-order harmonics of the laser beam.

## References

[1] Vorobyev, A. Y., & Chunlei, G. (2008). Colorizing metals with femtosecond laser pulses. Applied Physics Letters, 92(4). https://doi.org/10.1063/1.2834902

[2] Ali, S. A., Sharjeel, A. K., Rashid, A. G., & Emmanuel, S. (2019). Recent advances in femtosecond laser-induced surface structuring for oil-water separation, 9(8), 1554–1554. https://doi.org/10.3390/app9081554

[3] Xing, Y., Wu, Z., Yang, J., Wang, X., & Liu, L. (2020). LIPSS combined with ALD MoS2 nano-coatings for enhancing surface friction and hydrophobic performances. Surface &; Coatings Technology, 385, 125396. doi:10.1016/j.surfcoat.2020.125396

[4] Nykyruy, Y., Mudry, S., Shtablavyi, I., Borisyuk, A., Tsekhmister, Y., & Gnilitskyi, I. (2022). Formation of laser-induced periodic surface structures on amorphous fe- and co-based alloys and its impact on magnetic properties. Materials Chemistry and Physics, 287, 126317. doi:10.1016/j.matchemphys.2022.126317

[5] Sipe, J. E., Young, J. F., Preston, J. S., & Van Driel, H. M. (1983). Laser-induced periodic surface structure. I. Theory. Physical Review B, 27(2), 1141.

[6] Young, J. F., Preston, J. S., Van Driel, H. M., & Sipe, J. E. (1983). Laser-induced periodic surface structure. II. Experiments on Ge, Si, Al, and brass. Physical Review B, 27(2), 1155.

[7] Skolski, J. Z. P., Römer, G. R. B. E., Veld, A. J., Mitko, V. S., Obona, J. V., Ocelik, V., & De Hosson, J. T. M. (2010). Modeling of Laser Induced Periodic Surface Structures. Journal of Laser Micro/Nanoengineering, 5(3).

[8] Huang, M., Zhao, F., Cheng, Y., Xu, N., & Xu, Z. (2009). Origin of laser-induced near-subwavelength ripples: interference between surface plasmons and incident laser. ACS nano, 3(12), 4062-4070.

[9] Bonse, J., Rosenfeld, A., & Krüger, J. (2011). Implications of transient changes of optical and surface properties of solids during femtosecond laser pulse irradiation to the formation of laser-induced periodic surface structures. Applied Surface Science, 257(12), 5420-5423.

[10] Abdelmalek, A., Sotillo, B. E. L. E. N., Bedrane, Z., Bharadwaj, V., Pietralunga, S., Ramponi, R., ... & Eaton, S. M. (2017). Origin of femtosecond laser induced periodic nanostructure on diamond. AIP Advances, 7(10), 105105.

[11] Abdelmalek, A., Bedrane, Z., Amara, E. H., Ramponi, R., & Eaton, S. M. (2019, October). The efficacy of plasmonic model to calculate HSFL nanostructure period in Sapphire. In Journal of Physics: Conference Series (Vol. 1310, No. 1, p. 012017). IOP Publishing.

[12] Lu, M., Cheng, K., Qin, Z., Ju, J., Liu, J., & Huo, Y. (2022). Electromagnetic origin of femtosecond laser-induced periodic surface structures on GaP crystals. Optics Express, 30(6), 10152-10167.

[13] Dominic, P., Bourquard, F., Reynaud, S., Weck, A., Colombier, J. P., & Garrelie, F. (2021). On the insignificant role of the oxidation process on ultrafast high-spatial-frequency LIPSS formation on tungsten. Nanomaterials, 11(5), 1069.

[14] Khalil, B., Boltaev, G. S., Baker, A. A., Abbasi, N., & Alnaser, A. S. (2022). Controlling LIPSS formation on Ni surface using near-infrared laser beam and its low-order harmonics. Journal of Physics D: Applied Physics, 55(43), 435104.

[15] Petrović, S., Gaković, B., Zamfirescu, M., Radu, C., Peruško, D., Radak, B., ... & Mihailescu, I. N. (2017). Femtosecond laser processing of NiPd single and 5x (Ni/Pd) multilayer thin films. Applied Surface Science, 417, 16-22.

[16] Jalil, S. A., Yang, J., ElKabbash, M., Singh, S. C., & Guo, C. (2019). Maskless formation of uniform subwavelength periodic surface structures by double temporally-delayed femtosecond laser beams. Applied Surface Science, 471, 516-520.

[17] Höhm, S., Herzlieb, M., Rosenfeld, A., Krüger, J., & Bonse, J. (2016). Dynamics of the formation of laser-induced periodic surface structures (LIPSS) upon femtosecond two-color double-pulse irradiation of metals, semiconductors, and dielectrics. Applied Surface Science, 374, 331-338.

[18] Takenaka, K., Shinohara, N., Hashida, M., Kusaba, M., Sakagami, H., Sato, Y., ... & Tsukamoto, M. (2021). Delay times for ablation rate suppression by femtosecond laser irradiation with a two-color double-pulse beam. Applied Physics Letters, 119(23), 231603.

# Enhanced photocatalytic activities of $LaCoO_3$ perovskite nanostructures for degrading cationic and anionic dyes

**Muhammed Irshad**[1], **Soumya Columbus**[1*], **Javad B. M. Parambath**[1]

[1] Centre for Advanced Materials Research, Research Institute of Sciences and Engineering, University of Sharjah, Sharjah 27272, United Arab Emirates
[2] Department of Applied Physics and Astronomy, University of Sharjah, Sharjah 27272, United Arab Emirates

*Email: spal@sharjah.ac.ae

**Abstract**: Sustainable and cost-effective technologies are crucial for addressing environmental pollution issues. Perovskite photocatalysts have gained increased attention as a potential solution which could be accounted to their tuneable structural characteristics, flexible bandgap, and superior catalytic properties. In this study, we synthesized Lanthanum Cobaltite ($LaCoO_3$) perovskite nanostructures via co-precipitation for photocatalytic purposes. Morphological analysis revealed $LaCoO_3$ nanoparticles with an average size of ~33 nm whose crystalline characteristics were investigated as a function of annealing temperatures. We found that the pure and crystalline $LaCoO_3$ phase formed after the post-annealing process at 600° C. Photocatalytic studies showed that the degradation potential of $LaCoO_3$ was significantly improved when hydrogen peroxide ($H_2O_2$) was added as co-catalyst system. We achieved degradation efficiencies of up to 91% and 85% for MB and methyl orange (MO), respectively. $LaCoO_3/H_2O_2$ exhibited excellent photocatalytic potential for degrading both anionic and cationic organic dyes for environmental remediation applications.

## 1. Introduction

Advanced environmental protection strategies have gained immense research interest owing to the everlasting hike in global pollution. Pure water is the fundamental requirement for maintaining life on Earth. Unfortunately, freshwater resources are seriously affected by the release of harmful contaminants [1,2]. Industrial and other wastewater effluents mainly contain dyes, personal care products, fertilizers, pesticides, and organic additives [3-5]. Most organic dyes consist of multiple aromatic rings and are characterized by high thermal and chemical stability. The expulsion of these contaminants can cause an enhanced water pollution rate. The stability of the ecosystem can be thereby seriously affected, which necessitates the implementation of appropriate pollution control strategies.

Moreover, exposure to these dyes can drastically affect public health and well-being, as they are highly carcinogenic, mutagenic, and allergenic in nature [6]. Even trace levels of dyes can interfere with photosynthesis, thereby adversely influencing the aquatic environment as well. Methylene blue (MB) and methyl orange (MO) are common dyes that are stable and difficult to degrade [7]. These dyes are frequently employed in various industrial processes, which can be discharged into water bodies.

Therefore, regular monitoring and treatment of wastewater effluents are essential before disposal into the environment. Nevertheless, conventional water treatment techniques are inadequate to address the vast demands for achieving the goals of environmental sustainability, necessitating the implementation of economically viable water remediation tools for a greener environment [8].

Heterogeneous photocatalysis offers tremendous opportunities for eliminating micro-pollutants from wastewater [9]. Under UV or visible light irradiation, the catalysis of organic pollutants can be facilitated by the presence of an appropriate photocatalyst material. The absorption capability, charge recombination rate, and efficiency of solar light conversion are crucial in determining the functionality of the photocatalyst system. Semiconductor metal oxides such as titanium dioxide ($TiO_2$) and zinc oxide (ZnO) have been extensively employed for photocatalytic purposes [10,11]. However, their higher bandgap restricts solar light capture, preventing their use as visible-light-driven photocatalyst systems. Perovskite materials with the empirical formula $ABO_3$ can be designed with tunable properties for appropriate photocatalytic applications [12]. Lanthanum cobaltite ($LaCoO_3$) is an orthorhombic perovskite-like structure, which has been reported to have promising catalytic activities [13,14]. Owing to the suitable physiochemical properties, $LaCoO_3$-based photocatalyst systems have attracted much research interest. The large cationic size of Lanthanum has been found to improve the structural stability of the catalyst. It has also been reported that doping strategies can facilitate their absorption in the visible range, thus facilitating solar light-induced photocatalytic potentials [15-18].

In this study, we evaluated the photocatalytic efficiency of lanthanum cobaltite ($LaCoO_3$) nanoparticles synthesized via the co-precipitation technique. The structural and morphological properties of as-synthesized $LaCoO_3$ were characterized, including the effect of annealing temperature on the crystalline properties of $LaCoO_3$. The photocatalytic performance of $LaCoO_3$ with hydrogen peroxide ($H_2O_2$) as a co-catalyst was investigated for both anionic and cationic dyes. The mechanism of the photocatalytic action of $LaCoO_3/H_2O_2$ was illustrated. Our findings revealed that the $LaCoO_3/H_2O_2$ system is an excellent candidate for completely removing organic pollutants for environmental remediation applications.

## 2. Experimental

### *2.1. Synthesis and characterization of $LaCoO_3$ nanostructures*

Lanthanum cobaltite ($LaCoO_3$) nanostructures were synthesized using the co-precipitation method with equimolar amounts of Cobalt (II) acetate and Lanthanum (III) acetate (0.1 M each). After mixing the acetates, the solution was stirred for five minutes at ambient conditions. Next, sodium hydroxide (NaOH, 0.2 M) was added to precipitate the mixture, and the resultant solution was stirred at 80° C for 3 hours. After completing the reaction, the product was centrifuged and washed thoroughly with deionized water and absolute ethanol several times. Subsequently, the powder was dried at 70° C overnight. To study the effect of annealing on the structural properties of $LaCoO_3$, the powder was subjected to an annealing process from 300 to 600° C. The synthesized $LaCoO_3$ nanoparticles were analyzed for morphology and elemental confirmation using the Field Emission SEM/EDS (Thermo Scientific Apreo C). The crystalline properties were studied using an X-ray diffraction technique with an X-ray diffractometer (Bruker, D8 advance). The absorbance spectra of the dye moieties during photocatalysis were captured using the UV Shimadzu (model 2600) spectrometer.

### 2.2. *Photocatalysis studies*

As-synthesized $LaCoO_3$ nanoparticles were explored for degrading cationic and anionic dyes such as methylene blue and methyl orange, respectively. Degradation of dyes was carried out using a photochemical reactor (Techinstro, India) equipped with a 450 W UV lamp [19,20]. About 100 mL of aqueous solutions containing 0.02 mM dye were stirred with ~0.1 g of as-synthesized $LaCoO_3$, while the absorbance of dye solutions was monitored at regular intervals. Before UV-light exposure, the solution was kept in the dark to attain equilibrium between the adsorption and desorption of dye molecules on the catalyst surface. After the dark phase, ~50 μL of hydrogen peroxide was added as a co-catalyst before light exposure. After switching on the UV light, dye solutions were retrieved at 20-minute intervals, and absorbance spectra were measured until the completion of the degradation reaction.

## 3. Results and Discussion

$LaCoO_3$ nanoparticles have been successfully synthesized using a co-precipitation process whose crystalline properties were investigated as a function of annealing temperatures, as shown in Figure 1. XRD diffractograms revealed the subsequent evolution of crystallinity of as-synthesized nanoparticles by subjecting them to an annealing process from 300 to 600° C. One can see that crystalline peaks of $LaCoO_3$ were fully observed at 600° C. The peaks at 23.2 (012), 32.9 (110), 33.3 (104), 40.6 (202), 41.5 (006), 47.5 (024), 53.6 (122), 53.7 (116), 58.9 (214), 59.7 (018), 68.9 (220), 70.0 (028) were observed which confirms the formation of $LaCoO_3$ perovskite phase as per JCPDS file no. 25–1060 and 00-036-1392 [21].

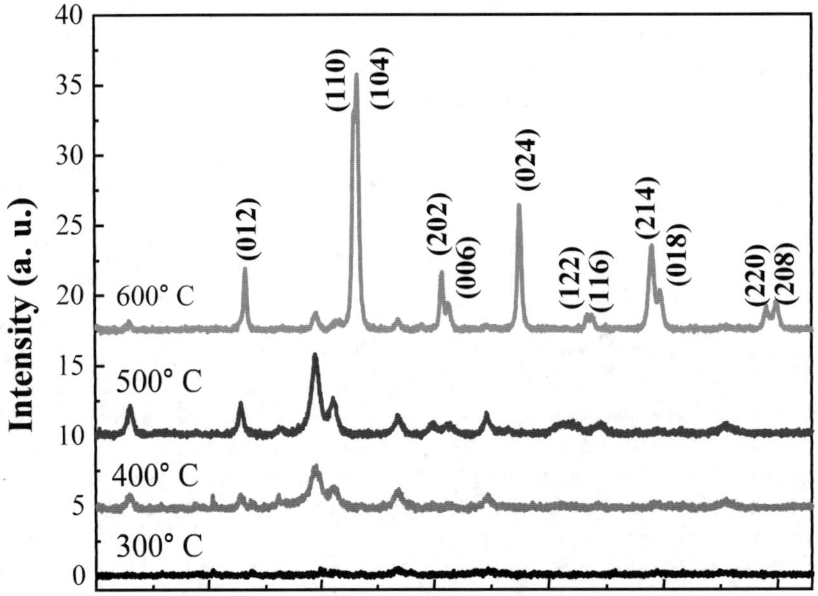


**Figure 1**: XRD pattern of as-prepared $LaCoO_3$ nanoparticles at different annealing temperatures from 300 to 600° C.

The morphology of as-synthesized lanthanum cobaltite nanoparticles has been demonstrated in Figure 2a. Homogenous and compact nanoparticle-sized aggregates could be seen from the FESEM image. Also, analysis using image J has revealed that the sizes of individual nanoparticles range from 20 to 40 nm, with an average size of 33 ± 0.69 nm. Furthermore, the elemental analysis using EDS confirmed the presence of elements such as La, Co, and O; thereby affirming the absence of any impurities within the as-synthesized powder. Meanwhile, the composition of $LaCoO_3$ is confirmed as 51.7% (La), 35.3% (O), and 13.1% (Co). The EDS mapping images demonstrated the distribution of elements as well (Figures 2d & 2e).

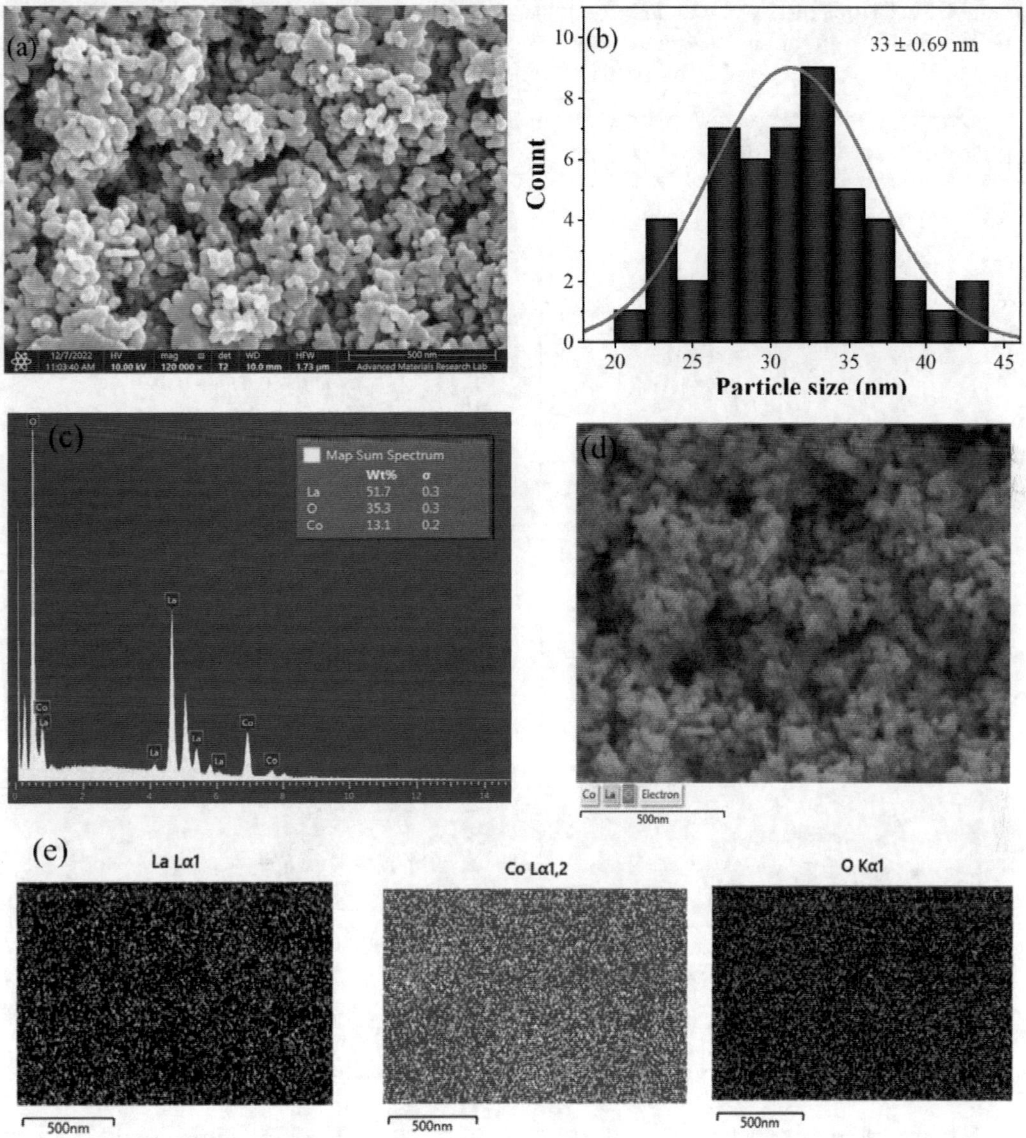


**Figure 2**: (a) FESEM image depicting surface morphology of as-synthesized $LaCoO_3$ nanoparticles; (b) corresponding particle size distribution; (c) & (d) EDS spectra and mapping image and (e) the constituent elements in the third row.

The photocatalytic performance of $LaCoO_3$ was evaluated using $H_2O_2$ as a co-catalyst for degrading both cationic and anionic dyes. It is clearly visible that the absorbance spectra of methylene blue decrease with increasing irradiation time (Figure 3a). After 120 minutes of light exposure, almost complete degradation of MB occurs. Figure 3b shows the plot of ln(A/A0) versus irradiation time with pseudo-first-order kinetics, where the slope of the linear fit provides the kinetic coefficient of the reaction. A kinetic coefficient (k-value) of $3.7 \times 10^{-2}$ was obtained, indicating superior photocatalytic properties of $LaCoO_3/H_2O_2$ photocatalysts.

The efficacy of photocatalysis was also studied using an anionic dye such as MO (Figure 3c). After 220 minutes of exposure, MO showed considerable decomposition. The degradation efficiencies for MB and MO were almost 91% and 85%, respectively (Figure 3d). Also, we investigated the photocatalytic activities of $LaCoO_3$ and $H_2O_2$ alone. We observed that ~20% and 18% degradation was observed within 120 min for $LaCoO_3$ and $H_2O_2$ respectively for degrading MB. However, $LaCoO_3/H_2O_2$ showed synergistic photocatalytic activity (91%) for MB decomposition.

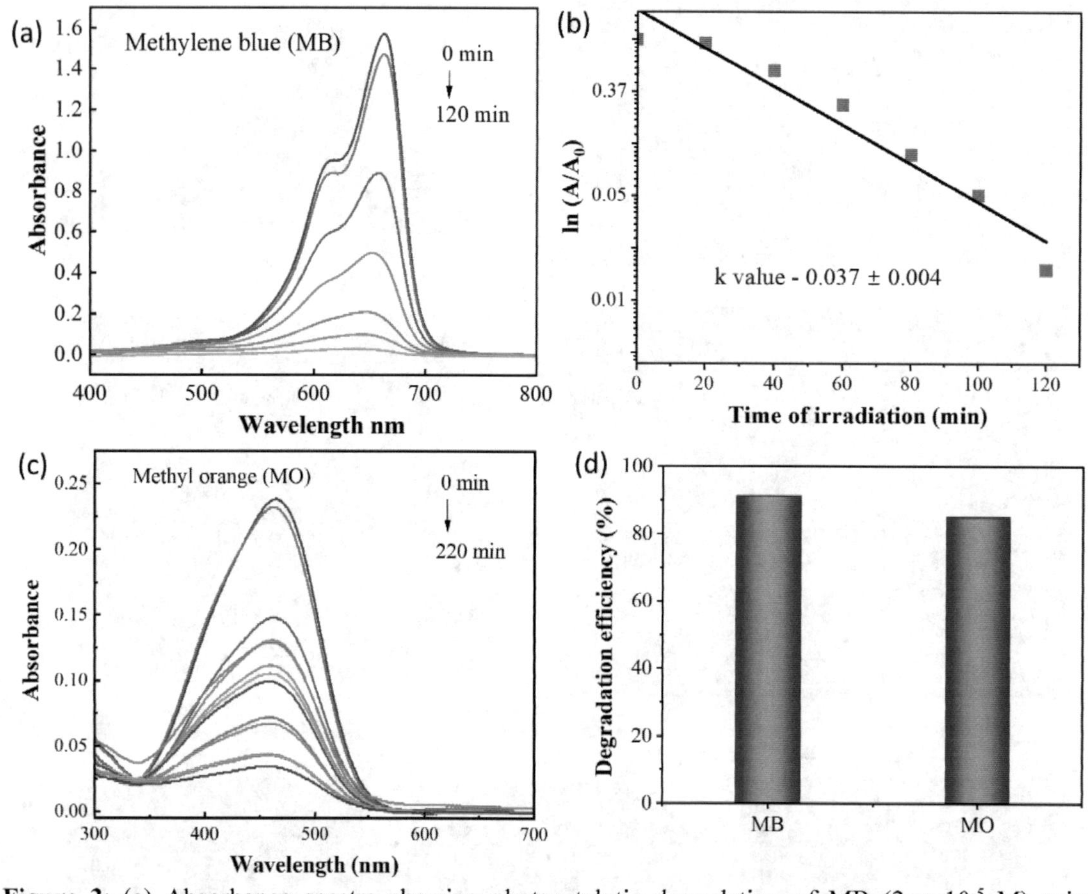


**Figure 3**: (a) Absorbance spectra showing photocatalytic degradation of MB ($2 \times 10^{-5}$ M) using $LaCoO_3/H_2O_2$; (b) plot of $\ln(A/A_0)$ versus time of irradiation; (c) Absorbance spectra of MO ($1 \times 10^{-5}$ M) and (d) degradation efficiencies of $LaCoO_3/H_2O_2$ for decomposing MB and MO.

The mechanism of photocatalytic degradation of organic pollutants assisted by $LaCoO_3/H_2O_2$ was also illustrated (Figure 4). Upon UV-light irradiation, photons can be absorbed by $LaCoO_3$, exciting valence band (VB) electrons to the conduction band (CB). The excited electrons can either be captured by $H_2O_2$ to form hydroxyl ($OH^·$) radicals or dissolve oxygen to undergo reduction, forming superoxide ($^·O_2^-$) radicals. The holes present in the VB can also react with water to form more hydroxyl radicals. Hydroxyl and superoxide radicals primarily attack the organic dye moieties, leading to their subsequent decomposition into carbon dioxide and water molecules [22,23].

Table 1 shows earlier studies that dealt with $LaCoO_3$-based photocatalysts for degrading MB and MO. In the present work, we attempted to investigate the photocatalytic efficiency of $LaCoO_3/H_2O_2$ for degrading both cationic and anionic dyes. Our findings showed that the proposed system is highly efficient for being used as a photocatalyst for environmental remediation applications.

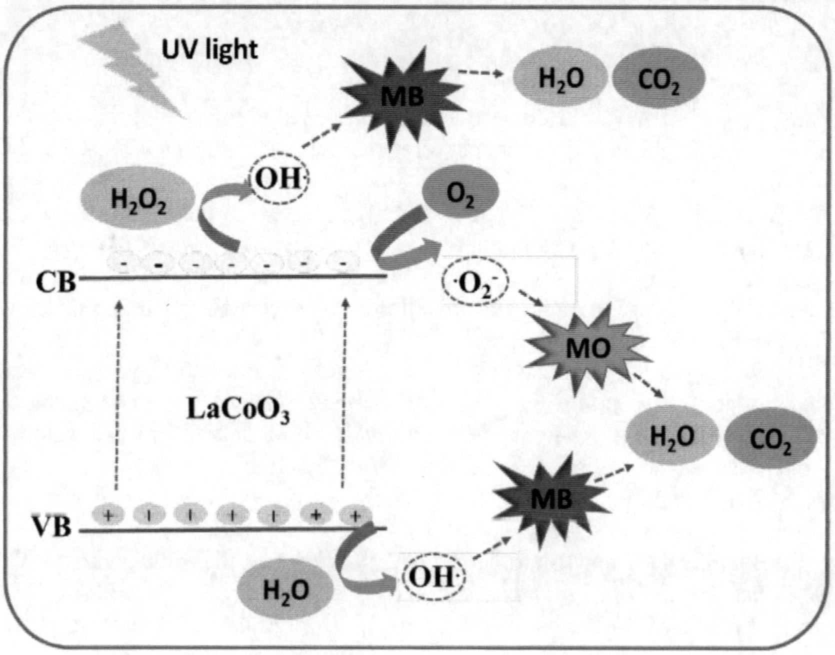


**Figure 4:** Photocatalytic mechanism of $LaCoO_3/H_2O_2$ for degrading MB

**Table 1:** Analysis of photocatalytic efficiencies of $LaCoO_3$-based systems for degrading dyes.

| Dyes | Photocatalysts | Dye conc. (mg/L) | Time (min) | Degradation efficiencies (%) | References |
|---|---|---|---|---|---|
| MB | $LaCoO_3/H_2O_2$ | 100 | 150 | 63 | [24] |
| | Sr doped $LaCoO_3$ | 30 | 150 | 12 | [25] |
| | $LaCoO_3/H_2O_2$ | 6.4 | 120 | 91 | This work |
| MO | $LaCoO_3$ | 10 | 240 | 100 | [26] |
| | $LaCoO_3$ hollow spheres | 10 | 60 | 87 | [27] |
| | $LaCoO_3/H_2O_2$ | 3.2 | 220 | 85 | This work |

## 4. Conclusions

Lanthanum cobaltite nanoparticles were synthesized successfully using the co-precipitation technique. The effect of annealing temperature on as-fabricated $LaCoO_3$ was investigated, which showed that 600° C was the optimum annealing temperature. The morphological analysis has shown that ~33 nm-sized $LaCoO_3$ particles have been formed. $LaCoO_3/H_2O_2$ system demonstrated 91 and 85% photocatalytic efficiencies for degrading methylene blue and methyl orange. The current investigation demonstrated that $LaCoO_3/H_2O_2$ could be exploited as an efficient photocatalyst for sustainably degrading organic pollutants.

**Acknowledgments**

The authors would like to acknowledge financial support from the University of Sharjah, United Arab Emirates (grant No 2102143096).

**References**

[1] Pal A, Gin K Y H, Lin A Y C, Reinhard M 2010 Impacts of emerging organic contaminants on freshwater resources: a review of recent occurrences, sources, fate, and effects *Sci. Total Environ.* **408** 6062-6069.

[2] Amin M T, Alazba A A and Manzoor U 2014 A review of removing pollutants from water/wastewater using different nanomaterials *Adv. Mater. Sci. Eng.* **2014** 1-24.

[3] Elgarahy A M, Elwakeel K Z, Mohammad S H and Elshoubaky G A 2021 A critical review of biosorption of dyes, heavy metals, and metalloids from wastewater as an efficient and green process *Clean. Eng. Technol.* **4** 100209.

[4] Mani S, Chowdhary P and Bhargava R N 2019 Textile wastewater dyes: toxicity profile and treatment approaches *Emerging and eco-friendly approaches for waste management* 219-244.

[5] Pavithra K G and Jaikumar V J J O I 2019 Removal of colorants from wastewater: A review on sources and treatment strategies. *J. Ind. Eng. Chem.* *75* 1-19.

[6] Padhi B S 2012 Pollution due to synthetic dyes toxicity & carcinogenicity studies and remediation. *Int. J. Environ. Sci.* **3** 940-955.

[7] Mathur N, Bhatnagar P and Sharma P 2012 Review of the mutagenicity of textile dye products. *Univers. j. environ. res. technol.* **2** 1-18.

[8] Hammouche J, Gaidi M, Columbus S and Omari M 2021 Enhanced photocatalytic performance of zinc ferrite nanocomposites for degrading methylene blue: effect of nickel doping concentration. *J Inorg Organomet Polym Mater.* **31** 3496-3504.

[9] Hammouche J, Daoudi K, Columbus S, Ziad R, Ramachandran K and Gaidi M 2021 Structural and morphological optimization of Ni doped ZnO decorated silicon nanowires for photocatalytic degradation of methylene blue. *Inorg. Chem. Commun.* **131** 108763.

[10] Ramachandran K, Ziad R, Columbus S, Daoudi K, Hammouche J, El Khakani M A, Chidambaram S and Gaidi M 2022 $TiO_2$/Si nanowires hybrid system for efficient photocatalytic degradation of organic dye. *Mater Sci: Mater Electron* **33** 9194–9203.

[11] Hammouche J, Daoudi K, Columbus S, Ziad R, Ramachandran K and Gaidi M 2021 Structural and morphological optimization of Ni-doped ZnO decorated silicon nanowires for photocatalytic degradation of methylene blue *Inorg. Chem. Commun.* **131** 108763.

[12] Kanhere P and Chen Z 2014 A review on visible light active perovskite-based photocatalysts *Molecules* **19** 19995-20022.

[13] Guo J, Li P and Yang Z. A 2019 2019 Novel Z-scheme g-$C_3N_4$/$LaCoO_3$ heterojunction with enhanced photocatalytic activity in degradation of tetracycline hydrochloride *Catal. Commun.* **122** 63-7.

[14] Wang L, Pang Q, Song Q, Pan X and Jia L 2015 Novel microbial synthesis of Cu doped $LaCoO_3$ photocatalyst and its high efficient hydrogen production from formaldehyde solution under visible light irradiation. *Fuel* **140** 267-74.

[15] Yao S, Zheng R, Li R, Chen Y, Zhou X, Ning X, Zhan L and Luo J 2020 $LaCoO_3$ acts as a high-efficiency co-catalyst for enhancing visible-light-driven tetracycline degradation of BiOI. *J. Am. Ceram. Soc.* **103** 1709-21.

[16] Nyamdavaa E, Uyanga E, Sevjidsuren G and Altantsog P 2017 Preparation and characterization of cerium doped $LaCoO_3$ perovskite. Journal of Surface Investigation: X-ray, *J. Synch. Investig.* **11** 366-70.

[17] Ling F, Anthony OC, Xiong Q, Luo M, Pan X, Jia L, Huang J, Sun D and Li Q 2016 PdO/$LaCoO_3$ heterojunction photocatalysts for highly hydrogen production from formaldehyde aqueous solution under visible light. Int. J. Hydrog. Energy. **41** 6115-22.

[18] Guo J, Dai Y Z, Chen X J, Zhou L L, Liu T H 2017 Synthesis and characterization of $Ag_3PO_4/LaCoO_3$ nanocomposite with superior mineralization potential for bisphenol A degradation under visible light. *J. Alloys Compd.***696** 226-33.

[19] Gaidi M, Daoudi K, Columbus S, Hajjaji A, El Khakani M A and Bessais B 2021 Enhanced photocatalytic activities of silicon nanowires/graphene oxide nanocomposite: Effect of etching parameters *J. Environ. Sci.* **101** 123-134.

[20] Ziad R, Columbus S, Elgamouz A, Daoudi K, Kawde A N, Ramachandran K and Gaidi M 2023 Multi-functional silver nanoprism-titanium dioxide hybrid nanoarrays for trace-level SERS sensing and photocatalytic removal of hazardous organic pollutants *Spectrochim. Acta - A: Mol. Biomol. Spectrosc.* **297** 122701.

[21] Ajmal S, Bibi I, Majid F, Ata S, Kamran K, Jilani K and Iqbal M 2019 Effect of Fe and Bi doping on $LaCoO_3$ structural, magnetic, electric and catalytic properties *J. Mater. Res. Technol.* **8** 4831-4842.

[22] Bahdaouia A, Amara S, Adnane L, Columbus S, Daoudi K and Mahieddine A 2022 Synthesis, characterization, and photo-electrochemical properties of novel $Cu(Cr_{1-x}, Al_x)_2O_4$ spinel oxides for enhanced photodegradation of organic dyes under UV light *Int. J. Hydrog. Energy.* **47** 9301-9318.

[23] Columbus S, Hammouche J, Ramachandran K, Daoudi K and Gaidi M 2022 Assessing the efficiency of photocatalytic removal of alizarin red using copper doped zinc oxide nanostructures by combining SERS optical detection *J. Photochem. Photobiol. A Chem.* **432** 114123

[24] Mocwana M L, Mokoena P P, Mbule P S, Beas I N, Kabongo G L, Ogugua S N and Tshabalala T E 2022 Photocatalytic Degradation of Methylene Blue and Ortho-Toluidine Blue: Activity of Lanthanum Composites LaxMOy (M: Fe, Co, Ni) *Catalysts.***12** 1313

[25] Trâm T L and Viễn T L 2015 Photocatalytic activity of Sr-doped LaCoO3 under UV illumination. *Tạp chí Khoa học và Công nghệ-Đại học Đà Nẵng*. **6** 42-5.

[26] Fu X, Chen Y, Yu GQ, Yu C, Wang HZ and Zhang GD 2011 Preparation, characterization and photocatalysis properties of ultrafine perovskite-type $LaCoO_3$ *Adv. Mater. Res.***197** 935-42.

[27] Fu S, Niu H, Tao Z, Song J, Mao C, Zhang S, Chen C and Wang D 2013 Low temperature synthesis and photocatalytic property of perovskite-type $LaCoO_3$ hollow spheres. *J. Alloys Compd.* **576** 5-12.

# A Study on Plasminogen-Ag Nanoparticles Interaction and its Application in Sensor Development

**A AbdelHamid[1], H Boukhatem[1], A Elgamouz[1*], and K Bajou[2]**

[1]Applied Chemistry Group, Department of Chemistry, College of Sciences, University of Sharjah, P.O. Box 27272, Sharjah, United Arab Emirates

[2]Department of Biology, College of Sciences, University of Sharjah, PO. Box 27272, Sharjah, United Arab Emirates

E-mail: aelgamouz@sharjah.ac.ae

**Abstract**. The interaction between plasminogen and Ag nanoparticles is studied using different techniques and applied for plasminogen sensor development. Ag nanoparticles are synthesized using an algal extract and their absorbance, emission, and electrochemical oxidation signals are detected and monitored as a function of plasminogen concentration. The variation in the optical and electrochemical properties of Ag nanoparticles is correlated with the changes in the hydrodynamic size of the bioconjugate at different plasminogen concentrations. A steady decrease in the absorbance and electrochemical oxidation peak of Ag nanoparticles is observed, while a threshold plasminogen concentration results in increasing the emission of Ag nanoparticles followed by a steady decrease in signal. The decrease in the optical and electrochemical oxidation signal of Ag nanoparticles agrees with the plasminogen-induced Ag nanoparticle agglomeration shown by dynamic light scattering. Calibration curves are established based on the absorbance, emission, and voltammetric studies obtaining a limit of detection as low as 0.740 nM with a wide linear range of 0.942–18.2 nM, which is a very promising analytical system for plasminogen detection, facilitating its progress as a biomarker for different biomedical applications.

## 1. Introduction

Plasminogen (Plg) is a plasma glycoprotein with a molecular weight of 92 kDa, present at a concentration of 2 μM with a half-life of around two days. It is a multifaceted zymogenic protein that needs to be activated to plasmin, a key fibrinolytic serine protease. The plasminogen-activating system is conventionally studied for its role in hemostasis. However, recently, there has been growing evidence of a much bigger role of this system in the human body. The multitude of Plg receptors allows it to exert other functions that affect the immune system, inflammatory processes, wound repair, protein aggregate clearance, and surprisingly, viral infection [1,2]. For example, it has been reported that a low level of Plg was correlated with inflammatory markers such as c-reactive protein (CRP) and interleukin-6 (IL-6), coagulation markers such as international normalized ratio (INR) and activated partial thromboplastin time (APTT), and organ dysfunction indicators such as high level of fasting blood glucose level and reduced glomerular filtration rate [3]. More interestingly, Plg has been found to play an important role in COVID-19 complications, with several reports on the correlation between low levels of Plg and increased mortality [3] and Plg's effectiveness in the treatment of associated lung lesions and hypoxemia [4]. Thus, Plg is a very promising COVID-19 prognosis biomarker and therapeutic agent.

The interaction between proteins and nanoparticles (NPs) is a complex process that has huge implications for diagnostics and drug delivery. Typically, proteins are adsorbed on the surface of the nanoparticles forming a "protein corona" with the reaction kinetics and dynamics determined by the type of interacting protein and NPs. Such interaction leads to changes in the functionality of the nanoparticles and also the protein structure, which exhibits conformational changes and unfolding processes with a major impact on its properties [5]. Protein-NP complexation has been studied in different systems and for different applications. For example, bovine serum albumin (BSA) has been used as a capping agent and stabilizer in the synthesis of Au [6] and Ag [7] NPs. Several types of NPs have been loaded with Plg activators for highly efficient and targeted delivery [8]. Recently, Boehmler *et al.* studied the reaction kinetics between Ag NPs and BSA, reporting an increase in the Ag dissolution rate constant upon BSA adsorption on its surface, which was dependent on the size of the Ag NPs and the BSA concentration. They explained this observation by BSA-$Ag^+$ complex formation and displacement by excess BSA [9].

Plg-NP complexation has wide implications in Plg detection and hemostasis management and needs to be well studied. Bychkova *et al.* studied the interaction between Plg and iron oxide NPs through non-specific binding and found ~70% decrease in Plg fibrinolytic activity upon conjugation [10]. Several studies have reported an inherent thrombolytic activity of Ag NPs that could be mediated through its interaction with Plg, triggering its activation, in addition to antagonizing its activation inhibitors [11-13]. Despite the evidence of major functionality changes upon Plg-NP interaction, studies on the dynamics and kinetics of such complexation, and its implications on conjugate complex properties and functionalities are very scarce.

In this work, we studied the interaction between Plg and Ag NPs, and characterized the Plg-Ag complexation at a wide range of Plg concentrations (0.942–18.2 nM), spectroscopically using UV-Vis and fluorimetric techniques, and electrochemically using square wave voltammetry (SWV). In addition, dynamic light scattering (DLS) was used to track complex size changes as the Plg concentration increased. Plg was found to impact the spectroscopic and electrochemical properties of Ag NPs significantly. Both UV-Vis spectroscopy and SWV showed a steady decrease in Ag NP response as the Plg concentration increased to 14.7 nM and 7.43 nM, respectively, followed by a slower decrease at higher concentrations. Fluorimetry showed a relatively stable Ag emission till 4.57 nM Plg, beyond which there was a sudden emission increase that decreased gradually upon adding more Plg. The steady and linear change of Ag NPs response as a function of Plg resulted in calibration curves with a wide linear range and low limits of detection (LOD) and quantification (LOQ). SWV showed the highest sensitivity with LOD and LOQ of 0.740 nM and 2.47 nM, respectively. Dynamic light scattering showed the appearance of larger particles as Plg concentration increased, which indicated Plg-induced Ag NP agglomeration that could have contributed to the change in Ag NP properties. The significant change in the optical, electrochemical, and structural properties of Ag NPs upon complexation with Plg shed light on the nature of their conjugation and demonstrated a robust methodology to develop a highly sensitive Plg sensor that could help in the development of Plg as a biomarker.

## 2. Experimental

### *2.1. Materials and Chemicals*

Plasminogen purified from human plasma using affinity column with Lysine-Sepharose beads, silver nitrate, sodium citrate dihydrate, and citric acid monohydrate were purchased from Sigma-Aldrich. *Noctiluca scintillans* algae were collected from the eastern coast of the United Arab Emirates. Graphite pencil lead (GPE) with a diameter of 0.5 mm (Pentel, HB) was used as the working electrode. Ultrapure water with a resistivity of 10–15 MΩ cm (Milli-Q Elix Essential® 5 ) was used in all experiments.

### 2.2. *Synthesis of Ag NPs*

Ag nanoparticles were synthesized using *Noctiluca scintillans* algae according to our earlier report [14] using 2.0% w/v algae extract, 0.1 M $AgNO_3$ at 80 °C, pH 7.2 for 2.5 h with an Ag:algal extract of 3:7 v/v.

### 2.3. *Instrumentation*

UV-Vis spectroscopy was carried out using Jenway 6850 spectrophotometer at a wavelength range of 300-800 nm and a resolution of 1 $nm^{-1}$. Transmission electron microscopy (TEM) was conducted using Titan Themis G2 Probe Cs Corrected Scanning Transmission Electron Microscope operating at 300 kV. Fluorimetry was conducted using a Shimadzu RF-6000 fluorimeter with an excitation wavelength of 434 nm and an emission wavelength range of 444–900 nm, a scan speed of 6000 nm/min, and a resolution of 1 $nm^{-1}$. Electrochemical measurements were conducted using CHI 1232c electrochemical workstation in a three-electrode cell using a graphite pencil working electrode with an immersion depth of 1 cm, Pt wire counter electrode, and Ag/AgCl (1 M KCl) reference electrode in a 0.1 M citrate buffer (pH 4) electrolyte. Dynamic light scattering was performed using Microtrac Nanotrac Wave II.

## 3. Results and Discussion

Metal NPs including Ag NPs show a unique optical phenomenon that arises from their nanoscale dimensions, which confine surface electrons. They can absorb electromagnetic radiation at a specific frequency that resonates with that of their surface electrons, which is mainly a function of the nanoparticle size. This phenomenon is known as surface plasmon resonance (SPR). The SPR absorbance band position and intensity are significantly affected by changes on the surface of the NPs, thus, SPR is widely applied for analytical applications [15,16]. We synthesized Ag nanoparticles using an algal extract according to our previous work [14]. The Ag NP absorbance peak was detected at 434 nm (Figure 1a), which indicated a size of ~60–80 nm [17]. This size range was confirmed using TEM, which showed Ag nanoparticles within the size range of 50–80 nm (Figure 1c). Plg was mixed with the Ag nanoparticles at increasing concentrations, and the effect on the Ag NPs SPR peak was monitored. It was observed that the peak intensity decreased steadily as the concentration of Plg increased from 0.942 nM to 14.7 nM, and then further decreased at a slower rate or almost stabilized at higher concentrations till 18.2 nM (Figure 1a–b). Both ranges were linear and we used the former as our main calibration curve from which we could calculate the LOD and LOQ to be 1.07 nM and 3.56 nM, respectively according to Equations (1–2) [18].

$$LOD = 3 \times \sigma/s \quad (1)$$

$$LOQ = 10 \times \sigma/s \quad (2)$$

Where $\sigma$ is the standard error of the intercept of the calibration curve and $s$ is the slope of the linear range of 0.942–14.7 nM. The effect of Plg on the SPR peak of Ag nanoparticles could be explained by its adsorption on the surface of the NPs [5], which disrupted the electron cloud oscillation and resulted in a change in SPR peak intensity. We were able to demonstrate that such change was concentration dependant and linear, and thus could be used for the quantitative detection of Plg.

Fluorescence is another optical phenomenon of metal nanoparticles that arises due to the quantum confinement effect that results in discrete energy levels [19]. Such discrete levels allow for the absorbance and emission of electromagnetic radiation, with several reported mechanisms, including the transition between the conduction band and holes in the d band, the presence of oligomeric metallic clusters, and electron interband transitions [20,21]. As with absorbance, metal NPs emissions are sensitive to surface changes that could form hybrid states at the interface of the two interacting moieties [19], and thus could be applied in electrochemical systems, especially given the inherently high sensitivity of fluorimetric analytical systems [22].

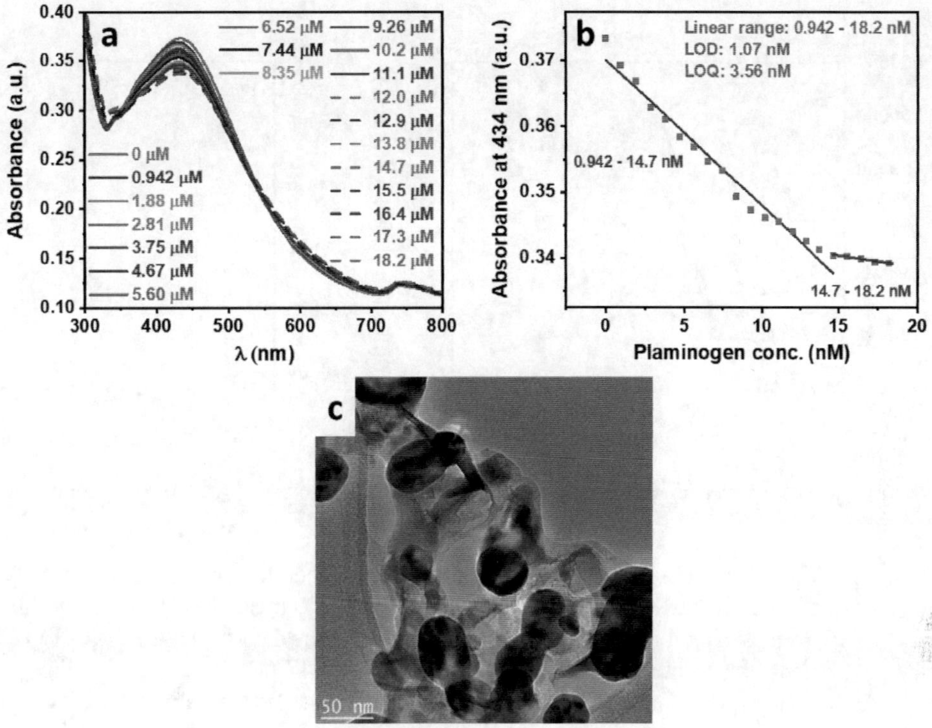


**Figure 1.** (a) UV-Vis absorbance spectra of Ag NPs at increasing Plg concentrations and (b) the corresponding calibration curve. (c) TEM image of the Ag nanoparticles.

To this end, the Ag nanoparticles were excited at 434.0 nm and the emission peak could be detected at 868.5 nm (Figure 2a), which was close to the values reported earlier [19]. The fluorimetry study revealed a more complex interaction between Ag NPs and Plg, where the emission peak intensity was stable upon Plg addition till 4.67 nm, which was a threshold Plg concentration upon which the emission intensity increased by ~32% and then decreased steadily thereafter (Figure 2a). A linear calibration curve could be established at the Plg concentration range of 5.60–18.2 nM, from which LOD and LOQ could be calculated to be 2.65 nM and 8.82 nM (Figure 2b). The sudden fluorescence intensity change is an intriguing phenomenon that could be explained by either Plg adopting certain conformational changes or Ag NPs structural rearrangement at the threshold concentration of 4.67 nM. This phenomenon is very interesting and will be further investigated in our future work. Although fluorimetry is inherently a very sensitive technique and showed a linear response after the threshold concentration, the system was not as sensitive as UV-Vis spectroscopy, which could be due to the nature of Plg effect on SPR and emission, which was more significant in the former than the latter.

Metal nanoparticles have excellent electrochemical properties due to their high conductivity and electrocatalytic activity, which is especially enhanced due to the large surface area of NPs [23]. The interaction of Ag NPs with BSA has been studied using an electrochemical voltammetric approach, where BSA was found to enhance the dissolution rate of Ag NPs [9]. In addition, electrochemical techniques are known to be among the most facile, cost-effective, and sensitive analytical systems, and thus could be very promising for point-of-care Plg detection [18,24]. We studied the interaction between Ag NPs and Plg on a graphite working electrode in a citrate buffer at pH 4 using SWV, recoding the oxidation peak of Ag NPs at ~0.4 V (Figure 3a).

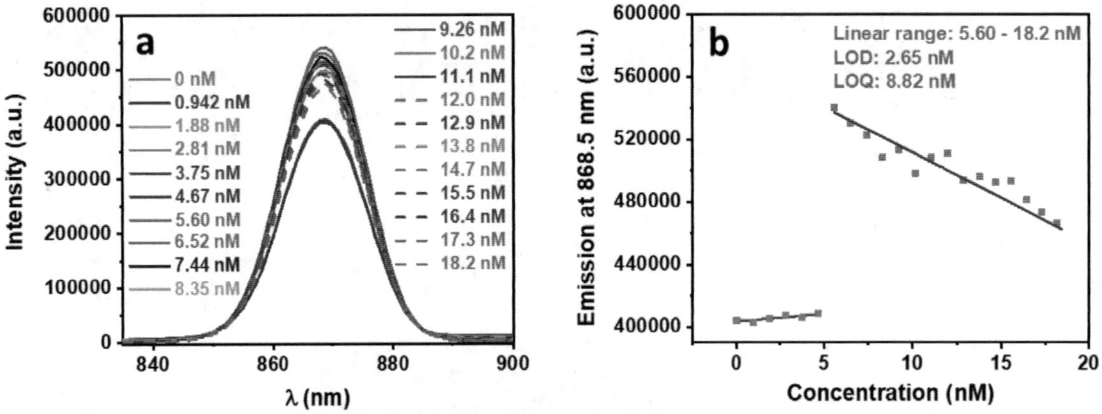


**Figure 2.** (a) Emission spectra of Ag NPs at increasing Plg concentrations and (b) the corresponding calibration curve.

Upon the addition of Plg, the Ag oxidation peak intensity started to decrease at a relatively high rate till 7.43 nM, after which the rate of change decreased but was still linear (Figure 3a). Two linear ranges could be derived from such calibration experiment, at 0.942–7.43 nM and 7.43–18.2 nM. LOD and LOQ were calculated using the former range to be 0.740 nM and 2.47 nM, respectively (Figure 3b). The decreasing Ag oxidation peak could be the result of two factors, the first is a Plg protective layer formed as a surface "protein corona" [5] and the second is an enhanced surface etching of Ag NPs as a result of Plg-Ag(I) complex formation, as reported earlier in the Ag-BSA system [9]. However, conclusive results would need further study of the system. The electrochemical approach achieved lower LOD and LOQ, as compared to spectroscopic approaches (Table 1), indicating its very high sensitivity and providing a guide for future studies on protein-metal NP interaction. The application of Ag-Plg complexation in Plg analytical detection using different techniques is summarized in Scheme 1.

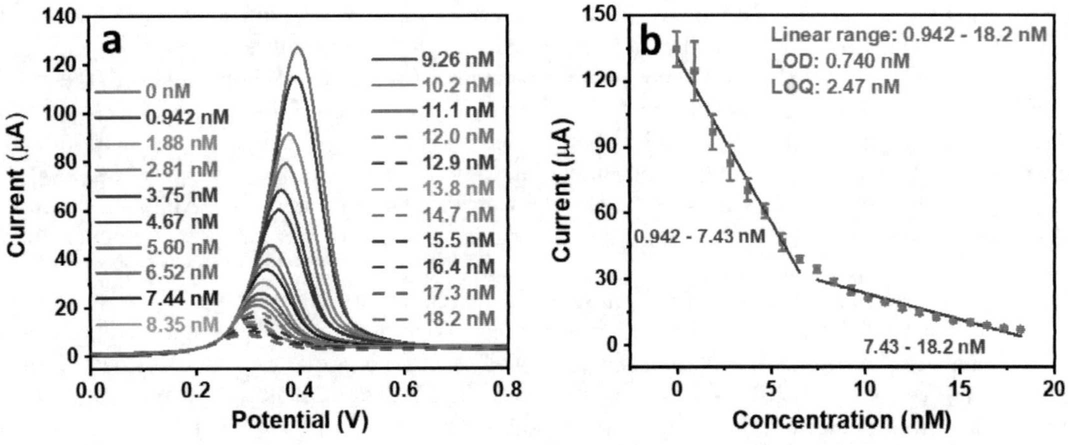


**Figure 3.** (a) SWV of Ag NPs at increasing Plg concentrations and (b) the corresponding calibration curve.

**Table 1.** Comparison between the Plg analytical parameters using spectroscopic and voltammetric techniques.

| Parameter | UV-Vis | Fluorimetry | SWV |
|---|---|---|---|
| **Linear ranges (nM)** | 0.942–14.7 | 5.60–18.2 | 0.942–7.43 |

| | | | |
|---|---|---|---|
| | 14.7–18.2 | | 7.43–18.2 |
| **LOD (nM)** | 1.07 | 2.65 | 0.740 |
| **LOQ (nM)** | 3.56 | 8.82 | 2.47 |

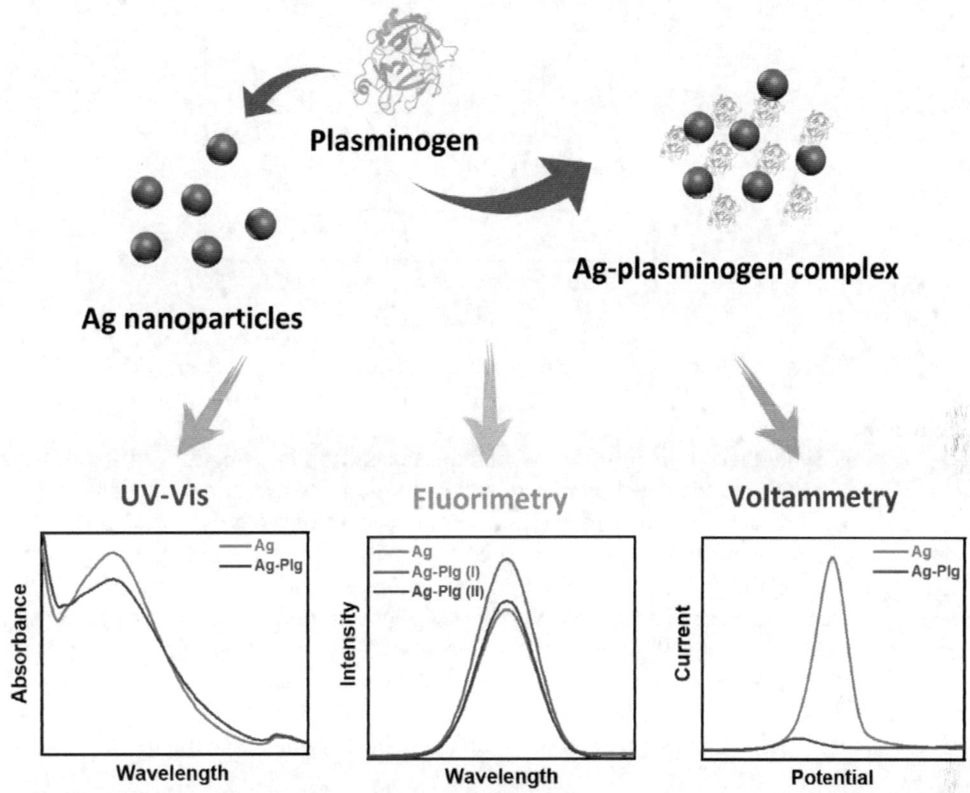


**Scheme 1.** Schematic displaying the application of Ag-Plg complexation in the analytical detection of Plg using different techniques.

DLS or photon correlation spectroscopy is a cost-effective and facile approach for nanoparticle sizing and monitoring size change over time or upon interaction with a specific reagent. It is based on the determination of diffusion coefficient based on the light scattering of particles, and then correlating the diffusion coefficient to their hydrodynamic radius using the Stokes-Einstein equation [25-27]. We used DLS to study the interaction between Ag nanoparticles and Plg to try to get an insight into the morphological changes taking place with increasing Plg concentration (Figure 4). The hydrodynamic size of Ag nanoparticles was in the range of 100–120 nm, which was not changed upon Plg addition till a concentration of 4.67 nM. The relatively larger size of Ag nanoparticles obtained using DLS, as compared to TEM, could be explained by the enclosure of Ag nanoparticles within residues of the algal extract (Figure 1c), which resulted in a larger hydrodynamic size. However, at higher Plg concentrations, a second peak appeared indicating the formation of larger size particles of ~240 nm and ~470 nm at Plg concentrations of 8.35 nM and 12.0/18.2 nM, respectively. The larger particles could have originated from Plg-induced Ag NP agglomeration, which could have increased the hydrodynamic size of the interacting particles resulting in the dual size range. This behavior could help to explain the reduction in absorbance, fluorescence, and electrochemical oxidation signals of Ag NPs as Plg concentration increased, where the NP agglomeration would have compromised the optical and electrochemical behavior of Ag NPs.

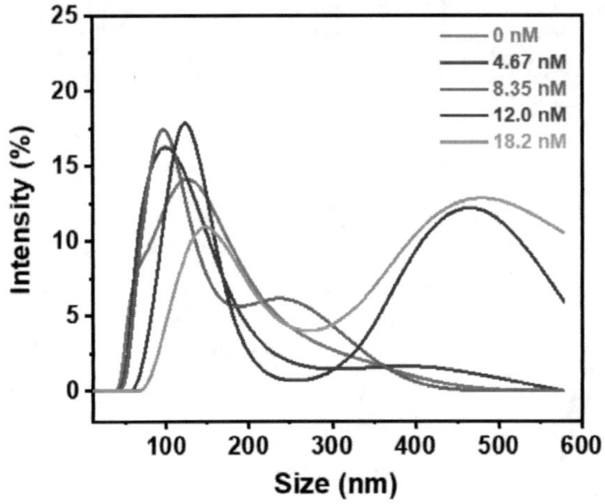


**Figure 4.** DLS size plots of Ag nanoparticles at increasing Plg concentrations.

## 4. Conclusion

We have studied the interaction between Ag nanoparticles and Plg using four different techniques; UV-Vis spectroscopy, fluorimetry, voltammetry, and dynamic light scattering, thus probing the effect of Plg on the SPR, emission, electrochemical, and size properties of Ag NPs. It was observed that the absorbance and electrochemical oxidation response of Ag NPs decreased steadily as Plg concentration increased, while that of fluorescence exhibited a more complex behavior, increasing after a threshold Plg concentration and then decreasing steadily at higher concentrations. The hydrodynamic size of Ag NPs showed a dual range upon Plg addition at concentrations higher than 4.67 nM indicating Plg-induced agglomeration. Calibration curves were established based on the spectroscopic and voltammetric studies, achieving very low LOD and LOQ of 0.740 nM and 2.47 nM, respectively, which opens the door for Plg detection as a biomarker in different biomedical applications.

**Acknowledgments**

The authors acknowledge the financial support received from the Research Institute of Science and Engineering (RISE), University of Sharjah, Sharjah, United Arab Emirates, Competitive Research Project (V.C.R.G. / R. 438/2021).

**Conflicts of Interest:** The authors declare no conflict of interest.

**References**

[1] Keragala C B, Medcalf R L 2021 *Blood* **137** 2881-2889
[2] Heissig B, Salama Y, Takahashi S, Osada T and Hattori K 2020 *Cell. Signal.* **75** 109761
[3] Della-Morte D *et al.* 2021 *Cell Death Dis.* **12** 773
[4] Wu Y, Wang T, Guo C, Zhang D, Ge X, Huang Z, Zhou X, Li Y, Peng Q and Li J 2020 *QJM* **113** 539-545
[5] Park S J 2020 *Int. J. Nanomedicine.* **15** 5783-5802
[6] Burt J L, Gutiérrez-Wing C, Miki-Yoshida M and José-Yacamán M 2004 *Langmuir* **20** 11778-11783
[7] Wang J, Zhao J and Ma G 2019 *Nano-Struct. Nano-Objects* **19** 100349
[8] Hassanpour S *et al.* 2020 *Small* **16** 2001647
[9] Boehmler D J, O'Dell Z J, Chung C and Riley K R 2020 *Langmuir* **36** 1053-1061

[10] Bychkova A V, Kostanova E A, Sadykova E Z, Biryukova M I, Muradova A G, Sharapaev A I, Degtyarev E N and Kovarski A L 2022 *Prep. Biochem. Biotechnol.* **52** 800-808
[11] Lateef A, Elegbede J A, Akinola P O and Ajayi V A 2019 *Pan African Journal of Life Sciences* **3** 157-182
[12] Azeez M A, Durodola F A, Lateef A, Yekeen T A, Adubi A O, Oladipo I C, Adebayo E A, Badmus J A and Abawulem A O 2020 *IOP Conf. Ser. Mater. Sci. Eng.* **805** 012043
[13] Lateef A, Akande M A, Azeez M A, Ojo S A, Folarin B I, Gueguim-Kana E and Beukes L S 2016 *Nanotechnol. Rev.* **5** 507-520
[14] Elgamouz A, Idriss H, Nassab C, Bihi A, Bajou K, Hasan K, Abu Haija M and Patole S P 2020 *Nanomaterials* **10** 1861
[15] Smitha S L, Nissamudeen K M, Philip D and Gopchandran K G 2008 *Spectrochim. Acta A Mol. Biomol. Spectrosc.* **71** 186-190
[16] Jana J, Ganguly M and Pal T 2016 *RSC Adv.* **6** 86174-86211
[17] Mulfinger L, Solomon S D, Bahadory M, Jeyarajasingam A V, Rutkowsky S A and Boritz C 2007 *J. Chem. Educ.* **84** 322
[18] AbdelHamid A, Elgamouz A, Khanfer M and Kawde A 2023 *Arab. J. Chem.* **16** 104674
[19] Kurdekar A D, Chunduri L A A, Chelli S M, Haleyurgirisetty M K, Bulagonda E P, Zheng J, Hewlett I K and Kamisetti V 2017 *RSC Adv.* **7** 19863-19877
[20] Abdullah A, Annapoorni S 2005 *Pramana* **65** 815-819
[21] Siwach O P, Sen P 2008 *Spectrochim. Acta A Mol. Biomol. Spectrosc.* **69** 659-663
[22] Sanchez J C, Toal S J, Wang Z, Dugan R E and Trogler W C 2007 *J. Forensic Sci.* **52** 1308-1313
[23] Ahmadi F, Raoof J B, Ojani R, Baghayeri M, Lakouraj M M and Tashakkorian H 2015 *Chinese J. Catal.* **36** 439-445
[24] Elgamouz A, Kawde A, Shehadi I A, Sayari S, Abdullah Mohammed S A, Abdelrazeq A, Nassab C N, AbdelHamid A A and Hasan K 2023 *ACS Omega* **8** 11880-11888
[25] Lim J, Yeap S P, Che H X and Low S C 2013 *Nanoscale Res. Lett.* **8** 381
[26] Khlebtsov B N, Khlebtsov N G 2011 *Colloid Journal* **73** 118-127
[27] Takahashi K, Kato H, Saito T, Matsuyama S and Kinugasa S 2008 *Part. Part. Syst. Charact.* **25** 31-38

# AUTHOR INDEX